Tay-Sachs Disease

Tay-Sachs
Disease

Serial Editors

Jeffery C. Hall
Waltham, Massachusetts

Jay C. Dunlap
Hanover, New Hampshire

Theodore Friedmann
La Jolla, California

Francesco Giannelli
London, United Kingdom

Tay-Sachs Disease

Edited by
Robert J. Desnick
Mount Sinai School of Medicine
of New York University
New York, New York

Michael M. Kaback
University of California, San Diego
School of Medicine
San Diego, California

ACADEMIC PRESS

A Harcourt Science and Technology Company

San Diego   San Francisco   New York
Boston   London   Sydney   Tokyo

This book is printed on acid-free paper.

Academic Press
A division of Harcourt, Inc.
525 B Street, Suite 1900, San Diego, California 92101-4495, USA
http://www.academicpress.com

Academic Press
Harcourt Place, 32 Jamestown Road, London NW1 7BY, UK
http://www.academicpress.com

International Standard Book Number: 0-12-017644-0

PRINTED IN THE UNITED STATES OF AMERICA
01   02   03   04   05   06   EB   9   8   7   6   5   4   3   2   1

This volume is dedicated to the patients, parents, and relatives who have suffered from Tay-Sachs disease. They motivated and inspired the physicians and scientists whose efforts to understand, prevent, and cure this devastating disease are described in this volume.

Contents

Contributors   xiii
Preface   xv

1 **Tay-Sachs Disease: From Clinical Description
   to Molecular Defect**   1
   Michael M. Kaback and Robert J. Desnick
     I.   Introduction   1
     II. 1880–1960: Clinical, Pathologic, and Genetic
          Advances   2
     III. 1960–1980: Lysosomes, Biochemical Defect, Prospective
          Prevention   3
     IV. 1980–Present: The Molecular Era and Therapeutic
          Horizons   5
     V.   Conclusion   6
          References   7

2 **Barney Sachs and the History of the Neuropathologic
   Description of Tay-Sachs Disease**   11
   Daniel P. Perl

3 **Early Epidemiologic Studies of Tay-Sachs Disease**   25
   Stanley M. Aronson

4 **Identification of the Accumulated Ganglioside**   33
   Lars Svennerholm
     I.   Substance X and Ganglioside   33
     II. N-Acetylgalactosamine is a Ganglioside
          Component   34
     III. Strandin   35
     IV. Chromatographic Separation of Gangliosides   36

V. Thin-Layer Chromatography—The Method
of Choice for Studies of Ganglioside Structure   37
VI. Tay-Sachs Ganglioside   38
References   40

5 **Discovery of the Hexosaminidase Isoenzymes   43**
Donald Robinson and John L. Stirling
I. Introduction   43
II. Fluorigenic Substrates   44
III. Mammalian Glycosidases   45
IV. Hexosaminidases   46
V. Differential Assay for Hexosaminidases A and B   47
VI. Structural Relationship between Hexosaminidases
A and B   48
References   48

6 **Tay-Sachs Disease: The Search for the
Enzymatic Defect   51**
Roscoe O. Brady
I. Historical Overview   51
II. Applications   58
References   59

7 **Discovery of β-Hexosaminidase A Deficiency
in Tay-Sachs Disease   61**
Shintaro Okada and John S. O'Brien
I. Introduction   61
II. John S. O'Brien's Recollection   62
III. Shintaro Okada's Recollection   62
References   66

8 **The G$_{M2}$-Gangliosidoses and the Elucidation
of the β-Hexosaminidase System   67**
Konrad Sandhoff
I. Amaurotic Idiocy   67
II. Glycolipid Analysis of Brains with Amaurotic
Idiocy   68

III.   Tay-Sachs Disease with Visceral Involvement
       (Variant 0)   70
IV.   Search for the Defect in Tay-Sachs Disease   71
V.   Variant AB and the G_{M2}-Activator Protein   76
VI.   Variant B1   80
VII.   Clinical and Biochemical Heterogeneity of G_{M2}
       Gangliosidosis—Degree of Enzyme Deficiency and
       Development of Different Clinical Syndromes   83
VIII. Addendum   85
       References   86

9 Subunit Structure of the Hexosaminidase Isozymes   93
Ernest Beutler

I.   Introduction   93
II.   Antibodies Against Hexosaminidase   95
III.   Unravelling the Subunit Structure
       Immunologically   96
IV.   Converting Hexosaminidase A to Hexosaminidase B   97
V.   Epilogue   99
       References   99

10 Molecular Genetics of the β-Hexosaminidase Isoenzymes: An Introduction   101
Edwin H. Kolodny

I.   Personal Recollections   102
II.   Biochemical Genetics of the Hexosaminidases   104
III.   Evolution of Molecular Biology   104
IV.   Analysis of DNA   107
V.   Classification of Mutations   108
VI.   Detection of Known Mutations   110
VII.   Screening for New Mutations   111
VIII. From Enzyme to Gene Structure   112
IX.   Mutations in the G_{M2} Gangliosidoses   117
X.   Genetically Engineered Animal Models   120
XI.   Conclusions   121
       References   121

11 Cloning the β-Hexosaminidase Genes   127
Richard L. Proia

12 The Search for the Genetic Lesion in Ashkenazi Jews with Classic Tay-Sachs Disease   137

Rachel Myerowitz

13 The β-Hexosaminidase Story in Toronto: From Enzyme Structure to Gene Mutation   145

Don J. Mahuran and Roy A. Gravel

    I.   Introduction   146
   II.   Structures of Hexosaminidase A and Hexosaminidase B   146
  III.   Isolation of cDNA Clones Coding for the α and β Chains   148
  IV.   Extensive Homology Between the Deduced α and β Primary Structures   149
   V.   Posttranslational Processing of the Pre-pro-α and Pre-pro-β Chains   149
  VI.   Structure–Function Relationships   154
 VII.   Molecular Heterogeneity in Tay-Sachs and Sandhoff Diseases   157
       References   160

14 Biosynthesis of Normal and Mutant β-Hexosaminidases   165

Elizabeth F. Neufeld and Alessandra d'Azzo

    I.   The Normal Biosynthetic Pathway   165
   II.   Biosynthesis of Mutant β-Hexosaminidases   169
       References   170

15 Recognition and Delineation of β-Hexosaminidase α-Chain Variants: A Historical and Personal Perspective   173

Kunihiko Suzuki

    I.   At the Beginning   173
   II.   Increasing Complexity   174
  III.   Era of Molecular Genetics   175
  IV.   Evolution of B1 Variant   176
   V.   Genotype–Phenotype Correlation   180
       References   182

16 **Late-Onset G_{M2} Gangliosidosis and Other Hexosaminidase Mutations among Jews**   185
Ruth Navon

     I.   Adult G_{M2} Gangliosidosis   185
    II.   Tay-Sachs Disease Among Moroccan Jews   192
   III.   Heat-Labile β-Hexosaminidase B and the Genotyping
        of Tay-Sachs Disease   194
        References   196

17 **Naturally Occurring Mutations in G_{M2} Gangliosidosis: A Compendium**   199
Barbara Triggs-Raine, Don J. Mahuran, and Roy A. Gravel

     I.   Introduction   199
    II.   β-Hexosaminidase A Mutations   201
   III.   β-Hexosaminidase B Mutations   210
    IV.   GM2A Mutations   215
     V.   Structure/Function Relationships of
        β-Hexosaminidase   215
        References   216

18 **Targeting the Hexosaminidase Genes: Mouse Models of the G_{M2} Gangliosidoses**   225
Richard L. Proia

19 **Molecular Epidemiology of Tay-Sachs Disease**   233
Neil Risch

     I.   Introduction   233
    II.   Mutations and Their Frequencies   236
   III.   The Demographic History of the Ashkenazim   244
    IV.   Statistical Modeling   246
     V.   Conclusion   249
        References   250

20 **Screening and Prevention in Tay-Sachs Disease: Origins, Update, and Impact**   253
Michael M. Kaback

     I.   Program Origins: The Place   253
    II.   The Events and the People   254

    III. The Program Is Conceived   256
    IV. From Baltimore to Jerusalem   257
     V. Results and Update   257
    VI. Impact and Conclusion   259
        References   260
        Appendix I   261

21 Not Preventing—Yet, Just Avoiding Tay-Sachs Disease 267

Charles R. Scriver

      I. Introduction   267
     II. Context   268
    III. The Patient with the Disease   268
    IV. Strategies to Avoid Tay-Sachs Disease   270
     V. Tay-Sachs Disease Carrier Testing: An Illustration of "Community Genetics"   271
    VI. Conclusion   272
        References   273

22 Experiences in Molecular-Based Prenatal Screening for Ashkenazi Jewish Genetic Diseases 275

Christine M. Eng and Robert J. Desnick

       I. Introduction   276
      II. Common Recessive Diseases in the Ashkenazim   277
     III. Sensitivity of Enzymatic and DNA-Based Carrier Screening   281
     IV. Experience with Multiple-Option Prenatal Carrier Screening   282
      V. Rationale for Multiple-Option Carrier Screening   283
     VI. Strategy for Multiple-Option Carrier Screening   284
    VII. Enzyme and DNA Testing   285
   VIII. Demographics and Test Acceptance   285
     IX. Frequency of Detected Carriers   286
      X. Detected Carrier Couples Choose Prenatal Diagnosis   287
     XI. Importance of Educational Intervention   287
    XII. Group Counseling Preferred   288
   XIII. Couple Screening Reduces Anxiety   288
   XIV. Acceptance and Selection of Prenatal Screening Tests   289

XV.   Confidentiality Issues   290
XVI.   Lessons Learned and Future Prospects   290
XVII. Type A Niemann-Pick Disease Detectability and Carrier
       Frequency in the Ashkenazi Population   291
XVIII. Canavan Disease Detectability and Carrier Frequency
       in the Ashkenazi Population   291
XIX.   Multiple-Option Carrier Screening for Five
       Disorders   292
XX.   Summary   293
       References   294

23 The Dor Yeshorim Story: Community-Based Carrier Screening for Tay-Sachs Disease   297
Josef Ekstein and Howard Katzenstein

I.   Introduction   298
II.   Understanding a Community at Risk   298
III. Early Efforts at Screening   301
IV.   Mechanics of the Premarital, Anonymous
      Screening Program   302
V.   Findings and Accomplishments   305
VI.   Research   308
VII. Can the Dor Yeshorim Model Be Applied to
      Other Communities?   308
VIII Analytical Laboratories   309
      References   310

24 Tay-Sachs Disease and Preimplantation Genetic Diagnosis   311
Christoph Hansis and Jamie Grifo

I.   Tay-Sachs Disease   311
II.   Preimplantation Genetic Diagnosis   312
      References   314

25 Treatment of G_{M2} Gangliosidosis: Past Experiences, Implications, and Future Prospects   317
Mario C. Rattazzi and Kostantin Dobrenis

I.   Introduction   317
II.   Early Enzyme Infusion Trials   318
III. Studies in G_{M2} Gangliosidosis Cats   321

IV. Cell Targeting of Hexosaminidase A   322
V. T TC-HEX A and Neuronal Storage   324
VI. Implications and Open Questions   326
VII. Bone Marrow Transplantation and Enzyme
Secretion   327
VIII. Delivery of Macromolecules to the Brain
Parenchyma   329
IX. CNS Gene Therapy   330
X. Conclusions   332
References   333

26 **Tay-Sachs Disease: Psychologic Care of Carriers and Affected Families**   341
Leslie Schweitzer-Miller

27 **Future Perspectives for Tay-Sachs Disease**   349
Robert J. Desnick and Michael M. Kaback
I. Introduction   349
II. Substrate Deprivation   350
III. Chemical Chaperones   351
IV. Stem Cells   351
V. Oligonucleotide Recombination   352
VI. Genetic Counseling and Psychosocial Support   352
VII. Prevention   353
References   354

Index   357

Contributors

Numbers in parentheses indicate the pages on which the authors' contributions begin.

Stanley M. Aronson[†] (25) Division of Biology and Medicine, Brown University, School of Medicine, Providence, Rhode Island 02912

Ernest Beutler (93) Department of Molecular and Experimental Medicine, The Scripps Research Institute, La Jolla, California 92037

Roscoe O. Brady (51) Developmental and Metabolic Neurology Branch, National Institute of Neurological Disorders and Stroke, National Institutes of Health, Bethesda, Maryland 20892

Alessandra d'Azzo (165) Department of Genetics, St. Jude's Children's Research Hospital, Memphis, Tennessee 38105

Robert J. Desnick (1, 275, 339) Departments of Human Genetics and Pediatrics, Mount Sinai School of Medicine of New York University, New York, New York 10029

Kostantin Dobrenis (317) Department of Neuroscience, Albert Einstein College of Medicine, New York, New York

Josef Ekstein (297) Dor Yeshorim, The Committee for Prevention of Jewish Genetic Diseases, Brooklyn, New York/Jerusalem, Israel

Christine M. Eng (275) Departments of Human Genetics and Pediatrics, Mount Sinai School of Medicine New York, New York 10029

Roy A. Gravel (145, 199) Departments of Cell Biology and Anatomy, and Biochemistry and Molecular Biology, University of Calgary, Calgary, Alberta, Canada T2N 1N4; and The Research Institute, The Hospital for Sick Children, and the Department of Laboratory Medicine and Pathobiology, University of Toronto, Toronto, Ontario, Canada M5G 1X8

Jamie Grifo (311) Department of Obstetrics and Gynecology, Program for IVF, Reproductive Surgery, and Infertility, New York Univesity School of Medicine, New York, New York 10016

Christoph Hansis (311) Department of Obstetrics and Gynecology, Program for IVF, Reproductive Surgery, and Infertility, New York Univesity School of Medicine, New York, New York 10016

[†]Deceased

Michael M. Kaback (1, 253, 349) Departments of Pediatrics and Reproductive Medicine, University of California, San Diego School of Medicine, San Diego, California 92123

Howard Katzenstein (297) Dor Yeshorim, The Committee for Prevention of Jewish Genetic Diseases, Brooklyn, New York/Jerusalem, Israel

Edwin H. Kolodny (101) Department of Neurology, New York University School of Medicine, New York, New York

Don J. Mahuran (145, 199) The Research Institute, The Hospital for Sick Children, and Department of Laboratory Medicine and Pathobiology, University of Toronto, Toronto, Ontario, Canada M5G 1X8

Victor A. McKusick (xix) McKusick–Nathans Institute of Genetic Medicine, Johns Hopkins University School of Medicine, Baltimore, Maryland

Rachel Myerowitz (137) St. Mary's College of Maryland, St. Mary's City, Maryland 20686

Ruth Navon (185) Department of Human Genetics and Molecular Medicine, Tel-Aviv University, Sackler School of Medicine, Ramat Aviv, Tel-Aviv 69978, Israel; and Laboratory of Molecular Genetics, Sapir Medical Center, Kfar Sava, 44281, Israel

Elizabeth F. Neufeld (165) Department of Biological Chemistry, UCLA School of Medicine, Los Angeles, California 90095

John S. O'Brien[†] (61) Department of Neurosciences, University of California San Diego School of Medicine, La Jolla, California 92093

Shintaro Okada (61) Department of Pediatrics, Osaka University Hospital, Yamada-oka, Suita, Osaka 565, Japan

Daniel P. Perl (11) Neuropathology Division, Mount Sinai Medical Center, New York, New York 10029

Richard L. Proia (127, 225) Genetics of Development and Disease Branch, National Institute of Diabetes and Digestive and Kidney Diseases, National Institutes of Health, Bethesda, Maryland 20892

Mario C. Rattazzi (317) Department of Human Genetics, NYS Institute for Basic Research in Developmental, Disabilities, Staten Island, New York

Neil Risch (233) Department of Genetics, Stanford University School of Medicine, Stanford, California 94305

Donald Robinson (43) Division of Life Sciences, King's College London, London SE1 9NN, United Kingdom

Konrad Sandhoff (67) Institut für Organische Chemie und Biochemie der Universität Bonn, D-53121 Bonn, Germany

Leslie Schweitzer-Miller (341) New York University Psychoanalytic Institute, New York University Medical School, New York, New York 10016

[†]Deceased

Charles R. Scriver (267) Department of Human Genetics, McGill University, and the deBelle Laboratory for Biochemical Genetics, McGill University—Montreal Children's Hospital Research Institute, Montreal, Quebec, Canada H3H 1P3

John L. Stirling (43) Division of Life Sciences, King's College London, London, SE1 9NN, United Kingdom

Kunihiko Suzuki (173) Neuroscience Center, Departments of Neurology and Psychiatry, University of North Carolina School of Medicine, Chapel Hill, North Carolina 27599

Lars Svennerholm[†] (33) Department of Psychiatry and Neurochemistry, Institute of Clinical Neuroscience, Göteborg University, Göteborg, Sweden

Barbara Triggs-Raine (199) Department of Biochemistry and Medical Genetics, University of Manitoba, Winnipeg, Canada R3E 0W3

[†]Deceased

Preface

Tay-Sachs disease is a paradigmatic genetic disorder for several reasons: It is a paradigm for the ethnic distribution of genetic disorders (as well as for the search for historical roots of that distribution and for the use to which the distribution can be put for screening and genetic counseling). It is a paradigm for disorders of a particular cytoplasmic organelle, the lysosome, including the pathogenesis and general clinical characteristic of lysosomal disorders. It is a paradigm for progress in understanding a genetic disorder, all the way from clinical description through identification of the material that accumulates in body tissues and finding the enzyme that is defective, to detecting mutations in the gene that encodes that enzyme. Ideally, it could be a paradigm for completing the circle of understanding from the mutations in the gene back to the distribution of the disorder at the population level and back to the individual patient for effective therapy.

This monograph is a marvelous recounting of the Tay-Sachs story by most of the major players. We are all indebted to the authors and editors who have created it. It leaves quite clear the unanswered questions that remain. Tackling those questions should be easier for scientists and clinicians alike, given the background provided here.

Victor A. McKusick
University Professor of Medical Genetics
McKusick–Nathans Institute of Genetic Medicine
Johns Hopkins University School of Medicine
Baltimore, Maryland

Acknowledgments

We gratefully acknowledge the following organizations and foundations who have generously supported research and educational efforts to prevent and cure Tay-Sachs disease. In the United States these include the National Foundation for Jewish Genetic Diseases; the National Tay-Sachs and Allied Disease Association, especially the New York Area Chapter; the Evan Lee Ungerleider Foundation; the Genetic Disease Foundation; and the March of Dimes Birth Defects Foundation. Their continuous dedication and encouragement has led to many of the accomplishments described in this volume. In particular, we wish to express our special thanks and appreciation to those who made this volume possible: Dr. Michael Katz, Vice President for Research, March of Dimes Birth Defects Foundation; George Crohn, President, National Foundation for Jewish Genetic Diseases; Stanley Michelman, and Marion Yanosky, Co-Presidents, National Tay-Sachs and Allied Diseases Association–New York Area Chapter; and Shari and Jeffrey Ungerleider, Co-Presidents, Evan Lee Ungerleider Foundation. We are most appreciative of their unrelenting personal and professional efforts to prevent the tragedy of this disease in the future.

R. J. Desnick
M. M. Kaback
July 2001

Tay-Sachs Disease: From Clinical Description to Molecular Defect

Michael M. Kaback*
Departments of Pediatrics and
Reproductive Medicine
University of California,
San Diego School of Medicine
San Diego, California 92123

Robert J. Desnick
Department of Human Genetics
Mount Sinai School of Medicine of New York University
New York, New York 10029

I. Introduction
II. 1880–1960: Clinical, Pathologic, and Genetic Advances
III. 1960–1980: Lysosomes, Biochemical Defect,
Prospective Prevention
IV. 1980–Present: The Molecular Era and Therapeutic Horizons
V. Conclusion
References

I. INTRODUCTION

In 1881, Warren Tay, a British ophthalmologist, first reported his observations of a peculiar yellowish macular degeneration in the fundi of an infant with symptoms and signs of progressive central nervous system degeneration (Tay, 1881). A few years later, he described two additional children similarly affected in the

*To whom correspondence should be addressed. E-mail: mkaback@ncsd.edu. Fax: (858) 279-8379. Telephone: (858) 495-7737.

Advances in Genetics, Vol. 44

same family (Tay, 1884). In 1886, Bernard Sachs, a neurologist in New York City, reported several additional infants with the same disorder, noting the familial nature of the condition, its occurrence in Jewish children of Eastern European ancestry, and a consistent pattern of early blindness, profound retardation, and death in early childhood (for which he introduced the name "amaurotic familial idiocy") (Sachs, 1896). In the following year, Sachs provided the first detailed description of the striking postmortem neuropathologic alterations in this condition (Sachs, 1897).

Named for these two clinicians who first described it, Tay-Sachs disease (TSD) is now also called GM2 gangliosidosis, type 1, reflecting the severe, early infantile form in which a glycosphingolipid (GM2 ganglioside) accumulates within neuronal lysosomes. Over the past century several hundred publications have appeared, expanding and furthering our understanding of Tay-Sachs disease and its pathogenesis (Gravel *et al.*, 1995). This volume chronicles many of these discoveries, most prominently those recent breakthroughs that have elucidated the fundamental biochemical and molecular/genetic basis of TSD. These discoveries have, in turn, provided the scientific foundations for the development of animal models of the GM2 gangliosidoses, for advances in preventive strategies, and, most recently, for possible therapeutic approaches to such disorders. A brief chronology of the most seminal contributions follows.

II. 1880–1960: CLINICAL, PATHOLOGIC, AND GENETIC ADVANCES

From the initial clinical descriptions of TSD at the end of the nineteenth century until the mid-1960s, major research contributions were predominantly in the areas of histopathology, lipid chemistry, genetic epidemiology, and ultrastructural analysis. Several early reports characterized the typical neuropathology of neuronal cells with "ballooned" lipid-filled cytoplasm in postmortem brain sections from children with TSD. Extensive studies on the composition of these lipids enabled Klenk and his colleagues in the late 1930s and early 1940s to identify the sphingolipids predominately accumulating in this disorder. He characterized them as gangliosides, a group of neuraminic acid-containing glycosphingolipids (Klenk, 1942). The structure of the specific ganglioside, GM2, was identified by Svennerholm in 1962 and further characterized by Ladeen and Salsman in 1965.

The earliest discussion of the autosomal recessive pattern of inheritance in TSD was presented by Slome in 1933, basing his conclusions on studies of 130 affected children in 80 families. This observation was confirmed in several subsequent epidemiological studies of more than 100 total families in the United States providing clear evidence for the ethnic predilection of TSD for Jewish children of Eastern and Central European ancestry, as well as presenting preliminary estimates of the gene frequencies for TSD among Ashkenazi Jews, Sephardic Jews, and non-Jews living in the United States (Aronson *et al.*, 1960; Myrianthopoulos, 1962; Aronson and Volk, 1962; Myrianthopoulos and Aronson, 1967).

III. 1960–1980: LYSOSOMES, BIOCHEMICAL DEFECT, PROSPECTIVE PREVENTION

The first characterization of intracytoplasmic membranous bodies in neurons from TSD brain was presented by Terry and Korey (1960). Subsequent publications revealed that these bodies contained ganglioside (Samuels *et al.*, 1962) and, by electron microscopic study, that the spirally wound membranous structures stained heavily with osmium (Terry *et al.*, 1962; Terry and Weiss, 1963) and were the site of acid phosphatase activity, indicating their probable lysosomal origin (Wallace *et al.*, 1967). It was during this period that subcellular "lysosomal particles" were first described by deDuve and co-workers (1955). It remained for Hers and colleagues to describe the underlying pathology in Type 2 glycogen storage disease (Pompe disease) and to propose the concept of lysosomal storage disorders as a category of human disease (Hers and Van Hoof, 1973).

The first two neuronal lipid storage disorders for which specific lysosomal hydrolase activities were demonstrated to be deficient were Gaucher disease and Niemann Pick disease (Brady *et al.*, 1965, 1966). These findings underscored the mechanism for intralysosomal glycolipid accumulation in these conditions; i.e., the sphingolipid substrate accumulates within the neuronal lysosome when the specific catabolic lysosomal hydrolase, required to catalyze the next step in substrate degradation, is deficient in activity. The findings of specific lysosomal acid hydrolase deficiencies in Gaucher disease, Niemann Pick disease, and metachromatic leukodystrophy (Mehl and Jatzkewitz, 1965) led to speculation that a deficiency of activity of lysosomal hexosaminidase, required to cleave the terminal N-acetylgalactosamine from GM2 ganglioside, was the specific defect in TSD (Brady, 1966).

Robinson and Stirling (1968) demonstrated that at least two human acidic hexosaminidase isoenzymes (hexosaminidase A and B) were present in normal human spleen. In the same year, Sandhoff and colleagues presented their findings in a child with apparent TSD but who had unusual clinical features: evidence of visceral storage. A complete absence of lysosomal hexosaminidase activity was found, in contrast to what appeared to be normal or near-normal hexosaminidase activities in tissues from children with typical TSD (Sandhoff *et al.*, 1968). This variant has been classified subsequently as GM2 gangliosidosis type 2, or Sandhoff disease.

In 1969, two laboratories reported the critical observation of deficient activity of one of the hexosaminidase isoenzymes (hexosaminidase A, or HexA) in tissues from children with TSD (Okada and O'Brien, 1969; Sandhoff, 1969).

These studies, employing synthetic hexosaminide substrates, were confirmed in the same year by investigations demonstrating deficient catabolism of radio labeled GM2 ganglioside (natural substrate) in TSD tissues as well (Kolodny *et al.*, 1969). The demonstration of deficient activity of hexosaminidase A in somatic tissues and body fluids from patients with TSD provided a highly sensitive, relatively simple, and inexpensive method for diagnosis of the disorder. These findings were then extended to cultivated skin fibroblasts derived from affected children and, shortly thereafter, were also demonstrated in cultured midtrimester amniocytes (Okada *et al.*, 1971). These discoveries provided the basis for prenatal detection of TSD in at-risk pregnancies (Schneck *et al.*, 1970; O'Brien *et al.*, 1971; Navon and Padeh, 1971). An additional and highly critical contribution during this period was the characterization of a partial deficiency of activity of this isoenzyme in serum, white blood cells, and cultured fibroblasts derived from obligate heterozygotes for the TSD gene (parents of afflicted children) (O'Brien *et al.*, 1970). Thus, carrier detection became feasible.

The aforementioned discoveries provided the basis for considering the "prospective prevention" of TSD (Kaback and O'Brien, 1971). Minor modifications of the serum and leukocyte assay method allowed first semiautomation (Kaback, 1973) and then complete automation (Lowden *et al.*, 1973) of the heat inactivation technique required for HexA quantification. This made it possible to screen accurately for TSD heterozygotes with large numbers of samples from child-bearing-age individuals in the increased-risk population (Ashkenazi Jews). It was possible, therefore, to identify individual carriers, and more specifically, couples at risk for TSD in their offspring (both parents carriers) before the birth of affected children. With the availability of antenatal diagnosis as a "positive" reproductive alternative for such couples, it was theoretically possible to prevent the birth of nearly all infants with TSD (assuming at-risk couples would choose to monitor each pregnancy by prenatal diagnosis and to terminate—abort—those in which the fetus was affected). In the same context, this availed all couples (even those at risk) a mechanism by which they could still have their own offspring, assured that they would be unaffected by TSD.

The initial effort at community education/heterozygote screening/genetic counseling was instituted in the Baltimore, Maryland/Washington, DC, communities in 1971. These efforts have been described in detail elsewhere (Kaback and Zeiger, 1972; Kaback, 1977). Other programs in cities throughout North America followed shortly thereafter. Subsequently, more than 100 cities spread over five continents initiated comparable genetic education, heterozygote screening, and counseling programs directed to the prevention of TSD. This international experience also has been reviewed recently (Kaback *et al.*, 1993). Suffice it to say that a greater than 90% reduction in the annual incidence of TSD has resulted in North America since such programs were initiated.

IV. 1980–PRESENT: THE MOLECULAR ERA AND THERAPEUTIC HORIZONS

Earlier biochemical studies identified the subunit structure of β-hexosaminidase isoenzymes and revealed important clues as to the structure of these catalysts (Beutler *et al.*, 1976; Geiger and Arnon, 1976). Immunologic techniques (combined with isotopic pulse-chase methods) were employed with cultured cells to define the sequential mechanisms of β-hexosaminidase biosynthesis and provided critical evidence for the heterogeneity of genetic defects in several types of GM2 gangliosidosis (Proia *et al.*, 1984; d'Azzo *et al.*, 1984).

Characterization of the subunit structure and elucidation of partial amino acid sequence of these enzymes made immunologic, genetic, and molecular methods applicable, first to the chromosomal localization of the α- and β-subunit genes (Chern *et al.*, 1977; Boedecker *et al.*, 1975), then to the isolation and cloning of the α-subunit cDNA, and subsequently to the definition of the complete genomic α-subunit nucleotide sequence on chromosome 15 (Myerowitz *et al.*, 1985; Proia and Soravia, 1987). An extensive homology (greater than 60%) is evident with the β-subunit gene on chromosome 5, suggesting that both α- and β-subunit genes are derived from a common ancestral gene (Triggs-Raine *et al.*, 1991).

With sophisticated methods of DNA-based mutation detection, more than 75 different mutations have been delineated in the α-subunit gene alone to date (Gravel *et al.*, 1995). These DNA alterations include all classes of mutations and span all 14 exons. Several intron–exon junction alterations also have been identified. Specific mutations associated with later-onset forms of GM2 gangliosidosis, including the B1, juvenile, and adult variants have been defined, usually in compound heterozygosity with one of the more common infantile alleles. Most mutations in the non-Jewish population have been detected in single families (private mutations) or only in a small number of patients where consanguinity or a high level of inbreeding has been evident.

Among Ashkenazi Jews, however, three mutations (two associated with infantile disease and one with adult onset) account for more than 96% of all mutations in this population (Paw *et al.*, 1990). An exon 1 deletion is prominent among non-Jewish French-Canadians (Myerowitz and Hogikyan, 1987), and an intron 9 alteration is frequently encountered among carriers of Celtic (Landels *et al.*, 1992) origin and among Cajun (McDowell *et al.*, 1992) and Pennsylvania Dutch carriers (Mules *et al.*, 1992).

The biochemical and molecular capabilities to identify TSD carriers and to define the mutations involved have provided the tools to characterize a variety of GM2 gangliosidosis variants. Not only have infantile, juvenile, and late-onset types been defined, but even within each classification a variety of mutations and compound mutation states have been described. Such distinctions are critical for genetic counseling and for genotype/phenotype assignments in this diverse group of disorders.

As to definitive treatment of the GM2 gangliosidoses, virtually no approach to date has shown any real evidence of effectiveness. This is so in spite of various previous attempts to intervene in or reverse the disease process. Purified enzyme infusions, cellular transfusions, and bone marrow transplantation—all have been attempted in afflicted children, but with virtually no evidence of even short-term benefit.

This dismal situation may, however, change dramatically in the future. With recombinant DNA techniques and homologous recombination methods, two groups have now produced mouse strains, either totally deficient in Hex A alone (TSD model) or deficient in both Hex A and Hex B (Sandhoff disease model) (Yamanaka *et al.*, 1994; Phaneuf *et al.*, 1996). Such models can provide critical opportunities to study disease pathogenesis as well as to assess various therapeutic modalities in the laboratory animal. Whether it be to study approaches to enzyme replacement in the central nervous system, targeted gene therapy, or techniques for removal of accumulated intralysosomal substrate, the availability of animal models for these diseases should prove invaluable in furthering the evaluation and development of new, safe, and effective treatments for affected human patients.

Obviously, such breakthroughs remain for the future. It is important to emphasize, therefore, that until such achievements are accomplished, education, carrier screening, and genetic counseling (with the option of prenatal diagnosis) are likely to continue as mainstays in the preventive management of these serious disorders.

V. CONCLUSION

In summary, TSD has served not only as a prototype of the neuronal lysosomal lipid storage disorders, but it also has been the model for population-based genetic education/heterozygote screening/genetic counseling efforts in the prospective control (prevention) of genetic disease. This volume recounts the seminal efforts that led to these achievements and the recent molecular genetic discoveries concerning TSD and related disorders. Importantly, it also describes the personal experiences of those who contributed to this process. It is emphasized that each discovery is derived from prior contributions and reflects, in the best sense, how multiple disciplines of biomedical science contribute to a growing foundation of knowledge on which new advances are made. Accordingly, this relatively rare and obscure disorder has served an importance far beyond its frequency, prevalence, or medical-economic impact. It is as a public health prototype, in the application of new genetic knowledge for the prevention and control of serious genetic disease, that the importance of TSD is best measured.

References

Aronson, S. M., and Volk, B. W. (1962). Genetic and demographic considerations concerning Tay-Sachs disease. In "Cerebral Sphingolipidoses" (S. M. Aronson and B. W. Volk, eds.), pp. 375–394. Academic Press, New York.

Aronson, S. M., Valsamis, M. P., and Volk, B. W. (1960). Infantile amaurotic family idiocy: Occurrence, genetic considerations and pathophysiology in the non-Jewish infant. *Pediatrics* **26**, 229–242.

Beutler, E., Yoshida, A., and Kuhl, W., *et al.* (1976). The subunits of hexosaminidase. *Biochem. J.* **159**, 541–549.

Boedecker, H. J., Mellman, W. J., and Tedesco, T. A., *et al.* (1975). Assignment of the human gene for Hex B to chromosome 5. *Exp. Cell Res.* **93**, 468–475.

Brady, R. O. (1966). The sphingolipidoses. *N. Engl. J. Med.* **275**, 312–318.

Brady, R. O., Kanfer, J. N., and Shapiro, D. (1965). Metabolism of glucocerebrosides. II. Evidence of enzymatic deficiency in Gaucher's disease. *Biochem. Biophys. Res. Commun.* **18**, 221–225.

Brady, R. O., Kanfer, J. N., Mock, M. B., and Frederickson, D. S. (1966). Metabolism of sphingomyelin. II. Evidence of enzymatic deficiency in Niemann-Pick disease. *Proc. Natl. Acad. Sci. (USA)* **55**, 366–369.

Chern, C. J., Kenneth, R., and Engel, E., *et al.* (1977). Assignment of the structural genes for the α-subunit of hexosaminidase A, mannose phosphate isomerase and pyruvate kinase to the region of 22q ter of human chromosome 15. *Somatic Cell Genet.* **3**, 533–542.

d'Azzo, A., Proia, R. L., Kolodny, E. H., Kaback, M. M., and Neufeld, E. F. (1984). Faulty association of alpha- and beta-subunits in some forms of beta-hexosaminidase A deficiency. *J. Biol. Chem.* **259**, 11070.

deDuve, C., Pressman, B. C., Gianetto, R., Wattiaux, R., and Appplemans, F. (1955). *Biochem. J.* **60**, 604.

Geiger, B., and Arnon, R. (1976). Chemical characterization and subunit structures of human N-acetylhexosaminidases A and B. *Biochemistry* **15**, 3484–3496.

Gravel, R., Clarke, J., Kaback, M., Mahuran, D., Sandhoff, K., and Suzuki, K. (1995). The G_{M2} gangliosidoses. In "The Metabolic and Molecular Basis of Inherited Disease" (C. Scriver, A. Beaudet, W. Sly, and D. Valle, eds.). McGraw-Hill, New York.

Hers, H. G., and Van Hoof, F. (eds.) (1973)[a]. Lysomes and Storage Diseases, p. 21. Academic Press, New York.

Kaback, M. M. (1973). Thermal fractionation of serum hexosaminidase: Applications to heterozygote detection and diagnosis of Tay-Sachs disease. In "Methods of Enzymology" (V. Ginsburg and E. Neufeld, eds.), Vol. 28, Academic Press, New York, pp. 862–867.

Kaback, M. M. (eds.) (1977). "Tay-Sachs Disease: Screening and Prevention," *Prog. Clin. Biol. Res.* Vol. 18.

Kaback, M. M., and O'Brien, J. S. (1971). The prevention of recessive genetic disease: Feasibility costs, and genetic impact. *Proc. Am. Ped. Soc. & Soc. Ped. Res.*, May, 283.

Kaback, M. M., and Zeiger, R. S. (1972). Heterozygote detection in Tay-Sachs disease: A prototype community screening program for the prevention of recessive genetic disorders. In "Sphingolipids, Sphingolipidoses, and Allied Disorders," Advances in Experimental Medicine and Biology, Vol. 19 (B. W. Volk and S. M. Aronson, eds.), pp. 613–632. Plenum Press, New York.

Kaback, M., Lim-Steele, J., Dabholkar, D., Brown, D., Levy, N., and Zeiger, K. (1993). Tay-Sachs disease: Carrier screening, prenatal diagnosis, and the molecular era. *J. Am. Med. Assoc.* **270**, 2307–2315.

Klenk, E. (1942). Uber die Ganglioside des Gehirns bei der infantilen amaurotischen Idiotie vom Typus Tay-Sachs. *Ber. Dtsch. Chem. Ges.* **75**, 1632–1636.

Kolodny, E. H., Brady, R. O., and Volk, B. W. (1969). Demonstration of an alteration of ganglioside metabolism in Tay-Sachs disease. *Biochem. Biophys. Res. Commun.* **37**, 526–531.

Landels, E. C., Green, P. M., Ellis, I. H., Fensom, A. H., and Bobrow, M. (1992). Beta-hexosaminidase splice site mutation has a high frequency among non-Jewish Tay-Sachs disease carriers from the British Isles. *J. Med. Genet.* **29**, 563.

Ledeen, R., and Salsman, K. (1965). Structure of the Tay-Sachs ganglioside. *I. Biochem.* **4**, 2225–2232.

Lowden, J. A., Skomorowski, M. A., Henderson, F., and Kaback, M. M. (1973). The automated assay of hexosaminidase in serum. *Clin. Chem.* **19**, 1345–1349.

McDowell, G. A., Mules, E. H., and Fabacher, P., *et al.* (1992). The presence of two different infantile Tay-Sachs disease mutations in a Cajun population. *Am. J. Hum. Genet.* **51**, 1071–1077.

Mehl, E., and Jatzkewitz, H. (1965). Evidence for genetic block in metachromatic leucodystrophy (ML). *Biochem. Biophys. Res. Commun.* **19**, 407–411.

Mules, E. H., Hayflick, S., Dowling, C., Kelly, T. E., Akerman, B. R., Gravel, R. A., and Thomas, G. H. (1992). Molecular basis of hexosaminidase A deficiency and pseudodeficiency in the Berks County Pennsylvania Dutch. *Hum. Mutat.* **1**, 298–302.

Myerowitz, R., and Hogikyan, N. D. (1987). A deletion involving Alu sequences in the beta-hexosaminidase alpha-chain gene of French Canadians with Tay-Sachs disease. *J. Biol. Chem.* **262**, 15396.

Myerowitz, R., Piekarz, R., Neufeld, E. F., Shows, T. B., and Suzuki, K. (1985). Human β-hexosaminidase α-chain: Coding sequence and homology with the β-chain. *Proc. Natl. Acad. Sci. (USA)* **82**, 7830–7838.

Myrianthopoulos, N. C. (1962). Some epidemiologic and genetic aspects of Tay-Sachs disease. *In* "Cerebral Sphingolipidoses" (S. M. Aronson and B. W. Volk, eds.), pp. 359–374. Academic Press, New York.

Myrianthopoulos, N. C., and Aronson, S. M. (1967). Reproductive fitness and selection in Tay-Sachs disease. *In* "Inborn Disorders of Sphingolipid Metabolism" (S. M. Aronson and B. W. Volk, eds.), pp. 431–441. Pergamon Press, Oxford, UK.

Navon, R., and Padeh, B. (1971). Prenatal diagnosis of Tay-Sachs genotypes. *Br. Med. J.* **4**, 17–20.

O'Brien, J. S., Okada, S., Chan, A., and Fillerup, D. L. (1970). Tay-Sachs disease: Detection of heterozygotes and homozygotes by serum hexosaminidase assay. *N. Engl. J. Med.* **283**, 15–20.

O'Brien, J. S., Okada, S., Fillerup, D. L., Veat, M. L., Adornato, B., Brenner, P. H., and Leroy, J. G. (1971). Tay-Sachs disease: Prenatal diagnosis. *Science* **172**, 61–64.

Okada, S., and O'Brien, J. S. (1969). Tay-Sachs disease: Generalised absence of a beta-D-N-acetylhexosaminidase component. *Science* **165**, 698–700.

Okada, S., Veat, M. L., Leroy, J., and O'Brien, J. S. (1971). Ganglioside GM2 storage diseases: Hexosaminidase deficiencies in cultured fibroblasts. *Am. J. Hum. Genet.* **23**, 55.

Paw, B. H., Tieu, P. T., Kaback, M. M., Lim, J., and Neufeld, E. (1990). Frequency of three Hex A mutant alleles among Jewish and non-Jewish carriers identified in a Tay-Sachs screening program. *Am. J. Hum. Genet.* **47**, 698–705.

Phaneuf, D., Wakamatsu, N., Huang, J. Q., Borowski, A., Peterson, A. C., Fortunato, S. R., Ritter, G., Igdoura, S. A., Morales, C. R., Benoit, G., Akerman, B. R., Leclerc, D., Hanai, N., Marth, J. D., Trasler, J. M., and Gravel, R. A. (1996). Dramatically different phenotypes in mouse models of human Tay-Sachs and Sandhoff diseases. *Hum. Mol. Genet.* **5**, 1–14.

Proia, R. L., and Soravia, E. (1987). Organization of the gene encoding the human β-hexosaminidase α-chain. *J. Biol. Chem.* **262**, 5677–5681.

Proia, R. L., d'Azzo, A., and Neufeld, E. F. (1984). Association of alpha- and beta-subunits during the biosynthesis of beta-hexosaminidase in cultured human fibroblasts. *J. Biol. Chem.* **259**, 3350.

Robinson, D., and Stirling, J. L. (1968). N-acetyl-beta-glucosaminidase in human spleen. *Biochem. J.* **107**, 321–327.

Sachs, B. (1896). A family form of idiocy, generally fatal, associated with early blindness. *J. Nerv. Ment. Dis.* **14**, 475–479.

Sachs, B. (1897). On arrested cerebral development with special reference to its cortical pathology. *J. Nerv. Ment. Dis.* **14**, 541.

Samuels, S., Korey, S. R., Gonatas, J., Terry, R. D., and Weiss, M. (1962). The membranous granules in Tay-Sachs Disease. *In* "Cerebral Sphingolipidoses" (S. M. Aronson and B. W. Volk, eds.), p. 304. Academic Press, New York.

Sandhoff, K. (1969). Variation of beta N-acetylhexosaminidase pattern in Tay-Sachs disease. *FEBS Lett.* **4**, 351–359.

Sandhoff, K., Andreae, U., and Jatzkewitz, H. (1968). Deficient hexosaminidase activity in an exceptional case of Tay-Sachs disease with additional storage of kidney globoside in visceral organs. *Life Sci.* **7**, 283–288.

Schneck, L., Friedland, J., Valenti, C., Adachi, M., Amsterdam, D., and Volk, B. W. (1970). Prenatal diagnosis of Tay-Sachs disease. *Lancet* **1**, 582–583.

Slome, D. (1933). The genetic basis of amaurotic family idiocy. *J. Genet.* **27**, 363–372.

Svennerholm, L. (1962). The chemical structure of normal human brain and Tay-Sachs gangliosides. *Biochem. Biophys. Res. Commun.* **9**, 436–441.

Tay, W. (1881). Symmetrical changes in the region of the yellow spot in each eye of an infant. *Trans. Opthal. Soc. UK* **1**, 55–57.

Tay, W. (1884). Third instance in same family of symmetrical changes in the region of yellow spot in each eye of an infant closely resembling those of embolism. *Trans. Opthal. Soc. UK* **4**, 158.

Terry, R. D., and Korey, S. R. (1960). Membranous cytoplasmic granules in infantile amaurotic idiocy. *Nature* **188**, 1000–1002.

Terry, R. D., and Weiss, M. (1963). Studies in Tay-Sachs disease. II. Ultrastructure of the cerebrum. *J. Neuropathol. Exp. Neurol.* **22**, 18–55.

Terry, R. D., Korey, S. R., and Weiss, M. (1962). Electron microscopy of the cerebrum in Tay-Sachs disease. *In* "Cerebral Sphingolipidoses" (S. M. Aronson and B. W. Volk, eds.), pp. 49–56. Academic Press, New York.

Triggs-Raine, B. L., Akerman, B. R., Clarke, J. T. R., and Gravel, R. A. (1991). Sequence of DNA flanking the exons of the HEX A gene and identification of mutations in Tay-Sachs disease. *Am. J. Hum. Genet.* **49**, 1041–1054.

Wallace, B. J., Lazarus, S. S., and Volk, B. W. (1967). Electron microscopic and histochemical studies of viscera in lipidoses. *In* "Inborn Disorders of Sphingolipid Metabolism" (S. M. Aronson and B. W. Volk, eds.), pp. 107–120. Pergamon Press, Oxford, UK.

Yamanaka, S., Johnson, M., Grinberg, A., Westphal, H., Crawley, J. W., Taniike, M., Suzuki, K., and Proia, R. L. (1994). Targeted disruption of the Hex A gene results in mice with biochemical and pathologic features of Tay-Sachs disease. *Proc. Nat. Acad. Sci. (USA)* **91**, 9975–9979.

2 Barney Sachs and the History of the Neuropathologic Description of Tay-Sachs Disease

Daniel P. Perl
Neuropathology Division
Mount Sinai Medical Center
New York, New York 10019

The history of the initial delineation of what is now referred to as Tay-Sachs disease is intimately associated with the early career of Bernard Sachs and his association with the Mount Sinai Hospital. Bernard Sachs was born in Baltimore on January 2, 1858, one of identical twins and the youngest of five children. His parents, Joseph and Sophia (Baer) Sachs, were German immigrants and his father, at the time, served as a teacher in a boarding school in that city. By the time the twins were one year of age, the family had moved to New York, where Bernard's father ultimately became headmaster of a preparatory school. Sachs' twin, Harry, died of scarlet fever at age 5. Although he grew up mostly in New York City, Barney, as he preferred to be called, spent two years of his youth living in Germany and thus became fluent in German. Many of the details of his childhood and early adult life are provided in a short but detailed autobiography which Sachs was writing at the time of his death, in 1944 at the age of 86. His family published a small number of copies of this volume posthumously (Sachs, 1949), one of which may be viewed in the history of medicine collection of the New York Academy of Medicine. He was the only one of the five children of Joseph and Sophia to become a physician, although his nephew, Ernest Sachs (son of Bernard's elder brother, Julius) later gained prominence as the first Professor of Neurosurgery at the Washington University School of Medicine in St. Louis.

Sachs entered Harvard College at the age of 16. While at Harvard he volunteered to read to the renowned psychologist and philosopher William James, who was then having problems with his eyesight. Following his interactions with Professor James, Sachs announced his intention to pursue a medical career and devote himself to "diseases of the mind." Based on a thesis entitled "Goethe as a man of science," Sachs graduated with honors in 1878. At graduation, he was also awarded the prestigious Bowdoin prize, which included an award of $50, for a dissertation entitled "A Comparison of the Fore and Hind Limb in Vertebrates." Upon his graduation from Harvard, Sachs enrolled to study medicine at the University of Strassburg. Under the treaty ending the Franco-Prussian War, Alsaice had come under the control of Germany. The University of Strassburg was developed under the German educational system and had only recently opened. The new medical school was considered the highlight of the university and boasted a prestigious faculty. For example, while a student there, Sachs studied anatomy under Waldeyer, pathology with von Recklinghausen, and medicine under Kussmaul. A semester in the fourth year of his medical studies was spent in Berlin, where he was exposed to pathology under Rudolph Virchow and further developed his interest in neurologic and psychiatric diseases under Carl Westphal.

Following receipt of his M.D. degree in June 1882, as was common practice at the time, Sachs elected to remain in Europe to take further training in the neurosciences. This period of training began with a one-year stay at the University of Vienna, where he studied in the laboratory of Theodore Meynert, then considered one of the world's leading experts in human neuroanatomy. At that time, a small number of recent graduates were also studying in Meynert's lab. Among these was the Viennese physician, Sigmund Freud, who had also recently graduated and was also interested in pursuing a career studying diseases of the brain. Sachs shared a bench with the young Dr. Freud and would later note that at that time there was nothing about Freud that would suggest the prominent role he would eventually play in the history of medicine. Based on their association in Meynert's laboratory, Freud and Sachs became close friends and maintained an active, life-long correspondence. Sachs demonstrated an interest, throughout his career, in aspects of psychiatric disease, but repeatedly cautioned that the field of psychiatry needed to rest on an organic basis. Late in life, Sachs would recall, "He (Freud) pegged away at anatomy, as we all did, and although his doctrines took him far afield, in his last letter to me, written only a few months before his death, he acknowledged he had never severed his relations with organic neurology. Evidently his early training left a deep impression on him. It would be well if some of his closest followers had a similar training" (Sachs, 1949).

Following his experience in Vienna, Sachs spent several months at the Salpêtrière in Paris, where he attended the clinical lectures and demonstrations of Jean Martin Charcot, then at the height of his fame. Charcot's abilities as a teacher and the dramatic flourishes he introduced into his weekly clinical presentations greatly impressed the young Sachs. He finished his clinical training at the London Hospital, where he had the opportunity to make daily clinical rounds with the leading British neurologist of that era, J. Hughlings Jackson. He considered Jackson to be a brilliant neurologist but complained that he spoke in such a soft voice that those standing at the foot of the bed on rounds could barely hear what he was saying.

With his clinical training completed, Bernard Sachs returned to New York in May 1884, where he joined the clinical practice of Dr. I. Adler. This practice was general in nature, but soon thereafter Sachs sought to confine his activities virtually exclusively to neurology and psychiatry. In 1886 he was elected to membership in the American Neurologic Association, and from 1886 to 1891 he was appointed editor of the *Journal of Nervous and Mental Disease*, the official publication of that organization. In 1894 he was selected to be President of the American Neurologic Association, and 38 years later he again held this prestigious position.

It was in 1887 that Sachs wrote his paper, "On Arrested Cerebral Development, with Special Reference to Cortical Pathology" (Sachs, 1887). The paper had initially been read before the American Neurologic Association in July of 1887 and was then published in the *Journal of Nervous and Mental Diseases*, of which he also was now the editor. In the manuscript Sachs graciously acknowledged the clinical assistance of his associate, Dr. Alder. Sachs and Adler did not get along well together; their dual practice was short-lived and their parting was apparently rather acrimonious. Nevertheless, Sachs acknowledged that the first case of Tay-Sachs disease that he examined and eventually reported was seen in a family that had been initially cared for by Dr. Adler. Although Sachs was later openly critical about how he had been treated by his senior associate, he remained grateful for what he recognized was perhaps the most important referral of his long and distinguished career.

Sachs also thanked Dr. Herman Knapp for his ophthalmologic evaluation of the patient and Dr. Ira van Gieson for cutting and staining the tissue sections derived from the autopsy which was performed following the child's death at the age of 2 years. The paper involves only a single case (referred to as patient "S"), but describes in detail the relentless clinical progression of the disorder as well as the characteristic ophthalmologic findings. The paper ends with a detailed description of the postmortem examination on the child. Sachs apparently acted as the neuropathologist for the report and noted that despite careful examination it was difficult to find any neurons with a normal histologic appearance. Indeed, he commented that "in my search throughout the brain, I have not come across more than a half dozen, if as many, pyramidal cells of anything like a normal appearance." He observed that most nerve cells were rounded and swollen, with a "detritus-like mass" which virtually obliterated the nucleus of the cell. The illustrations of the

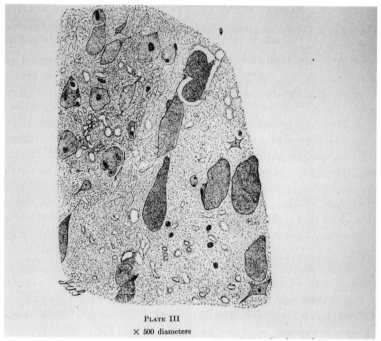

PLATE III
X 500 diameters

Figure 2.1. Camera lucida drawing of the microscopic appearance of the cerebral cortex appearing in the original publication by Sachs in 1887 on arrested cerebral development. The drawing was make by Dr. Ira van Gieson (note IVG signature), who assisted Sachs in preparing the tissue sections and staining the slides.

histologic features in the original paper were drawn by Dr. Ira van Gieson (all initialed "IVG") and clearly show the classic ballooned neurons of the cerebral cortex associated with Tay-Sachs disease (Figure 2.1). The development of the van Gieson method for staining elastic fibers in tissue sections represents a later contribution from this man, and this technique remains in active current use by pathologists.

In his paper of 1887, Sachs credits Dr. Knapp, the ophthalmologist who consulted on the child, for providing the description of the optic fundi. These observations include a description of a prominent cherry-red spot in the fovea. Sachs comments that late in the course of the illness the child became blind but that he could not pursue the nature of these clinical findings since he was not permitted to remove the eyes at autopsy. The optic nerves of the child were examined and were reported to be of normal appearance. Accordingly, he concluded, (from today's perspective with remarkable accuracy) that "blindness must therefore have been due either to the retinal changes or to a deficient cortical condition, or both." It should be remembered that ophthalmoscopes of that era were extremely primitive, employing a concave mirror to collect the light given off by an oil lamp.

It is apparent that at the time he wrote his initial description of the disease, Sachs was unaware of the independent publication in 1881 of ophthalmologic observations of a similar case by Warren Tay, an English ophthalmologist in practice at the London Hospital (Tay, 1881, 1884). It is of interest to note that upon visualizing these unique retinal changes, Dr. Tay sought two consultants to further evaluate the nonophthalmologic aspects of his case. The consultants were Jonathan Hutchinson (a noted clinical observer of his day who had described the characteristic abnormalities of the teeth associated with congenital syphilis) and Hughlings Jackson (Sachs' former mentor in neurology). Apparently neither consultant could offer additional light on the case. Jackson indicated that although the patient displayed weakness of the neck and limbs, he could find no further clinical evidence of cerebral involvement which could account for this sign. Curiously, Tay's patient died 6 months later, and at postmortem examination the only cerebral abnormality noted was a large cavity containing clotted blood which lay adjacent to the lateral ventricle. It is unclear if histologic studies were performed on the specimen. It was not until 1896 that Kingdon and Russell (1896) noted the similarity between the two published descriptions, one American and the other British, and concluded that the cases reported by both men were likely to be examples of the same disorder. Following this, Drs. Tay and Sachs began to be linked in an eponymic fashion when referring to the disease.

In the initial years of his practice in New York, Sachs had hospital appointments at the New York Polyclinical and the Montefiore Home. In 1893 it was announced to the staff of the Mount Sinai Hospital that "Dr. B. Sachs has been appointed consulting neurologist" but that "no separate ward or beds are assigned to this department, the object being mainly to afford consultation with this eminent specialist for the benefit of certain classes of patients in the Hospital." With this appointment the close professional association of Sachs with the Mount Sinai Hospital began, a relationship that was to continue for the next 52 years. In 1900 it was decided that a total of 12 beds (6 male and 6 female) were to be reserved for the use of the newly formed Neurology Division at Mount Sinai. These beds were placed under the supervision of the unit's first director, Dr. Sachs. In this way, the first distinct and separate neurologic service, consisting of specifically allocated beds, was established in a hospital in New York City.

Sachs continued to direct the neurology service at Mount Sinai until 1924, when, at age 66, he retired from his administrative duties. Despite his retirement as chief of the Neurology Service, Barney Sachs remained on the consulting staff of the Mount Sinai Hospital until his eighties, and continued to conduct his active consultant practice and contribute to the neurologic literature. Sachs was widely noted as a leader of American neurology. In 1885, soon after he returned to America, he completed an English translation of Meynert's textbook,

Psychiatry, A Clinical Treatise on Diseases of the Fore-Brain (Sachs, 1885). This volume served to introduce the contributions of this pioneering Viennese neuroanatomist and neuroscientist to the English-speaking scientific community. The volume in translation also served to quickly establish the scholarship of the young Sachs among his American neurology colleagues. In the introduction to his translation, Sachs acknowledged that this contribution was, at least in part, produced out of gratitude to his former mentor, Professor Meynert (Figure 2.2).

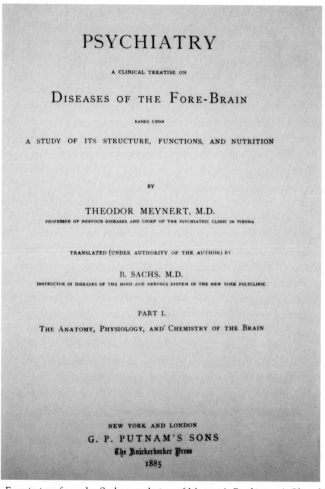

Figure 2.2. Frontispiece from the Sachs translation of Meynert's *Psychiatry, A Clinical Treatise on Diseases of the Fore-Brain.*

In 1895, on the advice of Dr. Charles Loomis Dana, his friend and professional counterpart at the Cornell University Medical College, Sachs wrote his volume entitled *A Treatise on the Nervous Diseases of Children* (Sachs, 1895). This is acknowledged as the first American textbook in the field of pediatric neurology. The original volume was over 600 pages long, went through three editions, was translated into German and Italian, and over the next 40 years was considered the definitive text in the discipline. Sachs twice served as the president of the New York Academy of Medicine and as its treasurer during the 1930s when that institution was undergoing serious financial difficulties related to the Great Depression. Among the many of Sachs' legacies to the New York medical community was his critical leadership role, which was considered largely responsible for saving this important institution from financial collapse. He also became the president of the New York Neurological Society and was the organizer of the First International Congress of Neurology (in 1931). Many of the most prominent citizens of New York were among his patients, and he was widely sought out as a consultant on difficult neurologic problems.

During his long career, Sachs continued to be interested in the disease that he had first described in the early years of his clinical practice. Further observations on additional cases emphasized the importance of the neuronal nature of the disease and its predilection for cerebral cortical involvement (Sachs, 1892, 1896, 1903). By 1896, he began to stress the familial nature of the disorder, having now identified the same condition in the sister of his initial case. By then he had collected a remarkable total of 19 cases, whom he noted were seen "amongst Hebrews and amongst them only." Sachs also commented on the high incidence of consanguinity among the parents of affected children. By 1898, with additional clinical observations, he defined the condition as "a heredo-degenerative form of disease occurring in infancy and characterized by a triad of manifestations: an arrest of all mental processes; a progressive weakness of all muscles of the body, terminating in generalized paralysis; and by rapidly developing blindness, associated with changes in the macula lutea, the cherry red spot and optic atrophy."

In 1905, Karl Schaefer, a Hungarian neuroanatomist and neuropathologist, recognized that not only were neurons affected in Tay-Sachs disease, but glial cells were also swollen with storage material. Schaefer maintained that the disease selectively involved those cells which were considered to be "neuroectodermal derivatives" and thus could be regarded as a "selective, germ cell disease." He too emphasized the marked swelling of the cytoplasm of affected cells, both neurons and glial cells. Using histochemical staining methods, Schaefer recognized that the cytoplasmic swelling within neurons and glia was in the form of what he referred to as "pre-lipoid granules," which he indicated reflected the abnormal accumulation of a lipid material, which he referred to as "lecithinoid."

In the early part of the 1900s, Sachs began collaborating with Dr. Isadore Strauss, the director of pathology at Mount Sinai. Together, they added further descriptions of patients with Tay-Sachs disease, including detailed neuropathologic observations (Sachs and Strauss, 1910). Sachs apparently hospitalized most of his patients with Tay-Sachs disease at the Mount Sinai Hospital, and its pathologists became familiar with the disorder through their detailed postmortem examinations (Figure 2.3).

It was not until the mid-1920s that a laboratory specifically devoted to neuropathology was established at the Mount Sinai Hospital, under the direction of Dr. Joseph Globus. Dr. Globus graduated from Columbia College in 1915 and received his M.D. degree from Cornell University in 1917. With the support of a Blumenthal Fellowship, funded by the Mount Sinai Hospital, Globus traveled to Hamburg, Germany, to receive 2 years of training in neuropathology under Alfons Jakob. In 1921, during the time that Globus was receiving his training, Jakob wrote the definitive description of three cases of the rapidly dementing disorder that would eventually be referred to as Creutzfeldt-Jakob disease.

In 1922 Globus returned to New York and shortly thereafter was asked to organize the neuropathology laboratory at Mount Sinai, the first such facility in New York City. Although the author cannot find any documentation of his direct involvement in initiating this advance, the development of capabilities in neuropathology at the Mount Sinai Hospital was clearly a priority for Sachs, the director of the Neurology Service. Although Globus and Sachs never published together (Sachs was 27 years older than Globus), they were professional colleagues at Mount Sinai Hospital for 22 years and clearly had discussed the nature of the entity that Sachs had originally described. Globus contributed detailed neuropathologic descriptions of a number of additional cases of Tay-Sachs disease and was the first to note the involvement of autonomic neurons in the mucosa of the bowel (Globus, 1942). Until the availability of direct enzymatic assays, this observation was routinely employed in establishing the clinical diagnosis of Tay-Sachs disease and related disorders through the use of morphologic studies of rectal biopsies.

In subsequent years relatively little additional information was added to the morphologic description of Tay-Sachs disease until the introduction of electron microscopy as a morphologic tool. In 1960, Terry and Korey first described the cytoplasmic membranous bodies in neuronal cytoplasm in a case of Tay-Sachs disease. This initial report was performed on a single postmortem brain specimen which had been fixed in formalin. Accordingly, the tissue was poorly preserved. Nevertheless, numerous "membranous granular bodies," measuring 0.5–2.0 μm in diameter, were identified within the neuronal cytoplasm. Similar structures were also reported in a specimen from a second case consisting of unfixed brain tissue which had been frozen immediately at autopsy and from a third case in which gradient centrifugation fractions from cerebral cortical specimens were available.

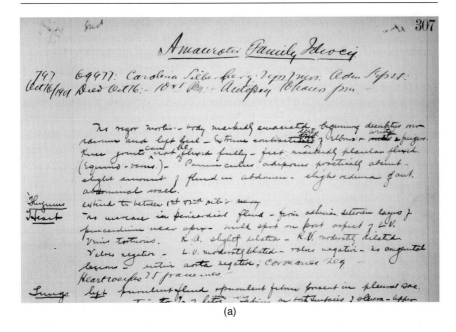

(a)

(b)

Figure 2.3. A portion of a handwritten Mount Sinai Hospital autopsy report on a case of amaurotic familial idiocy (Tay-Sachs disease). The child died on October 16, 1901, at age 2 years, 7 months, and was undoubtedly one of Sachs' patients. Of note is that the autopsy was performed by Emanuel Libman, then a newly appointed young assistant pathologist [(b) shows his signature at the end of the autopsy report]. Libman would also go on to an illustrious career in medicine, making notable contributions to our understanding of bacterial endocarditis. Libman's name endures through the term "Libman-Sachs endocarditis," a nonverucous form of endocarditis classically associated with systemic lupus erythematosis. However, the Sacks associated with the term was Benjamin Sacks (not Bernard Sachs), a cardiologist on the staff at Mount Sinai.

In 1963, Robert Terry and Martin Weiss (1963) published their classic ultrastructural studies on cerebral cortical biopsies from five cases of Tay-Sachs disease. In this study the biopsies were obtained neurosurgically and had been appropriately fixed for ultrastructural examination. The intracytoplasmic inclusions were now referred to by the authors as "cytoplasmic membranous bodies." The report is the second of multiple publications on Tay-Sachs disease all appearing in the same issue (volume 22, number 1) of the *Journal of Neuropathology and Experimental Neurology*. These seminal articles originated from the Departments of Neurology and Pathology of the newly formed Albert Einstein College of Medicine. The study involved brain biopsies of children with Tay-Sachs disease and combined detailed biochemical analyses performed under the direction of Dr. Saul Korey, with ultrastructural investigations led by Dr. Terry. This collaboration represented one of the first and best examples of collaborative investigations linking ultrastructural examination with the emerging discipline of analytic neurochemistry. Studies of this kind served as the beginning of our current understanding of the pathogenesis of inborn errors of metabolism affecting the central nervous system.

In their initial publication, Terry and Weiss indicate that the findings reported were synthesized from more than 2000 electron micrographs taken from the five available biopsy samples. These electron micrographs showed the presence of numerous cytoplasmic membranous bodies which remain the characteristic ultrastructural feature of the disease (Figure 2.4). The group recognized the lipid nature of the material accumulating within neurons, microglial cells, and pericytes and, based on their initial biochemical studies, suggested the ganglioside nature of the deposits. Dr. Terry, a man whose interests extended beyond the realm of science, was particularly pleased when one of the electron micrographs from this study was chosen for inclusion in an exhibit on art in science at the Museum of Modern Art in New York.

In large measure, I was introduced to the field of neuropathology by Dr. Stanley Aronson, who at the time was contributing to the epidemiology literature of Tay-Sachs disease (see Chapter 3). Early in his career, Dr. Aronson had spent time as a research fellow at Mount Sinai, working in the laboratory of Dr. Gregory Schwatzman. While at Mount Sinai, Aronson met Dr. Globus shortly before he died and described him as rather brusque, demanding man with little time for those whom he considered had only a superficial understanding of a clinical problem. In the beginning of my medical career, I could not imagine that some 20 years later, I would take over the chair in neuropathology that Globus had occupied in the hospital where Sachs made the observations on the disease that would bring him worldwide fame.

In the mid-1960s, as a medical student in my rotation in pediatrics, a visit was arranged to what was then called the Jewish Chronic Disease Hospital in Brooklyn (now the Kingsbridge Jewish Medical Center). This institution

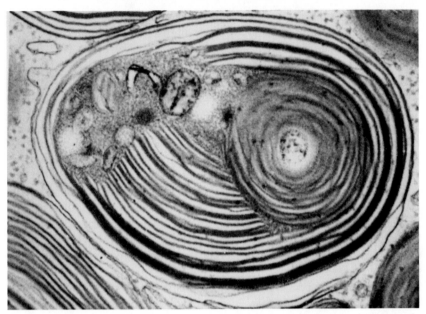

Figure 2.4. Electron photomicrograph of a cytoplasmic membranous body from a cerebral biopsy of a child with Tay-Sachs disease. This micrograph is from the classic ultrastructural study of Terry and Weiss (1963) and was exhibited at the Museum of Modern Art in New York.

was noted as a center for dealing with children suffering from inborn errors of metabolism affecting the nervous system, including cases of Tay-Sachs disease. During the day of our visit we were ushered into a room containing about 10 paralyzed marasmic infants, all in the terminal stages of Tay-Sachs disease or related conditions. We were cautioned to keep our voices down, as these children were sensitive to noise and easily startled. Each child was in a large metal crib, and we were asked to gently lower the sides so that we could examine the fundi for the characteristic cherry-red spot. One in our group fumbled with the heavy crib side, and it fell with a very loud metallic crash. In an instant, every child in the room showed the extreme hyperacusis which is characteristic of the disease, and soon almost all were displaying generalized seizures. It was an awful scene from the pages of Kafka or Dante, and one that I will never forget. I remember feeling utterly helpless; virtually nothing could be done to relieve the suffering of these children and their families.

Yet looking back on that day it is remarkable that in the time it took for me to obtain my M.D. degree and receive training in neuropathology, the ability to identify heterozygote Tay-Sachs disease carriers and achieve prenatal diagnosis through amniocentesis had been achieved. Using these advances, the network of Tay-Sachs screening programs grew and within a decade virtually eliminated the disease from its traditional Jewish setting. Based on this remarkable contribution, the only case of the disease that I have personally had the opportunity to examine as a practicing neuropathologist was the child of a non-Jewish couple living in northern Vermont. The infant had been diagnosed by Dr. Edwin Kolodny, then at the Fernald School, and at autopsy showed all of the neuropathologic features described by Sachs and his associates at Mount Sinai.

Archival tissue blocks from autopsies performed on cases personally examined by Sachs are no longer available at the Mount Sinai Hospital. Unfortunately, the lack of availability of space for long-term storage of such materials had forced their being discarded prior to the author's arrival. However, cases of Tay-Sachs disease that Dr. Globus had evaluated are still within the slide files of the Neuropathology Division. Most of these slides are in an excellent state of preservation and demonstrate many of the features which Globus illustrated in his publications on the disease. Although he must have been advanced in years, it is certainly possible that at least some of these autopsied cases had actually been evaluated clinically by Sachs, serving as a neurologic consultant in the later portion of his career.

Much has happened in recent years to more fully define the precise molecular basis of the disease originally described by Barney Sachs in 1887. However, from the perspective of our current understanding, looking back on his original morphologic descriptions and considering the nature of the tools available to this young, inquisitive man, only 5 years out of medical school and with little experience in carrying out investigations of this type, it is remarkable how precise and accurate Sachs had been. In his description of the single initial case in 1887, Sachs specifically noted the neuronal nature of the disorder, observed the accumulation of abnormal material within the nerve cells, and clearly pointed out that this storage represented the ultimate cellular expression of the disease. Although many of the details have now been added along the way, this still remains the essence of the pathology of the disorder.

References

Globus, J. (1942). Amaurotic family idiocy, introduction and a brief historical review. *J. Mount Sinai Hosp.* **9**, 451–503.

Kingdon, E. C., and Russell, J. S. R. (1896). Infantile cerebral degeneration with symmetrical changes in the macula. *Proc. Roy. Med. Chir. Soc.* **9**, 34.

Sachs, B. (1885). Translation of "Psychiatry, A Clinical Treatise on Diseases of the Fore-Brain" (by T. Meynert). Putman, New York.

Sachs, B. (1887). On arrested cerebral development with special reference to its cortical pathology. *J. Nerv. Mental Dis.* **14**, 541–553.

Sachs, B. (1892). A further contribution to the pathology of arrested cerebral development. *J. Nerv. Mental Dis.* **19**, 603–663.

Sachs, B. (1895). "A Treatise on the Nervous Diseases of Children." William Wood, New York.

Sachs, B. (1896). A family form of idiocy, generally fatal, associated with early blindness (amaurotic family idiocy). *J. Nerv. Mental Dis.* **21,** 475–479.

Sachs, B. (1898). Die amaurotische familiäre idiote. *Deutsche Mediz. Wochen.* **24,** 33–35.

Sachs, B. (1903). On amaurotic familial idiocy. A disease chiefly of the gray matter of the central nervous system. *J. Nerv. Mental Dis.* **30,** 1–13.

Sachs, B. (1949). Barney Sachs 1858–1944. Privately printed, New York.

Sachs, B., and Strauss, I. (1910). The cell changes in amaurotic family idiocy. *J. Exp. Med.* **12,** 685–695.

Schaefer, K. (1905). Zur pathogenesis der Tay-Sachs'schen Idiotie. *Neurol. Cbl.* **24,** 386–393.

Tay, W. (1881). Symmetrical changes in the region of the yellow spot in each eye of an infant. *Trans. Ophthal. Soc. UK* **1,** 55–57.

Tay, W. (1884). Third instance in the same family of symmetrical changes in the region of the yellow spot in each eye of an infant closely resembling those of embolism. *Trans. Ophthal. Soc. UK* **4,** 125–126.

Terry, R. D., and Korey, S. R. (1960). Membranous cytoplasmic granules in infantile amaurotic idiocy. *Nature* **188,** 1000–1002.

Terry, R. D., and Weiss, M. (1963). Studies in Tay-Sachs disease, II. Ultrastructure of the cerebrum. *J. Neuropathol Exp. Neurol.* **22,** 18–55.

3

Early Epidemiologic Studies of Tay-Sachs Disease

Stanley M. Aronson[†]
Division of Biology and Medicine
Brown University
Providence, Rhode Island 02912

Tay-Sachs disease (TSD), four decades ago, was known only to a modest number of physicians as an arcane storage disorder of infancy, a morbid footnote in the lengthier pediatric texts; to the general public, it was essentially unknown. The number of published cases were fewer than 300, and most of these were without competent clinical description or autopsy verification. Only two retrospective reviews of the disease stressing the genetic and epidemiologic features of the disorder had been undertaken: Slome, summarizing 135 cases (Slome, 1933); and Ktenides, reviewing an additional 99 cases (Ktenides, 1954).

In 1952, well before the emergence of DRGs, length-of-hospital-stay committees, and stringent admissions criteria, the Jewish Chronic Disease Hospital of Brooklyn dedicated much of its pediatric division to the extended and nonreimbursible care of infants and young children with "incurable brain diseases." There were no religious stipulations for admission to this unit, but because of the neighborhood where this facility was located, many of those hospitalized children were Jewish. Inevitably, then, a few infants with arrested cerebral development, blindness, hyperacusis, and that bizarre retinal change called the cherry-red spot, were brought to the unit for diagnostic evaluation and inpatient

[†]Deceased

Advances in Genetics, Vol. 44

care. Within a year, seven babies with Tay-Sachs disease had been admitted, to spend there the remainder of their brief lives.

To our knowledge, at that time this was the sole academically structured inpatient facility exclusively assigned for the study and continuing care of children with fatal metabolic or inherited diseases of the central nervous system. For this compassionate sheltering of dying youngsters alone, the clinical unit would have been remembered only by their anguished parents; but there emerged an additional factor in the form of a gifted, well-organized scientist and laboratory director. He was convinced that his hospital could not only provide excellent care for these helpless patients, but could also undertake clinical investigations which might yield medical insight into these relentlessly fatal diseases. The scientist was Bruno W. Volk, M.D., who became the intellectual and managerial force behind three decades of clinical, investigative, and educational activity in behalf of the Tay-Sachs children and their families.

Much of the early investigative efforts were necessarily primitive, diffuse, and at best marginal to the fundamental metabolic defect of TSD. A host of essentially noninvasive, serial measurements were recorded in order to achieve some notion of the evolution and trajectory of the disease. These studies were supervised by two fine pediatricians (Drs. Abram Kanof and Nathan Epstein), who invested much of their volunteer energies in this clinical service. These early studies included the weekly gathering of anthropometric data, as well as information concerning renal function, osseous growth, temperature homeostasis, and concentrations in serum and spinal fluid of enzyme and protein; the chemical studies were supervised by Dr. A. Saifer. (Kanof et al., 1959; Aronson et al., 1958; Saifer et al., 1959)

By 1954, 18 children with TSD (and many more with other neurologic disorders) had been hospitalized on this unit. Through accumulated experience, much was learned about improved ways of rendering nursing care, particularly in fulfilling the nutritional needs of unconscious children. Parents were frequently enlisted to assist in the daily care of their children. A special nursing education division was created by Nancy Valsamis, R.N., expressly to train nurses in the management of neurologically impaired infants and children.

At about this time, we were approached by a small cluster of families from North Philadelphia, all of whom shared in the tragedy of having, or having had, an infant with Tay-Sachs disease. We held lengthy meetings with them, from which was born the nucleus of the National Tay Sachs Association. From the beginning, this lay organization was dedicated resolutely to the impossible goal of eradicating the disease.

These committed parents fully appreciated the long path which stretched before them; they recognized that funds would have to be gathered and distributed, lobbying efforts undertaken, scientists persuaded to invest their creative energies into the complex problems of inborn disorders of sphingolipid metabolism, clinics supported, and educational campaigns mounted.

For a while the inpatient unit at the Jewish Chronic Disease Hospital remained the clinical concentration of this parental effort. It was in this sobering setting that the parents of TSD children met and consoled one another, and in the process found renewed strength with which to address the intrusive emotions which families experience when confronting a recessively transmitted fatal disease. Somehow never translated to the pages of medical texts are problems such as the anxiety about future pregnancies, parental depression, parental divorce, profound guilt experienced by the unaffected siblings, and ostracism by friends and relatives. Slowly, and initially without formal design, these parent discussion groups began to address the tragic psychosocial accompaniments of TSD. Aided by clergy and social workers, a continuing support structure evolved which offered social, psychiatric, and genetic counseling to family members, many of whom had previously believed their private burden to be unique. By means of these group sessions, their feeling of helplessness and isolation often diminished.

The expanding collection of cases of Tay-Sachs and allied diseases demanded a more extensive record-keeping system which could then be explored for genetic and epidemiologic insights. An archive was begun and a lengthy questionnaire was devised which sought information pertaining to three generations of maternal and paternal forebears of each proband, their religions, places of birth, occupations, fertilities, illnesses (including other cases of the sphingolipidoses), general causes of death, and ages at death. After appropriate permission, medical records, including autopsies, from other sources and institutions were incorporated into each TSD family record whenever possible.

The Sphingolipidosis Registry was thus formally established in 1956. It ultimately gathered data on 362 children afflicted with TSD and 44 with other allied disorders. An impressive amount of information accumulated, showing, incidentally, the remote and hitherto unknown relatedness of many of the TSD families, particularly those who were non-Jewish. Yet another observation was the relative ubiquity of the TSD gene as cases of TSD were found among blacks, Orientals, and other children of Asiatic origin. No cases have yet been ascertained in the Eskimo, Gypsy, or Mongolian populations.

The Registry contents were made available to numerous professional groups in New England, Quebec, Bethesda, Chicago, and Baltimore who were independently exploring various biologic, clinical, and genetic facets of TSD. The Registry also became the mechanism by which frozen autopsy-derived tissues and biological fluids, conserved by Dr. Volk and his associates, were shared with other investigators. Numerous laboratories in the United States and Europe were able either to start or accelerate their research activities on TSD by virtue of this tissue and fluid bank.

Simultaneously, Dr. N. C. Myrianthopoulos of the National Institutes of Health had been surveying the U.S. death certificates for cases of TSD. Our two bodies of information were joined and a fruitful collaboration was begun in order to compute the incidence rates and prevalence of the disease in population subsets defined by such criteria as religion, ethnic background, and geographic place of origin of the forebears. With the presumption of reasonably complete ascertainment of cases during the study interval, some TSD gene frequencies were then cautiously assigned to various ethnic and religious groups (Myrianthopoulos and Aronson, 1966).

The Registry data confirmed that the incidence rate of a number of the sphingolipidoses, notably Tay-Sachs, Niemann-Pick, and Gaucher diseases, was significantly elevated in the descendants of Ashkenazic Jews (in general, 100-fold higher than in non-Jewish Europeans or non-Ashkenazic Jews). In an effort to determine whether these storage disorders were distributed homogeneously among the entire Ashkenazic community, a closer study of the geographic origins of these families was undertaken. These labors were carried out in concert with V. McKusick and P. Brunt, who had been conducting a similar nationwide epidemiologic inquiry of familial dysautonomia or Riley-Day syndrome (RDS), a recessive autosomal disorder of the nervous system found almost exclusively in infants of Ashkenazic Jewish heritage.

In contrast to the clustering of sphingolipidosis forebears in the Baltic region and adjacent northeast provinces, the ancestry of children with RDS was more commonly traced to concentrations within the southern provinces of Eastern Europe, particularly Hungary, Rumania, Moldavia, Bucovina, Krakow, and Lwow. In general, these nationwide demographic studies portrayed significant parallels between the skewed distributions of the two ancestral groups (i.e., TSD and RDS) and the linguistic and cultural discontinuities which characterize Ashkenazic Jewry. It became evident that the historic events underlying the linguistic and cultural differentiation of Central and Eastern European Jewry (including such factors as the chronology of settlements and the degrees of contact or isolation among geographically discrete Jewish communities) helped to explain the existing heterogeneous frequencies of some of these inherited neurologic diseases (Aronson et al., 1967).

A further examination of the accumulated data within the Registry showed that the TSD ancestry exhibited an enhanced reproductive performance when compared with a carefully assembled control group; this advantage arose as a result of differential rates of survival, greatest for those who survived to reproductive age among the non-U.S. born, and diminishing or disappearing for those born in the United States.

Since lethal traits tend to disappear in time, some selective heterozygote advantage (perhaps of the order of about 1.25%) is needed to maintain the TSD gene at a stabile frequency despite absolute homozygote elimination. The Registry data derived only from the non-U.S.-born TSD ancestry exceeded this level, suggesting that the increase in gene frequency which had arisen in the historic past is now diminishing, at least in U.S. Jews.

It further seemed that the selective force(s) which allowed a strikingly augmented TSD gene frequency to develop in but one of many Jewish populations likely began to operate many centuries ago. To explain these differential frequencies we started with the assumption that the TSD gene frequency of the Jewish population prior to the Diaspora was homogeneous and about the same as that currently encountered in Sephardic Jews and in the West European non-Jewish population. In then seeking for an historic event which fundamentally altered the lives of some Jewish people but not others, we could not help but focus on the Roman conquest of Jerusalem in the first century. Although there had been Jewish dispersions before this Roman conquest, it was only after the Titus conquest that a significant European exodus of Jews had begun. We argued that the Jews who ultimately migrated to Eastern Europe developed a way of life which, in some manner, favored the rise of the TSD gene, while those who remained in the Middle East or Mediterranean region were subjected to a way of life which did not alter the baseline frequency of the TSD gene.

Our data led us to believe that heterozygote advantage expressed as a function of relative reproductive fitness rather than a founder effect was the genetic mechanism responsible for this selectively elevated TSD gene concentration. And while we could not initially identify the environmental agent which accounted for the reproductive advantage, we felt that it was probably adaptive or regulatory in character and likely to be subtle in its dynamics and manifestations. If it were a biological factor, it might involve selective resistance to a disease, and a disease which was probably endemic in distribution and chronic in character. We further hoped that this agent would become evident through a more rigorous analysis of the ecologic and epidemiologic forces uniquely influencing these dispersed Jewish communities. The evidence demonstrating a gradient of selective advantage further suggested that whatever environmental mechanism(s) operated to provide advantage for the TSD carriers in the northeastern European communities was no longer operative in this country.

When our data on the illnesses and causes of death of the TSD grandparents were more carefully inspected, a virtual absence of deaths caused by tuberculosis was seen, particularly when compared to the illnesses and causes of death in a control population of Jews without history of TSD in their families. Through the cooperation of the American Medical Center in Denver (founded in 1904 as the Jewish Consumptive's Relief Agency) we then reviewed the places of birth of 1466 foreign-born Jewish consumptives in a manner and time frame similar to those which we employed in our demographic studies of the TSD and control grandparents. It became readily evident that the TSD and tuberculosis data were derived from different populations. Accordingly, we suggested that these findings were consistent with the hypothesis that relative resistance to tuberculosis was the biologic mechanism by which the TSD gene imparted a selective advantage to its heterozygote carriers (Myrianthopoulos and Aronson, 1972).

The collaborative efforts between the TSD parents, the hospital, and the committed scientists led to yet another benefit. In April 1958 the first international symposium on Tay-Sachs disease was conducted on the grounds of the Jewish Chronic Disease Hospital. It lasted one day, attracted speakers and participants from four states and Germany (Professor E. Klenk), and was later published as a supplement to the *Journal of Diseases of Children*.

A second symposium followed in March 1961, two days in length and providing a podium for many more speakers from Sweden, the Netherlands, England, and the United States. During this symposium the nosology of sphingolipid diseases was placed on a more rational basis, and heretofore unknown sphingolipidoses were described. The papers presented were published as an independent text.

The third symposium took place in October 1965, also at the hospital, and included now virtually every scientist who had contributed to an understanding of these complex inborn disorders. There were participants from 11 foreign nations and many states. The 34 formal scientific presentations revealed again the accelerated growth of imaginative investigation pertaining to the hereditary diseases of sphingolipid metabolism.

The fourth and fifth international symposia on the sphingolipidoses were conducted in 1971 and 1975, again at the hospital, now called the Kingsbrook Jewish Medical Center. The proceedings of each of these meetings were published as texts.

The pediatric neurology service which, since 1951, has offered compassionate shelter for over 300 infants with TSD, is still in operation under the able direction of Dr. Larry Schneck (see current studies elsewhere in this volume.) The parallel laboratory facilities which studied these many children are still actively conducting creative research under the supervision of Dr. Masazumi Adachi.

References

Aronson, S., Lewitan, A., Rabiner, A., Epstein, N., and Volk, B. (1958). The megalencephalic phase of infantile amaurotic family idiocy. *Arch. Neurol. Psychiat.* **79**, 151–163.

Aronson, S., Herzog, M., Brunt, P., McKusick, V., and Myrianthopoulos, N. (1967). Inherited neurologic diseases of Ashkenazic Jewry: Demographic data suggesting non-random gene frequencies. *Trans. Am. Neurol. Assoc.* **92**, 117–121.

Kanof, A., Aronson, S., and Volk, B. (1959). Clinical progression of amaurotic family idiocy. *J. Dis. Child.* **97**, 656–662.

Ktenides, M. (1954). Au subjet de l'hérédité de l'idiotie amaurotique infantile (Tay Sachs). These No. 2264, L'Universitie de Genève.

Myrianthopoulos, N., and Aronson, S. (1966). Population dynamics of Tay Sachs disease. I. Reproductive fitness and selection. *Am. J. Hum. Genet.* **18**, 313–327.

Myrianthopoulos, N., and Aronson, S. (1972). Population dynamics of Tay Sachs disease. II. What confers the selective advantage upon the Jewish heterozygote? *In* "Sphingolipids, Sphingolipidoses & Allied Disorders" (B. Volk and S. Aronson, eds.), pp. 561–570. Plenum Press, New York.

Saifer, A., Volk, B., and Aronson, S. (1959). Neuraminic acid studies of biologic fluids in amaurotic family idiocy and related disorders. *J. Dis. Child.* **97,** 745–757.

Slome, D. (1933). The genetic basis of amaurotic family idiocy. *J. Genet.* **27,** 363–376.

4

Identification of the Accumulated Ganglioside

Lars Svennerholm[†]
Department of Psychiatry and Neurochemistry
Institute of Clinical Neuroscience
Göteborg University
Göteborg, Sweden

I. Substance X and Ganglioside
II. N-Acetylgalactosamine is a Ganglioside Component
III. Strandin
IV. Chromatographic Separation of Gangliosides
V. Thin-Layer Chromatography—The Method
of Choice for Studies of Ganglioside Structure
VI. Tay-Sachs Ganglioside
References

I. SUBSTANCE X AND GANGLIOSIDE

Landsteiner and Levine (1926) and Walz (1927) had found a new type of glyco-lipid in brain, kidney, and spleen that was water soluble and reacted with a pur-ple color when treated with orcinol-hydrochloric acid (Bial's reagent). During the 1930s, this new glycolipid was further studied in two European laboratories, that of Dr. Ernst Klenk in Cologne and that of Dr. Gunnar Blix in Uppsala. Klenk focused his work on the new glycolipid in the brains of children who died of Niemann-Pick (Klenk, 1935) and Tay-Sachs diseases (Klenk, 1939), while Blix examined the

[†]Deceased

Advances in Genetics, Vol. 44
Copyright © 2001 by Academic Press
All rights of reproduction in any form reserved.
0065-2660/01 $35.00

lipid in bovine brain (Blix, 1938). Klenk identified fatty acid, sphingosine, galactose, and the orcinol-positive substance and termed the new glycolipid Substance
X, while Blix identified acetylated hexosamine. Both Klenk and Blix used Bial's
reagent to monitor the isolation of the new glycolipid, and it is interesting to note
that as early as 1938, Blix suggested that the orcinol-reactive sugar was a member of a group of physiologically important compounds. In 1941 Klenk isolated
the orcinol-positive sugar from Substance X as a methoxy derivative he named
neuraminic acid, and the same year Klenk and Langerbeins (1941) published a
method for the determination of lipid-bound neuraminic acid.

Meanwhile, Klenk (1942) elaborated a procedure for purification of Substance X in which he indentified five components—fatty acid (mainly stearic
acid), sphingosine, galactose, glucose, and neuraminic acid. Because Klenk and
Langerbeins (1941) were unable to detect any ganglioside in cerebral white matter, and Feyrter (1939) had demonstrated that the cytoplasm of Purkinje cells and
other neurons in Tay-Sachs brain were loaded with a metachromatic lipid, Klenk
regarded Substance X as mainly or exclusively localized inside the ganglion cells.
With reference to its glycosidic nature, Klenk suggested the name *ganglioside* for
Substance X.

II. N-ACETYLGALACTOSAMINE IS A GANGLIOSIDE COMPONENT

When I began my research training in 1948, my teacher, Dr. Gunnar Brante,
suggested that I study whether ganglioside was localized only in the cytoplasm
of the neuronal cell body or also occurred in the nerve processes, axons, and
dendrites. It was quite clear that Klenk and Langerbein's (1941) method was
unsuitable for material that contained abundant myelin lipids. Therefore we chose
to study the presumptive white matter in the brains of newborn humans. This
unmyelinated tissue, which consisted of axons and glial cells, proved to have
an equally high or slightly higher ganglioside concentration than the adjacent
cerebral cortex. This finding suggested that ganglioside was a component of the
whole neuronal membrane and not limited to the neuronal cell body.

Klenk and Langerbein's (1941) ganglioside method, based on the determination of neuraminic acid, had several drawbacks: a complicated fractional
extraction method, dissolution in water of a crude ganglioside extract containing large amounts of water-insoluble lipids, and heating to $+142°C$ in an oil
bath. We therefore realized that it would be an advantage if we could use a different component than neuraminic acid for the determination of ganglioside, or
could supplement the neuraminic acid determination with the assay of another
component. Blix (1938) had previously indicated that it was probable that ganglioside contained hexosamine although Klenk (1942) had failed to demonstrate
this in his study of ganglioside composition. Elson and Morgan (1933) had also
elaborated a relatively simple and accurate hexosamine method that would not

require as extensive purification of a lipid extract as the neuraminic acid assay. My teacher, Gunnar Brante, demonstrated at the same time (1948), using a newly developed paper chromatographic method for sugars, that brain lipid extracts contained galactosamine. All these factors spoke in favor of a hexosamine method for the determination of ganglioside, assuming that Klenk (1942) was correct in claiming that brain contained only one ganglioside with homogenous carbohydrate composition. But before this approach could be accepted, it was necessary to isolate pure brain ganglioside, to demonstrate that N-acetylgalactosamine was one of its components, and to show that brain ganglioside contained a constant concentration of hexosamine. My first task was to isolate a large batch of ganglioside from bovine brain and examine it for hexosamine. The amino sugar was crystallized as the hydrochloride, and identified by its X-ray diffraction as galactosamine-HCl (Blix *et al.*, 1950, 1952). Blix (1938) had previously shown that submaxillary mucin and ganglioside showed the same reaction, using colorimetric methods. Galactosamine-HCl was also isolated from the mucin, and with mild acidic hydrolysis the natural orcinol-positive acidic sugar could be isolated. To differentiate it from the degraded methoxy derivative denoted neuraminic acid by Klenk (1941), we suggested the name sialic acid (Blix *et al.*, 1952) for the native compound.

III. STRANDIN

After the identification of galactosamine in ganglioside, I continued to isolate ganglioside from human brains, including a Tay-Sachs brain. The ratio between neuraminic acid and hexosamine showed marked variation in the various ganglioside preparations. Klenk (1951) had reinvestigated his ganglioside for hexosamine after our report appeared. He could verify our finding, and reported that hexose/hexosamine occurred in the ganglioside at a ratio of 5:1. Because he had previously suggested that ganglioside contained three hexoses per molecule, he proposed that in every second ganglioside molecule one hexose was replaced by a hexosamine. If Klenk was correct, the varying ratio between neuraminic acid and hexosamine was attributable to the variation in the hexosamine concentration of ganglioside. In most of my ganglioside preparations I found, like Klenk, a ratio between neuraminic acid and hexosamine of 2:1, but in the Tay-Sachs ganglioside the ratio was 1:1.

My continuing work was hampered by the low yields of ganglioside with the available ganglioside determination method and by the lack of a pure sialic acid standard. At this time, Folch *et al.* (1951) reported the isolation of a new high-molecular weight component of brain, which he named strandin. It was easy to demonstrate that brain strandin contained the same components as brain gangliosides, but strandin in water solution had a higher average molecular (micellar) weight than brain ganglioside isolated according to Klenk (1942), and it contained amino acids and sugars (trapped in the micell). Folch claimed that strandin and ganglioside could not be the same lipid because the brain concentration of strandin was several times higher than that of ganglioside as reported by Klenk. This discrepancy was due to the large losses of ganglioside in Klenk's isolation method (Svennerholm, 1957a), a fact Klenk never admitted. Therefore, it took several years before American biochemists accepted that strandin was a ganglioside preparation.

IV. CHROMATOGRAPHIC SEPARATION OF GANGLIOSIDES

Folch *et al.*'s (1951) study was also important because the group developed a new procedure, a solvent partition method, for the isolation of ganglioside, which markedly increased the recovery of a purer ganglioside fraction than that obtained using Klenk's procedure. Klenk considered brain ganglioside to have a homogenous carbohydrate composition, fatty acid–sphingosine–three hexoses–neuraminic acid, of which one hexose could be replaced by hexosamine. When I adapted Folch *et al.*'s method for the isolation of ganglioside, the proportion of neuraminic acid increased in comparison with ganglioside isolated using Klenk's (1942) method. This finding suggested that there was more than one ganglioside and that they differed in their numbers of neuraminic acids. We made several attempts to separate the gangliosides by countercurrent distribution using similar solvents as used by Folch *et al.* for the isolation of strandin, but the separation was disappointing. We achieved a much larger number of theoretical plates with chromatography on cellulose. The separation of ganglioside was monitored by the determination of hexosamine and N-acetylneuraminic acid (NeuAc)—the latter had recently been isolated from human brain (Svennerholm, 1955). Once I had obtained a pure standard of NeuAc, it was possible to develop new simple and accurate methods for the assay of sialic acid (Svennerholm, 1957b, 1957c); the resorcinol method (Svennerholm 1957c) is still the standard assay for gangliosides. We isolated gangliosides from brains of newborns and aged people and from Tay-Sachs brains. Using chromatography on cellulose, the gangliosides of normal human and bovine brain were found to be separated into three major fractions, whereas the gangliosides of Tay-Sachs brain were concentrated in the first, fast-migrating fraction. The results were reported at a symposium on cerebral lipidoses, held in Antwerp in July 1955 (Svennerholm, 1957d). Because the results were published only in the symposium volume, which may be difficult to obtain today, some data from the article are given in Table 4.1 in condensed form. It is clear from the table that in the ganglioside fractions the proportion of sialic acid increased and that of hexosamine diminished with aging. Gangliosides from Tay-Sachs brain contained more hexosamine and less sialic acid than gangliosides from normal brain, and the molar ratio of sialic acid/hexosamine was 1:1. When the gangliosides

Table 4.1. The Carbohydrate Composition of Human Brain Gangliosides

Brain tissue	Hexosamine	Hexose	NeuAc	NeuAc/Hexosamine molar ratio
	in percent			
Infantile brain	8.4	23.0	23.4	1.6:1
Senile brain	6.5	24.1	25.2	2.2:1
Tay-Sachs brain	10.2	22.7	18.8	1.0:1

were fractionated on cellulose, three major ganglioside fractions were obtained from normal brain. In the fastest-migrating and smallest fraction, 1, the sialic acid/hexosamine ratio was 1.0:1.0, in fraction 2 it was 1.7–2.1:1.0, and in fraction 3 it was 2.9–4.0, the last fraction containing free sialic acid in addition to gangliosides. When gangliosides from Tay-Sachs brain were fractionated, the fastest-migrating fraction was 90%, and the molar ratio of hexosamine:hexose: sialic acid was 1:2:1. This result led me to speculate that during the stepwise enzymatic degradation of gangliosides to cerebroside, the hexosamine had to be split off before the sialic acid. In Tay-Sachs disease, the hexosamine-degrading enzyme is lacking, which leads to the large accumulation of a ganglioside with terminal hexosamine. This hypothesis was considered interesting by Donald Fredrickson (1960), in his review on ganglioside lipidosis in the first edition of *The Metabolic Basis of Inherited Disease*.

It had been appropriate to continue the ganglioside studies with the elucidation of the structures of the partially separated gangliosides, and the possible difference between the normal fast-migrating gangliosides and the Tay-Sachs gangliosides. Our degradation studies showed the fast-migrating gangliosides of normal brain to consist of more than one ganglioside, but at this time we lacked simple and accurate analytical chromatographic methods for the assay of gangliosides and neutral complex glycolipids. Moreover, the capacity of the cellulose columns were too small for prepatory large-scale isolation of individual gangliosides in a simple way.

V. THIN-LAYER CHROMATOGRAPHY—THE METHOD OF CHOICE FOR STUDIES OF GANGLIOSIDE STRUCTURE

In our studies of plasma lipids (Hallgren *et al.*, 1960), we found silica gel column chromatography to have a markedly superior capacity for lipid separation to cellulose chromatography. In the late 1950s, the method for silica gel chromatography on glass plates (TLC) and an apparatus for the preparation of thin layers of fine silica gel on glass plates became available. This method revolutionized the possibilities of studying tissue gangliosides and neutral glycosphingolipids. Significantly better resolution than that achieved with the paper chromatographic methods was obtained, and the substances separated could be identified by color reagents containing mineral acids—e.g., the resorcinol assay. With the new tools it was simple to follow the stepwise degradation of gangliosides and allied gly-cosphingolipids by specific enzymes or weak mineral acids. Mammalian brain was shown to contain four major gangliosides, three of which were degraded by *Vibrio cholerae* sialidase to the fourth, which was a monosialoganglioside (Svennerholm, 1962), later designated GM1 (Svennerholm, 1963). This ganglioside was shown to consist of a sialidase-resistant sialic acid and four neutral sugars linked to the ceramide. The sialic acid was split off by weak acidic hydrolysis and isolated, crystallized, and characterized through its X-ray powder diagram (Svennerholm, 1956) as *N*-acetylneuraminic acid. The ceramide tetrasaccharide was then hydrolyzed with 10-fold–stronger HCl (0.1 M) and the ceramide saccharides formed were isolated and analyzed using TLC. This gave a relatively high yield of ceramide disaccharide, whereas that of ceramide trisaccharide was low; in the dialysate a disaccharide was the most abundant sugar, rather than monosaccharides. The disaccharide was hydrolyzed to galactose and *N*-acetylgalactosamine, and galactose was the only reducing sugar after borohydride treatment and acidic hydrolysis. The disaccharide gave a positive Morgan-Elson reaction (1934) even in the cold, and in light of the permethylation data the binding of galactose to C-3 of *N*-acetylgalactosamine was proposed. The ceramide disaccharide showed identical properties as lactosylceramide, and the following structure was suggested for the ceramide tetrasaccharide (Svennerholm, 1962): galactose β1→3 *N*-acetyl-galactosamine β1→4 galactose β1→4 glucose β1→1′ ceramide.

VI. TAY-SACHS GANGLIOSIDE

Gangliosides from Tay-Sachs brains were isolated with the same methods as used for the gangliosides from normal human brain. One single ganglioside constituted 75–90% of the total ganglioside content, depending on how advanced the disease process was. It was shown to migrate significantly faster than GM1 in several solvents. Like GM1, its sialic acid was resistant to hydrolysis with *V. cholerae* sialidase. After removal of the sialic acid with mild acid hydrolysis, the neutral ceramide saccharide showed identical migration in several solvents as the ceramide gangliotriaose of normal brain. The trisaccharide of normal and Tay-Sachs brain showed identical compositions, the same yields at stepwise degradation, and the same permethylated sugars. It was therefore suggested that the Tay-Sachs ganglioside had the same structure as the normal major monosialoganglioside, except that it lacked the terminal galactose (Svennerholm, 1962). The Tay-Sachs brain also showed a large accumulation of the neutral ceramide gangliotriaose, also known as asialo Tay-Sachs ganglioside. It was also demonstrated that the Tay-Sachs ganglioside occurred in brains of infants as well as in aged persons and was thus a normal brain ganglioside component. These findings further supported the hypothesis of a genetic lesion of a glycosphingolipid hexosaminidase in Tay-Sachs disease.

The structure of the basic monosialoganglioside of mammalian brain was established the following year in an elegant study by Kuhn and Wiegandt (1963), in which they demonstrated that N-acetylneuraminic was linked to the internal galactose in cis-configuration to C-3 of the terminal disaccharide. They also iden-tified a ganglioside G_0 in normal brain that differed from the major gangliosides in that it lacked terminal galactose. The structure of the Tay-Sachs ganglioside proposed by Kuhn and Wiegandt (1963) was confirmed by Makita and Yamakawa (1963) and by Ledeen and Salsman (1965).

A systematic study of this new class of lipids, gangliosides, began when Klenk (1935) detected them in the brain from a Niemann-Pick case, and some years later demonstrated large amounts of them in Tay-Sachs brain (Klenk, 1939). It took almost 30 years, however, until the exact structure of the Tay-Sachs gan-glioside was established. In this review I have tried to explain why it took such a long time. At the beginning we lacked suitable extraction methods for this new type of amphipatic molecule and chromatographic appliances for the separation of the closely related lipids. There were no commercial sources for sialic acids and no simple assay methods of them. All these problems had first to be solved before it was possible to attack the final goal—the structure of the Tay-Sachs ganglioside, whose beauty is shown in Figure 4.1.

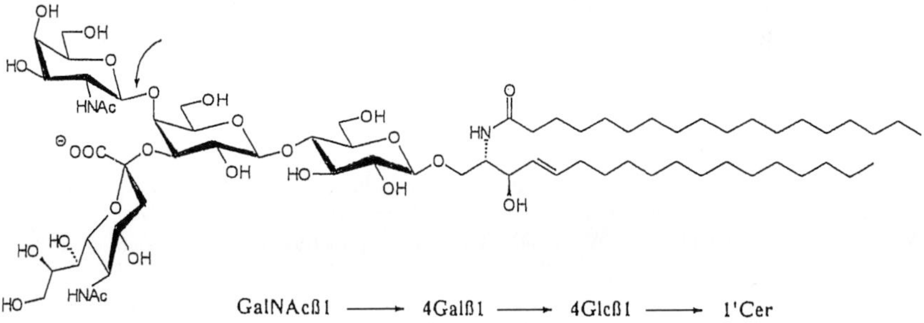

$$GalNAc\beta 1 \longrightarrow 4Gal\beta 1 \longrightarrow 4Glc\beta 1 \longrightarrow 1'Cer$$

$$\uparrow\ 3$$
$$\alpha$$
$$2$$

NeuAc

Figure 4.1. Structure of ganglioside GM2. The arrow indicates the bond to be cleared by hexosaminidase.

References

Blix, G. (1938). Einige Biobactungen über eine hexosaminhaltige Substanz in der Protagon-fraktion des Gehirns. *Scand. Arch. Physiol.* **80,** 46–51.

Blix, G., Svennerholm, L., and Werner, I. (1950). Chondrosamine as a component of gangliosides and of submaxillary mucin. *Acta Chem. Scand.* **4,** 717.

Blix, G., Svennerholm, L., and Werner, I. (1952). The isolation of chondrosamine from gangliosides and from submaxillary mucin. *Acta Chem. Scand.* **6,** 358–362.

Brante, G. (1948). Filter paper chromatography in lipid analysis. *Acta Soc. Med. Upsaliensis* **53,** 301–307.

Elson, L. A., and Morgan, W. T. J. (1933). *Biochem. J.* **27,** 1824–1828.

Feyrter, F. (1939). Zur Frage der Tay-Sachs-Schafferschen amaurotischen Idiotie. *Wirchows Arch.* **304,** 481–512.

Folch, J., Arsove, S., and Meath, J. A. (1951). Isolation of brain strandin, a new type of large molecular tissue component. *J. Biol. Chem.* **191,** 819–830.

Fredrickson, D. S. (1960). Infantile amaurotic family idiocy. *In* "The Metabolic Basis of Inherited Disease" (J. B. Stanbury, J. B. Wyngarden, and D. S. Fredrickson, eds.), pp. 553–596. McGraw-Hill, New York.

Hallgren, B., Stenhagen, S., Svanborg, A., and Svennerholm, L. (1960). Gas chromatographic analysis of the fatty acid composition of the plasma lipids in normal and diabetic subjects. *J. Clin. Invest.* **39,** 1424–1434.

Klenk, E. (1935). Über die Natur der Phosphatide und anderer Lipoide des Gehirns und der Leber bei der Niemann-Pickscher Krankheit. *Hoppe-Seyler's Z. Physiol. Chem.* **235,** 24–36.

Klenk, E. (1939). Beiträge zur Chemie der Lipoidosen. Niemann-Picksche Krankheit und amaurotische Idiotie. *Hoppe-Seyler's Z. Physiol. Chem.* **262,** 128–143.

Klenk, E. (1941). Neuraminsäure, das Spaltprodukt eines neuen Gehirnlipoids. *Hoppe-Seyler's Z. Physiol. Chem.* **268,** 50–58.

Klenk, E. (1942). Über die Ganglioside, eine neue Gruppe von zuckerhaltigen Gehirnlipoiden. *Hoppe-Seyler's Z. Physiol. Chem.* **273,** 76–86.

Klenk, E. (1951). Zur Kenntnis der Ganglioside. *Hoppe-Seyler's Z. Physiol. Chem.* **288,** 216–220.

Klenk, E., and Langerbeins, H. (1941). Über die Verteilung der Neuraminsäure in Gehirn. *Hoppe-Seyler's Z. Physiol. Chem.* **270,** 185–193.

Kuhn, R., and Wiegandt, H. (1963). Die Konstitution der Ganglio-N-tetraose und des Gangliosids G_I. *Chem. Ber.* **96,** 866–880.

Landsteiner, K., and Levene, P. A. (1926). On the heterogenetic hapten. *Proc. Soc. Exp. Biol. Med.* **24,** 343–344.

Ledeen, R., and Salsman, K. (1965). Structure of the Tay-Sachs ganglioside. *Biochemistry* **4,** 2225.

Makita, A., and Yamakawa, T. (1963). The glycolipids of the brain of Tay-Sachs disease. The chemical structures of a globoside and main ganglioside. *Jpn. J. Exp. Med.* **33,** 361–368.

Morgan, W. T. J., and Elson, L. A. (1934). A colorimetric method for the determination of N-acetylglucosamine and N-acetylchondrosamine. *Biochem. J.* **28,** 988–995.

Svennerholm, L. (1955). Isolation of sialic acid from brain gangliosides. *Acta Chem. Scand.* **9,** 1033.

Svennerholm, L. (1956). On sialic acid in brain tissue. *Acta Chem. Scand.* **10,** 694–696.

Svennerholm, L. (1957a). Determination of gangliosides in nervous tissue. *In* "Cerebral Lipidoses" (L. van Bogaert, J. N. Cumings, and A. Lowenthal, eds.), pp. 122–138. Blackwell Scientific Publications, Oxford, UK.

Svennerholm, L. (1957b). Quantitative estimation of sialic acids. I. A colorimetric method with orcinol-hydrochloric acid (Bial's) reagent. *Arkiv Kemi* **10,** 577–596.

Svennerholm, L. (1957c). Quantitative estimation of sialic acids. II. A colorimetric resorcinol-hydrochloric acid method. *Biochim. Biophys. Acta* **24,** 604–611.

Svennerholm, L. (1957d). The nature of the gangliosides in Tay-Sachs disease. *In* "Cerebral Lipidoses" (L. van Bogaert, J. N. Cumings, and A. Lowenthal, eds.), pp. 139–145. Blackwell Scientific Publications, Oxford, UK.

Svennerholm, L. (1962). The chemical structure of normal human brain and Tay-Sachs ganglioside. *Biochem. Biophys. Res. Commun.* **9,** 436–441.

Svennerholm, L. (1963). Chromatographic separation of human brain gangliosides. *J. Neurochem.* **10,** 613–623.

Walz, E. (1927). Über das Vorkommen von Kerasin in der normalen Rindermilz. *Hoppe-Seyler's Z. Physiol. Chem.* **166,** 210–222.

5

Discovery of the Hexosaminidase Isoenzymes

Donald Robinson and John L. Stirling*

Division of Life Sciences
King's College London
London SE1 9NN, United Kingdom

I. Introduction
II. Fluorigenic Substrates
III. Mammalian Glycosidases
IV. Hexosaminidases
V. Differential Assay for Hexosaminidases A and B
VI. Structural Relationship between Hexosaminidases A and B
References

I. INTRODUCTION

The investigation of the Tay-Sachs defect has pursued a classic course. The definition of the clinical symptoms and the realization that this was a familial disorder long preceded the laboratory analytical technology that was necessary for a precise identification of the biochemical lesion.

It was not until the early 1960s that developments in the structural analysis of gangliosides had reached a level at which a chemical identification of G_{M2} and asialo-G_{M2} as the characteristic storage products could be taken as confirmed (Svennerholm, 1962). By chance, this event coincided with a developing awareness of the function of the lysosome and the potential defects that could arise from enzymopathies of that structure after the identification of the enzyme defect in

*To whom correspondence should be addressed. E-mail: John.Stirling@kcl.ac.uk. Fax: 00 44 20848 4500. Telephone: 00 44 20848 4353.

Advances in Genetics, Vol. 44

Pompe's glycogenosis by H.-G. Hers (Hers, 1963). When this was integrated with the ongoing studies on the metabolic pathways of glycolipids by Gatt, Brady, and others, it was possible to make a logical prediction that an impairment of degradation at the terminal sugar moiety of the accumulating ganglioside was the basic cause of the disease (Brady, 1966). Electron microscope studies on affected tissues confirmed that the storage material was included in membrane-bound vesicles, as might be predicted for a lysosomopathy (Terry et al., 1962).

Since precise assay of enzyme activity on these hydrophobic or amphipathic natural substrates was an exacting task necessitating the preparation and use of radiolabeleled glycolipids, many workers including ourselves elected to study the enzymology of mammalian glycosidases using methods based on synthetic water-soluble substrates. However, whereas in other cases such as Gaucher's and Pompe's diseases the absence of a given activity correlated with disease symptoms might be diagnostically meaningful, the apparent abundance of hexosaminidase activity in samples from Tay-Sachs patients was uninformative or at least equivocal. There could be, some of us argued, a species of enzyme specifically evolved to interact with lipoid substrates and that might well not react with these simple synthetic products that had little physical resemblance to their naturally occurring counterparts. We now know in fact that nature has provided additional factors to assist this difficult interaction between an amphipathic substrate and an enzyme in an aqueous environment. Dr. Sandhoff's contribution here (see chapter 8) illustrates how skill and perseverance with these natural substrates led to the discovery of such an activating factor and also revealed, at a very early stage, the remarkable degree of genetic heterogeneity that could arise in what turned out to be a family of defects.

II. FLUORIGENIC SUBSTRATES

Undoubtedly, one of the factors that facilitated the study of lysosomal enzymes and led to discovery of the enzyme defect in Tay-Sachs disease was the availability of the 4-methylumbelliferyl (4-MU) fluorigenic substrates that enabled detection of the enzymes with very great sensitivity. Don had an active role in the synthesis and exploitation of these substrates and gives his personal view of these developments.

In 1961 I took up a post as Reader in Biochemistry at Queen Elizabeth College London, with the brief of introducing biochemistry into the undergraduate courses in the Faculty of Science. It was necessary at that time to set up a new field of research, and with Bob Price and Anna Furth, I began a study on mammalian glycosidases.

Previously I had been working on drug metabolism with the late R. T. Williams and co-workers. My contemporaries included K. S. Dodgson and Brian Spencer, who subsequently contributed much to our knowledge of the sulfatases, and our activities were concerned with feeding various examples of aromatic structures to animals and then identifying and quantitating the resultant metabolites. We were required to synthesize the potential metabolites, which were often not available commercially, and we were skilled in the isolation from urine of the various conjugates, particularly glucuronic acid derivatives and sulfate esters. The enzymes β-glucuronidase and aryl sulfatase were thus of interest to us as potential selective hydrolytic reagents for these metabolites, and the latter half of my doctoral thesis on the metabolism of aromatic nitro-compounds was devoted to various methods for the assay of sulfatases. The nitrocatechol method stems from this time (Robinson et al., 1951). The method was subsequently adapted by Roy (Roy, 1953) and exploited by Dodgson and Spencer when they established a new group in Cardiff in the late 1950s.

Thus, when Tony Mead and John Smith isolated the glucuronide of 7-hydroxy-coumarin and found that the conjugation suppressed the fluorescence of that compound, they naturally examined its possibility as a β-glucuronidase substrate (Mead et al., 1955). Theirs was the first description of the analytical potential of the 4-MU substrates that were to prove so crucial to the development of studies on lysosomal enzymes. Unfortunately, workers in the field were not ready to accept the new technique, and the method was rated "too exacting for general use" in the review of β-glucuronidase in the definitive tome on the enzymes (Levvy and Marsh, 1960).

In the meantime, R. T. Williams had become fascinated by the potential of fluorimetric analysis as a result of his close relationship with Sidney Udenfriend at the U.S. National Institutes of Health, and I, meanwhile becoming somewhat disenchanted with the prospect of putting yet another compound into the front end of a rabbit and spending the next 6 months determining what came out at the back, took every opportunity to examine the further potential of the elegant analytical method that I had seen on the next bench.

III. MAMMALIAN GLYCOSIDASES

By the time I was ready for the transition to Queen Elizabeth College and the setting up of a new group of my own, I had at hand a range of 4-MU substrates that I had tested against any convenient enzyme source that came to hand.

I had followed the story of the lysosome that was unfolding from the Louvain laboratory of de Duve, and therefore took as my new research target the characterization of mammalian glycosidases, for which I had the advantage of a particularly elegant and sensitive set of substrates and which were at that time so peripheral to the main thrust of biochemical research that I thought there was little possibility of serious competition for the fledgling department. I had not reckoned on Gery Hers' observation (Hers, 1963) that Pompe's disease was due to the lack of α-glucosidase (the substrate for which I had *not* synthesized), nor for his subsequent hypothesis regarding the fundamental lesion in lysosomal enzymes (Hers, 1965).

I had considered the possibility of genetic enzyme deficiencies in other glycosidases but had been told by an expert on mucopolysaccharidoses that the consensus was that these were aberrations of the biosynthetic pathway. Nevertheless, because I was already getting information on β-glucosidase from the work of Hope Abrahams on pig kidney (Abrahams and Robinson, 1969), where we had seen the value of starch gel electrophoresis for the identification of allelic forms, I proposed to examine spleens of Gaucher patients with a view to studying possible variations on β-glucosidase. However, while my files filled with promises from London pathologists of case material, no suitable samples had been located when the reports of Roscoe Brady and of Des Patrick (Brady *et al.*, 1965; Patrick, 1965) demonstrated the nature of the lesion for which we had at hand such an attractive diagnostic tool.

IV. HEXOSAMINIDASES

This was the third such study we had undertaken. Having established a foothold on β-galactosidase (Furth and Robinson, 1965) and β-glucosidase with the theses of Anna Furth and Hope Abrahams, the group was expanding. John Stirling came to the department as a Ph.D. student in October 1965 and continues the account.

I had spent the first few months as a research student learning techniques available in the lab for the assay and characterization of glycosidases and giving thought to a specific aspect to pursue in depth. These methods included starch gel electrophoresis, which was brought in by Norman Dance, who had just completed a thesis on variant hemoglobins with Huehns at University College. Isoenzymes of β-galactosidase, for example, could be located by flooding the gel with the 4-MU substrate, and after a short incubation, exposing them briefly to ammonia vapor to intensify the fluorescence. This worked best if the gels were sliced and the substrate applied to the cut surface, so that it was possible on any one occasion to use two different substrates on the top and bottom halves. David Leaback and Peter Walker had already synthesized the 4-MU N-acetyl-β-glucosaminide, having seen Don's original glycosides, and these compounds were now beginning to be commercially available as a result of the interest in lysosomal enzymes stimulated by the Louvain studies. I had tested a variety of these substrates without arousing my curiosity greatly, but on the occasion that I used the glucosaminide, it was immediately clear that there were two very active forms of the enzyme in roughly equal quantities. It was also clear that the characterization of the human hexosaminidase isoenzymes was going to provide excellent scope for my doctoral thesis. Since one of the isoenzymes ran to the anode and the other was slightly cathodic under our conditions, we thought of them as Acidic and Basic and called them A and B for this reason, although this was also a logical start for an alphabetical classification.

Walker and co-workers were already looking at the significance of this enzyme in serum, but the source from which the most convincing partial purification had been made was spleen as described by von Buddecke and Werries (1964). Our nodding acquaintance with Gaucher's disease also suggested this organ might be an important location of glycosidases, hence spleen was the chosen source.

V. DIFFERENTIAL ASSAY FOR HEXOSAMINIDASES A AND B

Heat denaturation was an obvious parameter to study, since we had previously used it to consider whether logarithmic plots of decay curves were mono- or biphasic as an indicator of whether more than one form of a given activity might coexist. From this point, a good deal of our effort went into finding the most efficient method of separating and concentrating the two forms and of devising a reliable differential assay.

Our previous lack of success at locating Gaucher specimens was repeated with a search for Tay-Sachs samples, and this, coupled with the reports that such specimens did not seem to lack activity, led us to give the diagnostic possibility little further thought. We inclined to the conclusion that the ganglioside-metabolizing hexosaminidase might be a specific enzyme that was not detected by the 4-MU substrates.

The announcement of the discovery of the two forms and the details of their characteristics were presented at a meeting of the Biochemical Society in September 1966 and published as a proceeding of the Biochemical Society in the *Biochemical Journal* in November the same year (Robinson and Stirling, 1966). It attracted no comment whatsoever. The full paper followed more than a year later (Robinson and Stirling, 1968), when we had had time to consider how these two forms might be related.

The rapidity with which the differential assay procedure was adopted for diagnostic purposes was a matter of some concern to us. It was after all an arbitrarily chosen set of conditions, and a slight variation of them could lead to different ratios, as could the presence of other species of enzyme such as Hex P, which we had found in pregnancy serum (Stirling, 1972). We compared our various criteria and concluded that we preferred to use ion exchange as the definitive method of distinguishing the two forms (Dance *et al.*, 1970).

VI. STRUCTURAL RELATIONSHIP BETWEEN HEXOSAMINIDASES A AND B

Why was the A form so much more negatively charged than the B form so that such a separation was possible? We felt it incumbent upon us to offer some hypothesis and considered protein phosphorylation and sialylation of glycoproteins as two likely causes. Finding enzyme sources that were not already contaminated with hexosaminidase to confuse the issue was difficult, but wheat germ acid phosphatase appeared to have no effect. Alkaline phosphatase was not tried because the pH necessary we supposed would denature our enzyme. We had seen reports of desialation of transferrin and other proteins and the effect this had on electrophoretic mobility and were delighted when, after trying a number of different supplies of neuraminidase, we found the Burroughs-Wellcome samples of "receptor destroying enzyme" produced the electrophoretic pattern we had hoped for. True, the effect was already noticeable in our so-called zero-time incubations, but the enzyme appeared to be so active that this could well be a result of highly susceptible species. After lecturing students on the value of "boiled enzyme" controls, we failed to use them in this case. Our hasty conclusion was that the effect could be attributed to Hexosaminidase A being a multisialated version of Hexosaminidase B. The ultimate reduction of electrophoretic mobility to that of Hexosaminidase B and the transient appearance of a B-like species during the heat denaturation of A, all supported a very close structural relationship. Only much later did we realize just how much egg was on our faces when both Rattazzi and Beutler (Carmody and Rattazzi, 1974; Beutler *et al.*, 1974) showed beyond doubt that the effect was a nonenzymic one due to merthiolate in the preparation. After all, when you buy an enzyme, you don't expect to be sold additives—or do you!

By 1972 we were wholly committed to a subunit model and said so in a letter to the Lancet (Robinson and Carroll, 1972), in the hope that we could catch the eye of clinical geneticists, and went on to try to demonstrate low-molecular-weight components (Robinson *et al.*, 1973; Carroll and Robinson, 1974). It was Beutler's group, however, with a comprehensive series of papers, who showed without doubt the fundamental basis for the two physical forms of the enzyme (Srivastava *et al.*, 1974), which was elegantly substantiated by Geiger and Arnon (1976).

References

Abrahams, H. E., and Robinson, D. (1969). β-D-Glucosidase and related enzyme activities in pig kidney. *Biochem. J.* **111,** 749–755.

Beutler, E., Villacorte, D., and Srivastava, S. K. (1974). Non-enzymic conversion of hexosaminidase A to hexosaminidase B by merthiolate. IRCS **2,** 1090.

Brady, R. O. (1966). The sphingolipidoses. *N. Engl. J. Med.* **275,** 312–318.

Brady, R. O., Kanfer, J. N., and Shapiro, D. (1965). Metabolism of glucocerebrosides. II. Evidence of an enzyme deficiency in Gaucher's Disease. *Biochem. Biophys. Res. Commun.* **18,** 211–225.

Carmody, P. J., and Ratazzi, M. (1974). Conversion of human hexosaminidase A to hexosaminidase B by crude *Vibrio cholerae* neuraminidase preparations. Merthiolate is the active factor. *Biochim. Biophys. Acta* **371,** 117–125.

Carroll, M., and Robinson, D. (1974). A low molecular-weight protein cross-reacting with human liver N-acetyl-β-D-hexosaminidase. *Biochem. J.* **137,** 217–221.

Dance, N., Price, R. G., and Robinson, D. (1970). Differential assay of human hexosaminidases A and B. *Biochim. Biophys. Acta* **222,** 662–664.

Furth, A. J., and Robinson, D. (1965). Specificity and multiple forms of β-galactosidase in the rat. *Biochem. J.* **97,** 59–66.

Geiger, B., and Arnon, R. (1976). Chemical characterisation and subunit structure of human N-acetylhexosaminidases A and B. *Biochemistry* **15,** 3484–3493.

Hers, H.-G. (1963). α-Glucosidase deficiency in generalised glycogen storage disease (Pompe's Disease). *Biochem. J.* **86,** 1–6.

Hers, H.-G. (1965). Inborn lysosomal diseases. *Gastroenterology* **48,** 625–633.

Levvy, G. A., and Marsh, C. A. (1960). β-Glucuronidase *In* "The Enzymes," 2nd ed. (P. D. Boyer, H. Lardy, and K. Myrback, eds.), pp. 397–407. Academic Press, New York and London.

Mead, J. A. R., Smith, J. N., and Williams, R. T. (1955). The biosynthesis of the glucuronides of umbelliferone and 4-methylumbelliferone and their use in fluorimetric determination of β-glucuronidase. *Biochem. J.* **61,** 569–574.

Patrick, A. D. (1965). A deficiency of glucocerebrosidase in Gaucher's Disease. *Biochem. J.* **97,** 17C–19C.

Robinson, D., and Carroll, M. (1972). Tay-Sachs disease: Interrelation of hexosaminidases A & B. *Lancet* **i,** 322.

Robinson, D., Smith, J. N., and Williams, R. T. (1951). Colorimetric determination of arylsulphatase. *Biochem. J.* 49: lxxiv.

Robinson, D., and Stirling, J. L. (1966). N-acetyl-β-glucosaminidase in human spleen. *Biochem. J.* **101,** 18P.

Robinson, D., and Stirling, J. L. (1968). N-acetyl-β-hexosaminidases in human spleen. *Biochem. J.* **107,** 321–327.

Robinson, D., Stirling, J. L., and Carroll, M. (1973). Identification of a possible sub-unit of hexosaminidase A and B. *Nature* **243,** 415.

Roy, A. B. (1953). The sulphatases of ox liver: 1 The complex nature of the enzyme. *Biochem. J.* **53,** 12–15.

Srivastava, S. K., Yoshida, A., Awasthi, Y. G., and Beutler, E. (1974). Studies on human β-D-N-acetylhexosaminidases II. Kinetic and structural properties. *J. Biol. Chem.* **249,** 2049–2053.

Stirling, J. L. (1972). Separation and characterisation of N-acetyl-β-glucosaminidase A and P from maternal serum. *Biochim. Biophys. Acta* **271,** 154–162.

Svennerholm, L. (1962). The chemical structure of normal brain and Tay-Sachs gangliosides. *Biochem. Biophys. Res. Commun.* **9,** 436–441.

Terry, R. D., Korey, S. R., and Weiss, M. (1962). Electron microscopy of the cerebrum in Tay-Sachs disease. *In* "Cerebral sphingolipidoses" (S. M. Aronson and B. W. Volk, eds.), pp. 49–56. Academic Press, New York.

von Buddecke, E., and Werries, E. (1964). Reinigung und eigenschaften einer β-D-acetyl-D-hexosaminidase aus Rindermiltz. *Z. Naturforsch.* **19,** 798–800.

6

Tay-Sachs Disease: The Search for the Enzymatic Defect

Roscoe O. Brady
Developmental and Metabolic Neurology Branch
National Institute of Neurological and Communicative Disorders and Stroke
National Institutes of Health
Bethesda, Maryland 20892

I. Historical Overview
II. Applications
    A. Diagnosis of Patients
    B. Enzyme Replacement Trials
    References

I. HISTORICAL OVERVIEW

The search for the enzymatic defect in Tay-Sachs disease began just over 40 years ago with a simple galactose tolerance test in a patient with Gaucher disease. It was known that patients with this disorder accumulated a sphingoglycolipid called glucocerebroside and that was in contrast with the major sphingolipid of brain, which is galactocerebroside. Since nothing whatsoever was known about the metabolism of sphingolipids, one of the first investigations that appeared reasonable was to determine whether Gaucher patients had a metabolic abnormality that prevented the synthesis of galactocerebroside. This was initially approached with the galactose tolerance test in a young woman with Gaucher disease. It was found that the disposition of galactose by this individual was perfectly normal. This observation led in turn to examining the synthesis of glucocerebroside and galactocerebroside in surviving slices of human spleen tissue obtained at operation. It was learned that the synthesis of both cerebrosides occurred normally. Furthermore, there was no excessively rapid formation of glucocerebroside. These findings led to the postulate that something was wrong with a catabolic enzyme required for the degradation of the accumulating glycolipid (Trams and Brady, 1960).

Although numerous attempts were made to assess glucocerebroside catabolism using (1) unlabeled natural material, (2) radioactive glucocerebroside prepared biosynthetically, and (3) ^{3}H-glucocerebroside prepared by the Wilzbach isotope-exchange procedure, no significant progress was made until this compound was chemically synthesized with ^{14}C in specific portions of the molecule. Using ^{14}C-glucocerebroside labeled in the glucose moiety, an enzyme was discovered in human tissues that catalyzed the hydrolytic cleavage of the hexose, and the products of the reaction were glucose and ceramide. The activity of this enzyme was examined in human tissues obtained from control sources and from patients with Gaucher disease. The activity of this glucosidase was found to be greatly diminished in the preparations obtained from Gaucher patients (Brady et al., 1965, 1966).

Emboldened by this discovery, I began to look at potential sites of enzymatic deficiencies in other disorders, including Tay-Sachs disease. A dilemma was immediately apparent. Since ganglioside G_{M2} is branched in the terminal portion of its molecule, its catabolism could theoretically begin either through the action of a hexosaminidase-cleaving N-acetylgalactosamine or by a neuraminidase (sialidase) that catalyzed the cleavage of N-acetylneuraminic acid. In fact, as will be seen, most mammalian tissues, including brain, contain both of these enzymes. In reviewing my laboratory notebooks for the preparation of this manuscript, I found an interesting entry concerning an experiment performed on May 24, 1965 (Figure 6.1). I was examining hexosaminidase activity in a human brain enzyme preparation that had been fractionated by ammonium sulfate and then applied to a TEAK ion-exchange chromatography column using p-nitrophenyl-β-D-N-acetylgalactosminide as substrate.

I noted on the top of the second page that there appeared to be two hexosaminidase enzymes. One of these passed through the ion-exchange column and the other was adsorbed. Further evaluation carried out on the following day indicated that both appeared to have optimal catalytic activity around pH 4.0. Would that I had then performed such an experiment with a preparation of Tay-Sachs brain tissue! A number of enzymatic experiments were then carried out with unlabeled ganglioside G_{M2}. The addition of G_{M2} inhibited the hydrolysis of the artificial substrate using the 30.5 to 43.5 ammonium sulfate fraction that contained both enzymes. However, no hydrolysis of the ganglioside itself was detected by colorimetric analysis. The reaction was carried out in the presence of detergent, which should have obviated the necessity for an activator protein (Li and Li, 1984).

In other experiments performed in 1965, Julian Kanfer, David Shapiro, and I synthesized sphingomyelin labeled in the choline portion of the molecule.

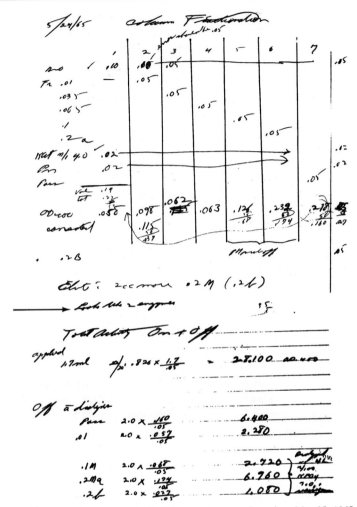

Figure 6.1. Photocopy from my notebook of an experiment performed on May 25, 1965, on the purification of human brain hexosaminidase. A 30.5–43.5% saturated ammonium sulfate fraction of a high-speed supernatant solution was applied to and fractionally eluted from a TEAE ion-exchange column. (Arrow added for this chapter.)

Using this compound, we discovered that there was a drastic reduction of sphingomyelinase activity in tissues obtained from patients with Types A and B Niemann-Pick disease (Brady *et al.*, 1966). The discoveries of the enzymatic deficiencies in Gaucher disease as well as Niemann-Pick disease led to the prediction of the specific metabolic abnormalities in Fabry disease, generalized (G_{M1}) gangliosidosis, and the assumption that insufficient activity of a hexosaminidase that catalyzed the hydrolysis of N-acetylgalactosamine from ganglioside G_{M2} might be the enzymatic defect in Tay-Sachs disease (Brady et al., 1966).

It was at this point that the difficulties began. Lars Svennerholm, who had initiated his extensive work on ganglioside isolation, came to my laboratory in October 1965, and we began a more intensive examination of the enzymatic hydrolysis of unlabeled gangliosides. It was quite easy for us to demonstrate β-galactosidase activity in mammalian tissues that catalyzed the hydrolysis of the terminal molecule of galactose from ganglioside G_{M1}. However, we had no success in detecting the cleavage of N-acetylgalactosamine from unlabeled ganglioside G_{M2}. We obtained, however, suggestive evidence for the enzymatic hydrolysis of N-acetylneuraminic acid from this glycolipid. Because there was no method available for the chemical synthesis of ganglioside G_{M2} [its synthesis was reported much later (Sugimoto et al., 1986)], I decided to try to label it with ^3H by the Wilzbach procedure that had been used by Andrew Gal to prepare radioactive ceramidetrihexoside which we used to demonstrate the enzymatic defect in Fabry disease (Brady et al., 1967). We were never able to obtain ganglioside G_{M2} by this procedure which was generally labeled with ^3H that had a sufficiently low radioactive background for meaningful metabolic studies.

In 1967, Edwin Kolodny joined my laboratory, and we began a comprehensive new series of investigations on Tay-Sachs disease. At this juncture, I was convinced that we had to have specifically labeled ganglioside G_{M2} to follow its metabolism, and the only way that I could envision obtaining it was biosynthetically. At the time of these investigations, I believed it critical to try to assess which portion of the ganglioside G_{M2} molecule we should first label, since this was likely to be a long endeavor. To decide whether the defect might be insufficient hexosaminidase activity, I measured N-acetylgalactosaminidase activity in unfractionated human brain extracts using the nitrophenyl-β-D-galactosylpyranoside substrate. Not only was there ample hexosaminidase activity present in the brain of Tay-Sachs patients measured by this procedure, but it was actually increased over that in control human brain specimens. This experiment led to a crucial false step in this investigation. I surmised that because hexosaminidase activity was not deficient, the enzymatic defect in Tay-Sachs disease might be due to insufficient neuraminidase activity. The experiments with the chromogenic substrate performed in May 1965 that indicated the possibility of two hexosaminidases in human brain were not seriously reconsidered at this point, principally because I had not been able to show the cleavage of N-acetylgalactosamine from unlabeled ganglioside G_{M2}.

Dr. Kolodny and I therefore began experiments to label ganglioside G_{M2} in the N-acetylneuraminic acid moiety by intracerebral injection of ^3H-N-acetylmannosamine, a specific precursor of N-acetylneuraminic acid. For these

Table 6.1. Metabolism of [^{3}H]-Tay-Sachs Ganglioside Specifically Labeled in the N-Acetylneuraminyl Moiety

Patients	Sialic acid hydrolyzed (pmol/mg protein/h)
Controls (4)	389.
Tay-Sachs disease (4)	351.

experiments, we used neonatal rats, since it was known that ganglioside formation in the brain occurred at its maximal rate during this period. After a year and a half of work, he succeeded in isolating pure ganglioside G_{M2} labeled in the N-acetylneuraminic acid portion of the molecule (Kolodny et al., 1970). We substantiated the earlier indication that mammalian brain contained an enzyme that catalyzed the hydrolysis of N-acetylneuraminic acid from ganglioside G_{M2}. However, the rate of the reaction was extremely slow compared with other sphingolipid hydrolases. We were therefore reluctant to use frozen tissues for the examination of this enzyme. It was difficult to obtain fresh samples of control human brain tissue. Because L. Schneck and co-workers had demonstrated an accumulation of ganglioside G_{M2} in heart muscle of Tay-Sachs patients, and because muscle biopsy specimens could be obtained relatively easily, I asked Dr. Bruno Volk at the Kingsbrook Medical Center in Brooklyn if he would assist us in determining whether the activity of this neuraminidase could be detected in human muscle tissue. Using ganglioside G_{M2} labeled with ^{3}H in the sialic acid moiety, we found that normal human muscle tissue did, in fact, contain this enzyme. We then assayed the activity of this enzyme in fresh muscle biopsy preparations obtained from four human controls and from four patients with Tay-Sachs disease. As may be seen in Table 6.1, the activity of this enzyme was completely normal in the samples from the Tay-Sachs patients.

It was therefore apparent that one had to investigate the catabolism of ganglioside G_{M2} labeled in the N-acetylgalactosaminyl moiety as the possible site of the enzymatic defect in Tay-Sachs disease.

At this time, I made a second error. Because I assumed that only Kolodny and I and our co-workers had sufficient expertise to prepare ganglioside G_{M2} specifically labeled in the aminosugar portion of the molecule with any degree of expediency, in December 1968 I submitted an abstract to the American Society of Biological Chemists describing our observations with sialic acid–labeled ganglioside G_{M2} (Kolodny et al., 1969a). This abstract was published in March and the data were reported at the April 1969 meeting of this society. Because of the interest in this investigation, we were asked to prepare a verbatim copy of the presentation for the press conference that took place the next day. The first page of this presentation, on which we indicated that neuraminidase is not the enzymatic defect in Tay-Sachs disease, was as follows:

FEDERATION OF THE AMERICAN SOCIETIES FOR EXPERIMEN-
TAL BIOLOGY, 53rd ANNUAL MEETING

Subject . . . . Studies on the Metabolism
. . . . of Tay-Sachs Ganglioside
Authors . . . . Edwin H. Kolodny, M.D.; Roscoe
. . . . O. Brady, M.D.; Jane M. Quirk, M.S.;
. . . . Julian N. Kanfer, Ph.D.
Address . . . . Laboratory of Neurochemistry,
. . . . National Institute of Neurological
. . . . Diseases and Stroke,
. . . . National Institutes of Health,
. . . . Bethesda, Maryland 20014
Time . . . . 4:00 p.m., Wednesday, April 16, 1969
Place . . . . Penn 1, Haddon Hall, Atlantic City,
. . . . New Jersey
Program . . . . Phospholipids
Convention Address Haddon Hall

RELEASE TIME
a.m. April 17

Tay-Sachs disease produces severe central nervous system damage early in infancy and leads to death before the age of four. It is
due to an inborn error of metabolism characterized by the accumulation of Tay-Sachs ganglioside within nerve cells. This compound is a
lipid-carbohydrate complex containing sialic acid. Through the use of
radioactively-labeled Tay-Sachs ganglioside, our studies demonstrate the
presence of an enzyme in rat and human tissues which removes sialic acid
from Tay-Sachs ganglioside. The enzyme has been characterized and the
products of the reaction identified. In order to learn whether a deficiency of this enzyme could account for the accumulation of Tay-Sachs
ganglioside in patients with Tay-Sachs disease, the activity of this enzyme
was measured in human skeletal muscle obtained at biopsy from control
subjects and from patients with Tay-Sachs disease. The activity of the
enzyme in muscle preparations from Tay-Sachs patients did not differ
from the activity in controls. We therefore feel that the metabolic defect
in Tay-Sachs disease cannot be attributed to a deficiency of this enzyme.

We then undertook the labeling of ganglioside G_{M2} with ^3H in the
N-acetylgalactosaminyl moiety through the intracerebral injection of N-acetyl-

Table 6.2. Enzymatic Hydrolysis of Tay-Sachs Ganglioside-[^{3}H] Labeled in the N-Acetylgalactosaminyl Moiety

Patients	Sialic acid hydrolyzed (pmol/mg protein/h)
Controls (4)	304.
Tay-Sachs disease (3)	0.

Source: From Kolodny *et al.*, 1969b.

[^{3}H]-D-galactosamine. Using this aminosugar-labeled G_{M2} as substrate, we found that normal human muscle contains an enzyme that catalyzes the cleavage of N-acetylgalactosamine from ganglioside G_{M2}. This enzyme was completely lacking in similar muscle biopsy preparations obtained from three patients with Tay-Sachs disease (Table 6.2).

The metabolic defect in Tay-Sachs disease was thereby established using authentic labeled ganglioside G_{M2}. The report of this finding was published on October 22, 1969. However, as will be apparent in the next chapter, Okada and O'Brien had published two months earlier on August 15, 1969 that a specific hexosaminidase isozyme was in absent in conventional patients with Tay-Sachs disease. Although disappointed by the sequence of events surrounding the establishment of the metabolic defect in this disorder, we continued our investigations with the two labeled ganglioside G_{M2} preparations. We characterized an enzyme in mammalian intestinal tissue that catalyzes the cleavage of sialic acid from

Table 6.3. Ganglioside Catabolism by Human Brain Lysosomes

	Enzyme		
	Hexosaminidase		Sialidase
	Substrate		
	^{14}C-G_{M2} pmol/mg protein/h	4-MU-GlcNAc nmol/mg protein/h	^{3}H-G_{M2} pmol/mg protein/h
Tissue source			
Controls (7)	140 ± 53	134 ± 11	217 ± 33
Tay-Sachs disease			
1. Classic	0	513	225
2. Classic	0	1416	n.d.
3. Classic	6	3	232

n.d. = not determined.
Source: From Tallman *et al.*, 1972.

ganglioside G_{M2} (Kolodny et al., 1971). We carried out extensive metabolic studies with ganglioside G_{M2} that we subsequently labeled in the N-acetylgalactosaminyl moiety by enzymatic synthesis in vitro (Quirk et al., 1972). We found that rat and human brain lysosomes contained both ganglioside G_{M2} N-acetylneuraminidase and N-acetylgalactosaminidase activities (Tallman et al., 1971; Tallman and Brady, 1972), and that ganglioside G_{M2} hexosaminidase was specifically lacking in lysosomes obtained from the brain of Tay-Sachs patients (Table 6.3).

II. APPLICATIONS

A. Diagnosis of patients

Most testing for Tay-Sachs homozygotes and heterozygotes is performed with artificial substrates such as 4-methylumbelliferyl-β-D-N-acetylglucosaminide. However, as will also be brought out later in this book, certain Tay-Sachs variants show completely normal activity with this material and with chromogenic substrates. Therefore, it is still occasionally necessary, as in the activator-deficient AB variant form of Tay-Sachs disease (Conzelmann and Sandhoff, 1978), to use labeled ganglioside G_{M2} to establish the diagnosis. An interesting converse of this situation was found in hexosaminidase A-deficient adults when labeled ganglioside G_{M2} catabolism was examined in leukocyte preparations. In these patients, adequate catabolism of the ganglioside was found despite the deficiency of Hex A as determined with artificial substrates (Tallman et al., 1974a).

B. Enzyme replacement trials

There is an extraordinarily large discrepancy between the activity of purified hexosaminidase A with artificial substrates and the natural ganglioside G_{M2}, even in the presence of detergents (Tallman et al., 1974b). It is therefore imperative to determine the activity of all enzyme preparations that are being considered for enzyme replacement trials with the natural ganglioside. Thus, in the initial trial of enzyme replacement in a patient with the O-variant form of Tay-Sachs disease, the activity of hexosaminidase isolated from human urine was checked against labeled G_{M2} (Johnson et al., 1973). In this investigation, we found that none of the intravenously injected hexosaminidase reached the brain. However, by 4 h after infusing the enzyme, a 40% reduction in the level of globoside in the recipient's plasma was observed. This important finding that exogenous enzyme can reduce accumulated sphingolipid has been confirmed in many other studies (summarized by Brady, 1984). It was postulated that the rapid fall in plasma sphingolipid may be the result of catabolism within circulating leukocytes (Brady et al., 1973).

Because of the lack of effect of intravenous hexosaminidase A and the fact that direct intracerebral injections of enzymes exerted no beneficial effects, it was apparent that enzyme replacement would not be feasible for Tay-Sachs disease unless some method could be developed that permitted intravenously injected enzyme to reach the substance of the brain. Accordingly, my colleagues and I altered the blood–brain barrier by intracarotid infusion of hyperosmolar sugar solutions.

Although this procedure permitted delivery of measureable hexosaminidase to the brain, only a small fraction of the administered enzyme gained access to the central nervous system using this procedure (Barranger *et al.*, 1979; Neuwelt *et al.*, 1981). Therefore, enzyme replacement therapy for Tay-Sachs disease has been in abeyance for several decades. If a safe procedure is developed that allows the blood–brain barrier to be opened more extensively, or for a longer period of time, then one might anticipate additional attempts using this approach. This strategy seems reasonable, since purified human placental hexosaminidase A appears to be a mannose-terminated glycoprotein and cholinergic neurons have a mannose lectin on their surface that will assist in targeting the enzyme to neurons (Scmueler, U. and Brady, R. O., unpublished observations). These observations, combined with the fact that the exogenous hexosaminidase A can reach lysosomes (Neuwelt *et al.*, 1984) where ganglioside G_{M2} is stored in the form of membranous bodies (Tallman *et al.*, 1971), provide a degree of hope for the eventual treatment of Tay-Sachs disease by enzyme replacement.

References

Barranger, J. A., Rapoport, S. I., Fredericks, W. R., Pentchev, P. G., MacDermot, K. D., Steusing, J. K., and Brady, R. O. (1979). Modification of the blood-brain barrier: Increased concentration and fate of enzymes entering the brain. *Proc. Natl. Acad. Sci. (USA)* **76**, 481–485.

Brady, R. O. (1966). The sphingolipidoses. *N. Engl. J. Med.* **275**, 312–318.

Brady, R. O. (1984). Enzyme replacement in the sphingolipidoses. *In* "The Molecular Basis of Lysosomal Storage Disorders" (J. A. Barranger and R. O. Brady, eds.), pp. 481–494. Academic Press, Orlando, FL.

Brady, R. O., Kanfer, J. N., and Shapiro, D. (1965). Metabolism of glucocerebrosides. II. Evidence of an enzymatic deficiency in Gaucher's disease. *Biochem. Biophys. Res. Commun.* **18**, 221–225.

Brady, R. O., Kanfer, J. N., Bradley, R. M., and Shapiro, D. (1966). Demonstration of a deficiency of glucocerebroside cleaving enzyme in Gaucher's disease. *J. Clin. Invest.* **45**, 1112–1115.

Brady, R. O., Gal, A. K., Bradley, R. M., Martensson, E., Warshaw, A. L., and Laster, L. (1967). Enzymatic defect in Fabry's disease. Ceramidetrihexosidase deficency. *N. Engl. J. Med.* **276**, 1163–1167.

Brady, R. O., Tallman, J. F., Johnson, W. G., Gal, A. K., Leahy, W. E., Quirk, J. M., and Dekaban, A. S. (1973). Replacement therapy for inherited enzyme deficiency: Use of purified ceramidetrihexosidase in Fabry's disease. *N. Engl. J. Med.* **289**, 9–14.

Conzelmann, E., and Sandhoff, K. (1978). AB variant of infantile G_{M2} gangliosidosis: Deficiency of a factor necessary for stimulation of hexosaminidase-A catalyzed degradation of ganglioside G_{M2}. *Proc. Natl. Acad. Sci. (USA)* **75**, 3979–3983.

Johnson, W. G., Desnick, R. J., Long, D. M., Sharp, H. L., Krivit, W., Brady, B., and Brady, R. O. (1973). Intravenous injection of purified hexosaminidase A into a patient with Tay-Sachs disease. In "Enzyme Therapy in Genetic Diseases" (R. J. Desnick, R. W. Bernlohr, and W. Krivit, eds.), Birth Defects Original Article Series, IX, pp. 120–124.

Kolodoy, E. H., Brady, R. O., Quirk, J. M., and Kanfer, J. N. (1969a). Studies on the metabolism of Tay-Sachs ganglioside. Fed. Proc. **28,** 596.

Kolodoy, E. H., Brady, R. O., and Volk, B. W. (1969b). Demonstration of an alteration of ganglioside metabolism in Tay-Sachs disease. Biochem. Biophys. Res. Commun. **37,** 526–531.

Kolodny, E. H., Brady, R. O., Quirk, J. M., and Kanfer, J. N. (1970). Preparation of radioactive Tay-Sachs ganglioside labeled in the sialic acid moiety. J. Lipid. Res. **11,** 144–149.

Kolodny, E. H., Kanfer, J. N., Quirk, J. M., and Brady, R. O. (1971). Properties of a particle-bound enzyme from rat intestine that cleaves sialic acid from Tay-Sachs ganglioside. J. Biol. Chem. **246,** 1426–1431.

Li, Y.-T., and Li, S.-C. (1984). The occurrence and physiological significance of activator proteins essential for the enzymic hydrolysis of G_{M1} and G_{M2} gangliosides. In "The Molecular Basis of Lysosomal Storage Disorders" (J. A. Barranger and R. O. Brady, eds.), pp. 79–91. Academic Press, Orlando, FL.

Neuwelt, E. A., Barranger, J. A., Brady, R. O., Pagel, M., Furbish, F. S., Ouirk, J. M., Mook, G. E., and Frenkel, E. (1981). Delivery of hexosaminidase-A to the cerebrum following osmotic modification of the blood-brain barrier. Proc. Natl. Acad. Sci. (USA) **78,** 5838–5841.

Neuwelt, E. A., Barranger, J. A., Pagel, M., Quirk, J. M., Brady, R. O., and Frankel, E. P. (1984). Delivery of active hexosaminidase across the blood-brain barrier in rats. Neurology **34,** 1012–1019.

Okada, S., and O'Brien, J. S. (1969). Tay-Sachs disease: Generalized absence of a β-D-N-acetylhexosaminidase component. Science **165,** 698–700.

Quirk, J. M., Tallman, J. F., and Brady, R. O. (1972). The preparation of trihexosyl- and tetrahexosyl-gangliosides specifically labeled in the N-acetylgalactosaminyl moiety. J. Labeled Compounds **VIII,** 483–494.

Sugimoto, M., Numata, M., Koike, K., Nakahara, Y., and Ogawa, T. (1986). Total synthesis of gangliosides G_{M1} and G_{M2}. Carbohydrate Res. **156,** C1–C5.

Tallman, J. F., and Brady, R. O. (1972). The catabolism of Tay-Sachs ganglioside in rat brain lysosomes. J. Biol. Chem. **247,** 7570–7575.

Tallman, J. F., Brady, R. O., and Suzuki, K. (1971). Enzymatic activities associated with membranous cytoplasmic bodies and isolated brain lysosomes. J. Neurochem. **18,** 1775–1777.

Tallman, J. F., Johnson, W. G., and Brady, R. O. (1972). The metabolism of Tay-Sachs ganglioside: Catabolic studies with lysosomal enzymes from normal and Tay-Sachs brain tissue. J. Clin. Invest. **51,** 2339–2345.

Tallman, J. F., Brady, R. O., Navon, R., and Padeh, B. (1974a). Ganglioside catabolism in hexosaminidase A deficient adults. Nature **252,** 254–255.

Tallman, J. F., Brady, R. O., Quirk, J. M., Villalba, M., and Gal, A. E. (1974b). Isolation and relationship of human hexosaminidases. J. Biol. Chem. **249,** 3489–3499.

Trams, K. G., and Brady, R. O. (1960). Cerebroside synthesis in Gaucher's disease. J. Clin. Invest. **39,** 1546–1550.

7

Discovery of β-Hexosaminidase A Deficiency in Tay-Sachs Disease

Shintaro Okada*
Department of Pediatrics
Osaka University Hospital
Yamada-oka, Suita, Osaka 565, Japan

John S. O'Brien†
Department of Neurosciences
University of California San Diego School of Medicine
La Jolla, California 92093

I. Introduction
II. John S. O'Brien's Recollection
III. Shintaro Okada's Recollection
   References

I. INTRODUCTION

In this discussion, Shintaro Okada and John S. O'Brien present some thoughts and recollections about how they discovered the deficiency of hexosaminidase A in Tay-Sachs disease nearly 30 years ago.

There are five levels of sophistication in understanding the molecular basis of a hereditary disease such as Tay-Sachs disease. The first level is the description of the phenotype, including the clinical features, progression, and pathology of the disease. The second level is determination of the mode of genetic transmission. The third level is the discovery of the biochemical defect responsible for the

*To whom correspondence should be addressed. E-mail: shin@ped.med.osaka-u.ac.jp.
Fax: +81-6-6879-3939. Telephone: +81-6-6879-3930.
†Deceased

pathophysiology of the disorder. The fourth level is the elucidation of the enzyme, catalytic, or structural protein defect that underlies the biochemical defect. The fifth level is the definition of the genetic defect at the nucleic acid level (O'Brien *et al.*, 1971a).

II. JOHN S. O'BRIEN'S RECOLLECTION

My first contact with patients with Tay-Sachs disease came as a National Institutes of Health postdoctoral fellow at the City of Hope Medical Center in Duarte, California. In 1960 there were many patients with Tay-Sachs disease on a ward specializing in the care of children with catastrophic illnesses. I received a tremendous emotional jolt when I saw Tay-Sachs patients in all stages of their illness. Especially moving was the experience of following a 6-month-old child who was admitted in the early stages of the disease, the second offspring of a young couple whose older sister had Tay-Sachs disease. I followed the progression of his illness from a smiling, outgoing, happy, cherubic baby boy who lost intellectual function and reached a state of decerebrate rigidity by the end of the fellowship 18 months later. That experience was the main impetus for me to work on Tay-Sachs disease. I found when I gave the Bernard Sachs Memorial Lecture to the Child Neurology Society in 1982 and learned more about Bernard Sachs, that he had made the statement, "All true scientific research in medicine stems from the bedside." In my case this was completely true. Of course, Bernard Sachs was responsible for describing the clinical and pathological phenotype of Tay-Sachs disease and determining the mode of genetic transmission by pedigree analysis, as is ably documented in this symposium.

III. SHINTARO OKADA'S RECOLLECTION

When John O'Brien and I met for the first time in Japan in 1965, we talked about the newly described lipid storage disorder, generalized gangliosidosis (O'Brien *et al.*, 1965). He emphasized that ganglioside stored in this disease had a different structure from that of Tay-Sachs ganglioside, and had normal structure as G_{M1}. At that time, I could not understand why storage of normal G_{M1} ganglioside resulted in such a severe progressive disease as generalized gangliosidosis. In 1967, I joined John's group in Los Angeles. Discussing my first research project in the United States, John and I decided to test the degradation of G_{M1} ganglioside using brain and liver from a patient with generalized gangliosidosis. The tissues were from the first described case by John and Dr. Landing a few year before (O'Brien *et al.*, 1965). By that time, 1967, from the first to the third levels of sophistication for many hereditary lipidoses had been achieved, and further study to clarify some diseases such as Gaucher disease was successfully completed by Dr. Brady and his group

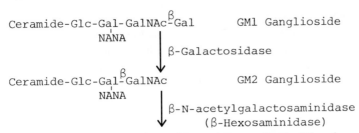

Figure 7.1. Structure and metabolic pathway of G_{M1} and G_{M2} gangliosides (Glc = glucose; Gal = galactose; GalNAc = N-acetylgalactosamine; NANA = N-acetylneuraminic acid).

(1965). We were about to enter the time of enzymological study of lysosomes for understanding the pathogenesis of hereditary lipidoses. Considering the structure of G_{M1} ganglioside with β-galactoside residue at the terminal position, I decided to check β-galactosidase (Figure 7.1). Luckily enough, I soon found a complete deficiency of acid β-galactosidase in the tissues of G_{M1} gangliosidosis using an artificial p-nitrophenyl substrate. But as John told me that impaired degradation of G_{M1} ganglioside in the disease should be investigated, it took another several months for me to complete our project. We finally discovered the primary defect in this disease and published it in *Science* (Okada and O'Brien, 1968). We then moved from Los Angeles to La Jolla, California.

When the work on G_{M1} gangliosidosis was finished successfully, we started considering Tay-Sachs disease. Our hope was that the same enzymological strategy might also apply. The correct structure of accumulated ganglioside in Tay-Sachs disease had been elucidated by Professors Makita and Yamakawa in Tokyo (1963); a terminal nonreducing N-acetyl-β-D-galactosamine residue was evident (Figure 7.1). We decided to focus on a defect of ganglioside N-acetyl-β-D-galactosaminidase. For our initial experiment we used asialo G_{M2} for assaying this enzyme, but the sensitivity of the colorimetric method was not satisfactory. Then using synthetic β-D-N-acetyl-galactosamine or glucosamine substrates, we found no significant reduction in activity in patients with Tay-Sachs disease. The key to understanding the defect was the discovery by Professor Donald Robinson and Dr. John Stirling (1968) showing that two different hexosaminidase forms were demonstrated in human tissue by starch gel electrophoresis, one of which, hexosaminidase A, was more acidic than the other, hexosaminidase B. We quickly turned to starch gel electrophoresis to determine the isozyme patterns in Tay-Sachs tissues. We were then working in a small wooden laboratory behind the beautiful Salk Institute buildings. One autumn evening in 1968 we found the absence of hexosaminidase A in frozen tissues from several patients. Professor Robinson's group had determined that the A form was physically more labile than the B form, and we were concerned that artifactual denaturation of hexosaminidase may have occurred even though pathological control tissues did not reveal any diminution. Nonetheless, we drove throughout the Los Angeles Basin collecting samples from Tay-Sachs patients and relatives. That evening, after returning to our lab in San Diego, we isolated fresh leukocytes while looking at the beautiful sunset off La Jolla, and ran electrophoresis on the freshly obtained pellets overnight. Once again, hexosaminidase A was absent in the samples from Tay-Sachs patients and the heterozygotes appeared to have reduced activities on the gels consistent with a gene dosage effect.

On the way to Los Angeles to visit patients, John drove his car while I sat next to him as guide. But I was surprised at a map he gave me. Though we were driving on the freeway, there was no new freeway on his old map of Los Angeles. Some freeways were indicated only by dotted lines, as either under construction or still planned. John kept asking me which exit to use, and I could only guess. I must say that this was one of the most difficult tasks in our Tay-Sachs disease project.

We published our results in the August issue of *Science* in 1969 (Okada and O'Brien), and this paper has become a citation classic (1981). We next turned our attention to prenatal diagnosis of Tay-Sachs disease, since in several of the families with which we had made contact, there were pregnancies at risk for an affected child. Professor Jay Seegmiller, who was being recruited to the University of California San Diego (UCSD) faculty and had been deeply involved with prenatal diagnosis of Lesch-Nyhan disease, was kind enough to send us amniotic cell lines grown in his laboratory. We quickly demonstrated that hexosaminidase A was present in normal amniotic cells at levels comparable with those in most tissues. We then obtained cells from at-risk pregnancies and monitored a total of 15 pregnancies in the first month of testing. We determined that six fetuses were affected and confirmed these by analysis of tissues from each affected fetus (Figure 7.2) (O'Brien *et al.*, 1971b). A long period of anxiety occurred, since the first affected fetus was not detected until the fifth in the series; we were concerned that we were missing the diagnosis. Soon, a sufficient number of prenatal diagnoses had been made to establish the reliability of the test in families in which one or more affected children had been born.

We also began work on a test for detection of Tay-Sachs disease heterozygotes. It was clear from a consideration of Slome's data that 80% of all Tay-Sachs births were the first affected children in the family. To effectively reduce the incidence of the disease by prenatal diagnosis, at-risk carrier couples had to be identified before procreation. Once again, Professor Robinson pointed to the differential susceptibility of hexosaminidase A and B to thermal inactivation at 50°C (Robinson and Stirling, 1968). We decided to determine whether this property could be exploited to quantify hexosaminidase A in human serum as a test for carriers, since we had found reduced hexosaminidase A in leukocytes of obligate heterozygotes. We worked out conditions for thermal inactivation of hexosaminidase A at 50°C in serum from normals and Tay-Sachs disease patients to establish the extremes, and applied the assay to as many obligate heterozygotes as we could obtain. We also assayed a large number of random samples from patients

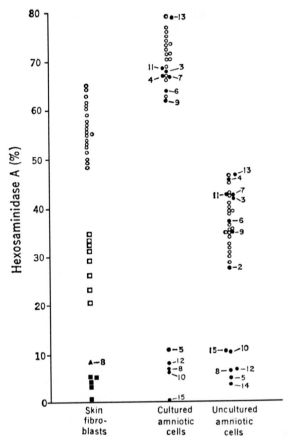

Figure 7.2. Hexosaminidase A activity in high-risk pregnancies expressed as percentage of total hexosaminidase. Filled circles: 15 high-risk pregnancies; open circles: controls; open squares: parents of Tay-Sachs disease patients closed squares: patients with Tay-Sachs disease.

in the hospital and found large variations in the percentage of hexosaminidase A activity, mostly decreasing, in subjects with diabetes, trauma, heart disease (especially coronary infarction), and liver diseases (Figures 4 and 6 in O'Brien *et al.*, 1970). Nonetheless, the test appeared to be a reasonably reliable carrier detection method using serum or plasma. We published our method and findings in the *New England Journal of Medicine* in 1970 (O'Brien *et al.*). This paper has also become a citation classic (1985) and so has Professor Robinson's paper on two hexosaminidases (1985). Our investigations were fully supported by all members of John's laboratory, including Arnie Miller, Mae Wan Ho, and others.

The rest of the story is now well known. Several laboratories, especially those of Professors J. Alexander Lowden and Michael Kaback, polished the test, automated it, worked out additional problems including the situation where birth control pills and pregnancy give false positive values, polished the leukocyte assay, and carried out mass screening in at-risk populations (Kaback *et al.*, 1974; Lowden *et al.*, 1973, 1974).

Hopefully, the brave new world of genetic engineering and gene therapy will obviate the need for the Tay-Sachs prevention program. Let us hope that this is the case. For us, this work has been certainly one of the most gratifying efforts of our own scientific careers. It is really our pleasure that there are numerous young adults throughout the world who have been born free of Tay-Sachs disease because of the prevention program. We are proud to show you our first prenatal diagnosis case for Tay-Sachs disease in 1969: this baby girl was born normal. We also have a collection of pictures of normal California children after prenatal diagnosis for Tay-Sachs disease. We believe that if Dr. Bernard Sachs were alive today, he would also be pleased.

References

Brady, R. O., Kanfer, J. N., and Shapiro, D. (1965). Metabolism of glucocerebrosides. II. Evidence of an enzyme deficiency in Gaucher's disease. *Biochem. Biophys. Res. Commun.* **18**, 221–225.

Citation Classic (1981). Current Contents, Life Science 24:16. (Okada & O'Brien, 1969).

Citation Classic (1985). Current Contents, Clinical Practice 13:18. (O'Brien *et al.*, 1971a).

Citation Classic (1985). Current Contents, Life Science 28:23. (Robinson & Stirling, 1968).

Kaback, M. M., Zeiger, R. S., Reynolds, L. W., and Sonneborn (1974). Approaches to the control and prevention of Tay-Sachs disease. *In* Steinberg, A., Bearn, A. (eds.): "Progress in Medical Genetics," Vol. 10 (M. M. Kaback and R. S. Zeiger, eds.), pp. 103–134. Grune & Stratton, New York.

Lowden, J. A., Skomorowski, M. A., Henderson, F., and Kaback, M. M. (1973). Automated assay of hexosaminidases in serum. *Clin. Chim.* **19**, 1345–1349.

Lowden, J. A., Zuker, S., Wilensky, A. J., and Skomorowski, M. A. (1974). Screening for carriers of Tay-Sachs disease: A community project. *Can. Med. Assoc. J.* **111**, 229–233.

Makita, A., and Yamakawa, T. (1963). The glycolipids of the brain of Tay-Sachs disease. The chemical structure of globoside and main ganglioside. *Jpn. J. Exp. Med.* **33**, 361–368.

O'Brien, J. S., Stern, M. S., Landing, B. H., O'Brien, J. K., and Donnell, G. M. (1965). Generalized gangliosidosis. *Am. J. Dis. Child.* **109**, 338–346.

O'Brien, J. S., Okada, S., Chen, A., and Fillerup, D. L. (1970). Tay-Sachs disease. Detection of heterozygotes and homozygotes by serum hexosaminidase assay. *N. Engl. J. Med.* **283**, 15–20.

O'Brien, J. S., Okada, S., Ho, M. W., Fillerup, D. L., Veath, M. L., and Adams, K. (1971a). Ganglioside storage diseases. *Fed. Proc.* **30**, 956–969.

O'Brien, J. S., Okada, S., Fillerup, D. L., Veath, M. L., Adornato, B., Brenner, P. H., and Leroy, J. G. (1971b). Tay-Sachs disease prenatal diagnosis. *Science* **172**, 61–64.

Okada, S., and O'Brien, J. S. (1968). Generalized gangliosidosis: β-galactosidase deficiency. *Science* **160**, 1002–1004.

Okada, S., and O'Brien, J. S. (1969). Tay-Sachs disease: Generalized absence of a β-D-N-acetylhexosaminidase component. *Science* **165**, 698–700.

Robinson, D., and Stirling, J. L. (1968). N-acetyl-β-glucosaminidase in human spleen. *Biochem. J.* **107**, 321–327.

8

The G$_{M2}$-Gangliosidoses and the Elucidation of the β-Hexosaminidase System

Konrad Sandhoff

Institut für Organische Chemie und
Biochemie der Universität Bonn, D-53121 Bonn, Germany

I. Amaurotic Idiocy
II. Glycolipid Analysis of Brains with Amaurotic Idiocy
    A. Tay-Sachs Ganglioside (TSG, G$_{M2}$) and Ganglioside
       G$_{M1}$ as Storage Compounds
III. Tay-Sachs Disease with Visceral Involvement (Variant 0)
IV. Search for the Defect in Tay-Sachs Disease
V. Variant AB and the G$_{M2}$-Activator Protein
VI. Variant B1
VII. Clinical and Biochemical Heterogeneity of G$_{M2}$
     Gangliosidosis—Degree of Enzyme Deficiency
     and Development of Different Clinical Syndromes
VIII. Addendum
    A. Hexosaminidases
    B. G$_{M2}$ Activator
    C. Pathogenesis
    Acknowledgment
    References

I. AMAUROTIC IDIOCY

Warren Tay (Tay, 1881) was the first to describe the clinical picture of infantile amaurotic idiocy. He observed a cherry-red spot surrounded by a white halo in the macula of the retina of a 1-year-old child with mental and physical retardation. The term familial amaurotic idiocy was coined more than 100 years ago by the neurologist Bernhard Sachs (Sachs, 1887, 1896a, 1896b) when he described

Advances in Genetics, Vol. 44
Copyright © 2001 by Academic Press
All rights of reproduction in any form reserved.
0065-2660/01 $35.00

the clinical picture and the morphological feature in 19 cases of the disease: distended cytoplasm of the neurons and ballooning of their dendrites. Thereafter, the classical, infantile form of amaurotic idiocy (iai) became known as Tay-Sachs disease (TSD). In addition, late infantile (Jansky-Bielschowsky), juvenile (Vogt, Spielmeyer), and adult (Kufs, Hallervorden) forms with similar morphologic features but different clinical symptomatology have been described. There was little progress in the understanding of the "disease" until the development of chemical, biochemical, and refined ultrastructural means of investigation. In the late 1930s the biochemist Ernst Klenk (1937, 1939, 1940), from the University of Köln, detected a new group of acid glycosphingolipids as storage material in the brain of patients with amaurotic idiocy. For their high concentration in normal ganglion cells (Klenk, 1942a, 1942b), they were named "Ganglioside" (gangliosides) and their acidic sugar component "Neuraminsäure" (neuraminic acid). The first ganglioside structure, that of ganglioside G_I (G_{M1}), was elucidated by Kuhn and Wiegandt (1963). The main neuronal storage compound in TSD was identified by Svennerholm (1962), and its structure was elucidated by Makita and Yamakawa (1963) and Ledeen and Salsman (1965). In the electron microscope, Terry and co-workers (Terry and Korey, 1960; Terry and Weiss, 1963; Samuels et al., 1963) observed storage granules (membranous cytoplasmic bodies) in the cytoplasm of the distended neurons that resemble pathologically modified lysosomes incapable of degrading their storage material.

II. GLYCOLIPID ANALYSIS OF BRAINS WITH AMAUROTIC IDIOCY

A. Tay-Sachs ganglioside (TSG, G_{M2}) and ganglioside G_{M1} as storage compounds

During my thesis work with Horst Jatzkewitz, an acetone-fixed brain with iai arrived in the laboratory around June 1962. A lipid analysis, which I, as a beginner, was certainly not supposed to do, and which I therefore undertook secretly on a small piece of brain, revealed two storage compounds, A' (G_{M2}) and B' (G_{A2}). I purified them quickly but stopped the structural analysis when Lars Svennerholm published the identification of the main storage compound in TSD, ganglioside G_{M2}, at the end of 1962 (Svennerholm, 1962). A screening of 12 formalin-fixed brains with amaurotic idiocy and appropriate controls identified A' (G_{M2}) and B' (G_{A2}) as storage compounds in 11 cases. However, the twelfth one did not show any accumulation of these lipids at all. This case (J.K.), who died at the age of $7\frac{3}{4}$ with late iai, stored ganglioside A (G_{M1}) and its sialic acid-free residue (G_{A1}) (Jatzkewitz and Sandhoff, 1963) (Figure 8.1).

       Though all major gangliosides of appropriate control brains were converted by the storage in formalin to G_{M1} and G_{A1}, it became clear from the

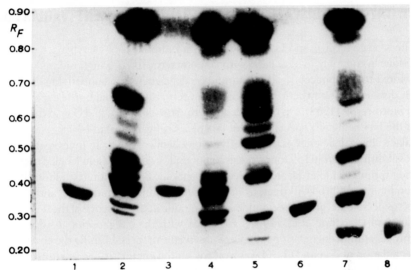

Figure 8.1. Thin-layer chromatogram of lipid extracts of brain tissue from two usual infantile cases and from one special form of late-infantile amaurotic idiocy (Jatzkewitz and Sandhoff, 1963) Adsorbent, 400-μm-thick layer of Kieselgel G, Merck; solvent system, propanol–conc. ammonia–water (6:2:1); height of the solvent front, 15 cm; detection, anisaldehyde, sulfuric acid in acetic acid (reagent of Kägi-Miescher). 1, 20 μg of neuraminic acid-free residue of ganglioside Tay-Sachs (B′, now named G_{A2}, R_F 0.33); 2, 250 μg of total lipid extract of the brain cortex of a case of iai (fresh tissue); 3, 20 μg of ganglioside Tay-Sachs (A′, G_{M2}, R_F 0.37); 4, 250 μg of total lipid extract of the brain cortex of a case of iai, preserved in formalin for 26 years; 5, 250 μg of total lipid extract of normal brain cortex, preserved in formalin for 26 years; 6, 20 μg of neuraminic acid-free residue of ganglioside A (G_{M1}, R_F 0.30); 7, 250 μg of total lipid extract of the brain cortex of a special form of late iai, preserved in formalin for 26 years; 8, 20 μg of neuraminic acid-free residue of ganglioside A (G_{A1}, R_F 0.26).

quantitative analysis that the brain of J.K. had an original accumulation of G_{A1} and consequently also of G_{M1}: the total ganglioside and G_{A1} content was 2.9-fold elevated and the ratio of G_{A1}/G_{M1} was 0.7 instead of 0.35 in appropriate controls. This clearly indicated that about 50% of the G_{A1} and more than 60% of the G_{M1} found in the brain of this patient resulted from an original storage process. A simultaneous accumulation of G_{A1} and a di- or oligosialylganglioside could be excluded for enzymatic reasons; it would imply the simultaneous defect of two different enzymes. Therefore, I still think that our 1963 paper (Jatzkewitz and Sandhoff, 1963) was probably the first description of G_{M1} gangliosidosis.

The quantitative analysis of further brains revealed a much reduced storage of G_{M2} in Niemann-Pick disease and in juvenile and adult cases of ai as compared with those with the infantile form of the disease (Jatzkewitz *et al.*, 1965).

III. TAY-SACHS DISEASE WITH VISCERAL INVOLVEMENT (VARIANT O)

The storage of G_{M2} and G_{A2}, which have a terminal β-N-acetylgalactosamine residue in common, in most cases with iai (later termed G_{M2} gangliosidosis) clearly pointed to a β-N-acetyl-galactosaminidase deficiency as an underlying cause of the disease. A biosynthetic block was almost certainly ruled out by the experiment of Burton et al. (1968). On the other hand, the storage of G_{M1} and G_{A1} in the brain of the biochemically special form of iai (J.K.; G_{M1} gangliosidosis) suggested a β-galactosidase deficiency as the underlying biochemical cause. To investigate these possibilities, I purified all four storage compounds in 1963, tritium labeled them in their sphingoid moieties, and studied their enzymic degradation (Sandhoff, 1965; Sandhoff et al., 1964). Extracts from mammalian kidney degraded G_{A1} and G_{A2} to lactosylceramide, glucosylceramide, and ceramide in the presence of the detergent Cutscum (Sandhoff et al., 1964). However, with the gangliosides as substrates, only very minor degradation rates were observed. After establishing the enzymatic degradation of G_{A1} and G_{A2}, I was working on the enzymatic degradation of G_{M2} while waiting for fresh autopsy tissues of patients with ganglioside storage in order to investigate the suggested metabolic blocks in these diseases.

In October 1966, deep-frozen autopsy material arrived from an infantile case (A.J.) with ai. The glycolipid analysis soon demonstrated differences from the cases I had studied before. Besides the storage of G_{M2}, the neuronal storage of G_{A2} was much more pronounced; in addition, globoside was accumulated in the visceral organs and, most important, hexosaminidase activity was almost completely absent (Figure 8.2). This defect was demonstrated with four different substrates (p-nitrophenyl-β-D-N-acetylglucosaminide, p-nitrophenyl-β-D-N-acetylgalactosaminide, glycolipid [^{3}H]G_{A2}, and [^{3}H]globoside) in four different organs.

Deep-frozen tissues from two further cases arrived shortly thereafter. One (C.B., from H. Moser, Boston) was later identified as the first case of AB variant, and the second (I.S., from Munich) was later identified as my first case with variant B. In these two cases total hexosaminidase levels were normal or even increased, especially in the brain tissues. Sialidase activity appeared to be normal in all three cases, catalyzing a slow conversion of G_{M2} to G_{A2}. Degradation of G_{M2} to G_{M3} was extremely slow even in the presence of detergents and could only be demonstrated qualitatively by crude and concentrated hexosaminidase preparations. Though the normal catabolism of G_{M2} and the defect in the classical TSD remained unclear, these data, including the total hexosaminidase deficiency in the case of A.J., were presented in July 1967 at the second Symposium on Cerebral Lipidoses in Coimbra, Portugal, and in the proceedings of the symposium (Sandhoff et al., 1968).

The respective manuscript was rejected by Biochemical and Biophysical Research Communications for being outside the scope of the journal.

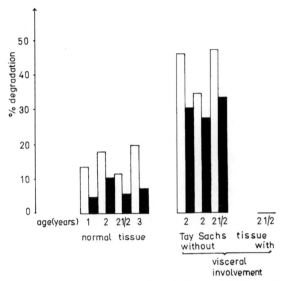

Figure 8.2. Hexosaminidase activity in normal and Tay-Sachs brain tissues (Sandhoff *et al.*, 1968). Substrates: □ *p*-nitrophenyl-β-D-N-acetylglucosaminide; ■ asialo residue of Tay-Sachs ganglioside. The incubation volume (2.1 ml) contained either 1 μmol *p*-nitrophenyl-β-D-N-acetylglucosaminide or 0.01 μmol asialo residue of Tay-Sachs ganglioside G_{A2} with 3 mg of sodium taurocholate and 2 ml of 0.4% brain cortex homogenate in 0.1 M citrate phosphate buffer, pH 4.5. The mixtures were incubated either for 30 min or for 3 h at 37°C. The ordinate shows the product formed, expressed as a percentage of the substrate used.

Klaus Harzer, who joined the laboratory, later analyzed hexosaminidase levels in blood samples of two families with visceral involvement and total hexosaminidase deficiency (Harzer *et al.*, 1971). The legal father of the first case (J.V.G.) had normal levels. This problem was resolved after a Belgian judge ordered a blood sample to be taken from a soldier who was accused by the mother of being the father. He indeed had hexosaminidase levels reduced to about 50% of normal. After being confronted with these data by the judge, he confessed to being the father.

IV. SEARCH FOR THE DEFECT IN TAY-SACHS DISEASE

In 1963, Korey and Stein presented a gangliosidase system in which they assayed the disappearance of ganglioside sialic acid in the presence of rat brain homogenates at pH 7. Using tritium-labeled gangliosides G_{M1} and G_{M2}, I could not find any degradation products under these conditions. However, homogenates of rat kidney or spleen degraded $[^3H]G_{M2}$ slowly between pH 4 and 5.4 (4.5% of 140 μg of G_{M2} within 3 h by homogenate of 0.5 g spleen). Ceramide, glucosyl-ceramide, and glycolipid G_{A2} were identified as degradation products.

In 1967 I resumed purification of hexosaminidases and found a variety of different hexosaminidase (Hex) species in different mammalian species after preparative gel electrophoresis or isoelectric focusing (Sandhoff, 1968). There-fore, the separation of human hexosaminidase into two major species was not so impressive to me. It was described before in an abstract by Robinson and Stirling (1966) (which unfortunately I read only later) and in full length in 1968 (Robinson and Stirling, 1968).

Isoelectric focusing of hexosaminidase of brain tissue of the two available cases with iai gave contradictory results in the spring of 1968. The brain tissue of C.B. (later designated variant AB) contained both Hex A and Hex B activity: even at elevated levels, the brain tissue of the second case (later designated vari-ant B) showed increased activity of Hex B and was deficient in Hex A activity (Figure 8.3).

Hex A is more thermolabile than Hex B. Therefore, an artifact could not be ruled out completely. Furthermore, at that time I could not demonstrate that Hex A and not Hex B is involved in G_{M2} catabolism. Since the case with both hexosaminidases A and B active (C.B.) and a clinical diagnosis typical for TSD and confirmed storage of G_{M2} and G_{A2} had no Hex A deficiency, data on Hex A deficiency were kept in the drawer until the summer 1969. In April 1969 we got frozen brain tissue from a new case with iai and thereafter tissue of still another one. Both showed again increased levels of Hex B and defective Hex A activity. These data were published in the summer of 1969 (Sandhoff, 1969). Then I learned that J. S. O'Brien also did not find Hex A activity in several patients with TSD (Okada and O'Brien, 1969). We could then confirm the Hex A deficiency in deep-frozen tissue samples of eight ad-ditional cases with the Jewish form of TSD, which, Horst Jatzkewitz got from A. C. Crocker (Boston) in July 1969. These specimens were included in a thorough study on lipid storage in the three enzymatic variants of TSD found so far (Sandhoff et al., 1971).

Though the Hex A deficiency was a reproducible finding in the clas-sical form of TSD and was also confirmed by Hultberg (1969) in Sweden, the significance of Hex A for the normal catabolism of G_{M2} remained obscure. So far degradation of G_{M2} had been demonstrated only with extracts from rat tissue homogenates (Sandhoff, 1965), calf brain (Frohwein and Gatt, 1967), and human muscle (Kolodny et al., 1969). Therefore, I purified Hex A and Hex B from post-mortem human liver 3000- and 1000-fold, respectively, and studied their speci-ficity against the storage compounds. Both isoenzymes were active on $[^3H]G_{A2}$ and on $[^3H]$globoside in the presence of a crude sodium taurodeoxycholate preparation

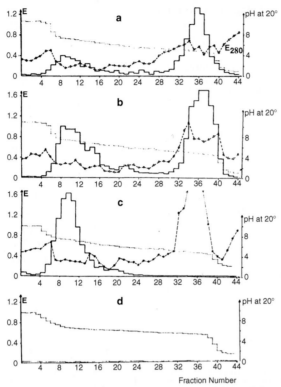

Figure 8.3. Variation of β-N-acetylhexosaminidase pattern in the brain tissue of cases with TSD (Sandhoff, 1969). (a) Normal human pattern. (b) Unaltered pattern in conventional Tay-Sachs case (now named variant AB). (c) Partially defective pattern in a conventional Tay-Sachs case (now named variant B). (d) Defective pattern in the special Tay-Sachs case with visceral storage of kidney globoside (now named variant 0 or Sandhoff disease). Curves were obtained by isoelectric focusing. ———, β-N acetylhexosaminidase activity measured at 410 nm; (–o–o–o–), extinction at 280 nm; (•–•–•–•), pH values.

as detergent. On the other hand, only Hex A converted [^{3}H]G$_{M2}$, labeled in its sphingoid moiety, to a compound identical with G$_{M3}$, as evidenced by thin-layer chromatographic analysis (Sandhoff, 1970; Sandhoff and Wässle, 1971). However, the reaction rate (0.1 nmol G$_{M2}$/min/mg) was very slow and could hardly account for the ganglioside turnover in normal brain. Presumably, the *in vitro* incubation conditions using some randomly chosen detergent were not very efficient and could be improved.

Several major questions remained:

How fast could Hex A hydrolyze G_{M2} under more suitable *in vitro* conditions?

What is the relationship among Hex A, Hex B, and Hex S?

What is the defect in patient C.B.? Is his Hex A mutated though still active on synthetic substrates and G_{A2}, or is his disease caused by the absence of still another, so far unknown factor involved in G_{M2} degradation?

Are the enzymic variants observed in TSD in biochemical terms really three different diseases?

To answer the last question, we first developed thin-layer chromatographic methods for the quantitative determination of the main glycolipids and phospholipids. These methods were used to quantitate the glycolipid storage and the lipid pattern parallel to enzyme alterations in tissue of one case without Hex deficiency (AB variant), in three cases with deficiency of Hex A and B (0 variant), and in the eight cases with Hex A and Hex S deficiency (variant B), obtained mostly from A. C. Crocker (Boston) (Sandhoff *et al.*, 1971) (Figure 8.4).

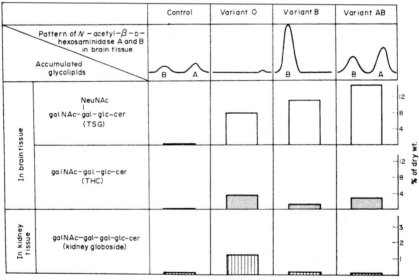

Figure 8.4. Correlation between the alteration or the hexosaminidase pattern and the glycolipid accumulation in three variants of TSD (Sandhoff *et al.*, 1971). Glycolipid content is expressed as percentage of dry weight; each column represents the mean value of three cases (variant AB: one case), each case representing the average of three determinations.

The three enzymatic variants clearly differed in their storage pattern: brains with variant B accumulated besides G$_{M2}$ only minor amounts of G$_{A2}$, whereas the variant AB brain showed in comparison a markedly increased accumulation of G$_{A2}$ and even higher G$_{M2}$ levels. Variant 0 brains had in comparison the highest G$_{A2}$ and the lowest G$_{M2}$ levels. In addition, the visceral organs showed globoside storage. Myelin lipids like cerebrosides and sulfatides were drastically reduced in the brains of all three variants, in variant 0 more than in variant B and here more than in variant AB. β-Glucosidase activity was increased in the brains of all three variants but to a different degree: 14-fold in variant 0, 9-fold in variant B, and only 7-fold in variant AB. Hexosaminidases still active in the

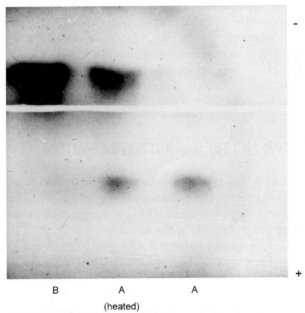

B   A   A
(heated)

Figure 8.5. Conversion of the acidic N-acetyl-β-D-hexosaminidase A to a basic form (Sandhoff, 1973). N-acetyl-β-D-hexosaminidase A, purified about 4,000-fold from human liver, was heated for 30 min at 50°C in a 0.01 M phosphate buffer, pH 5.0, 0.1 M ammonium sulfate, protein concentration 10–20 μg/ml. An aliquot (15 μl) of the enzyme solution was subjected to starch-gel electrophoresis in a 0.04 M phosphate buffer, pH 7.0, for 3 h, according to the procedure of Robinson *et al.* (1967). The enzyme bands were visualized according to the method of Hayashi (1965) using the N-acetyl-β-D-glucosaminide of 7-bromo-3-hydroxy-2-naphthoaniside as substrate and fast garnet GBC at pH 4.4 as diazonium salt. B = N-Acetyl-β-D-hexosaminidase B, purified from human liver; A, heated = N-acetyl-β-D-hexosaminidase A, purified from human liver and heated for 30 min at 50°C as described above; A = N-acetyl-β-D-hexosaminidase A, purified from human liver.

brains with variant AB (Hex A and Hex B) and variant B (Hex B) exhibited normal substrate specificities, pH profiles, and thermolabilities. The relationship between hexosaminidases A and B (and S) remained obscure. An experiment I presented at the Sarasota meeting in 1972 (Sandhoff, 1973) showed that highly purified Hex A could be converted almost quantitatively to Hex B as assayed by starch gel electrophoresis, in the absence of added sialidase or merthiolate. During this conversion, total Hex activity against synthetic substrate dropped only slightly (Figure 8.5).

The explanation for this experiment was forwarded by Srivastava and Beutler (1973) when they introduced immunochemical techniques to analyze the interconversion of purified hexosaminidases by freezing and thawing. They concluded that both major hexosaminidases have a common subunit, Hex A being a heteropolymer $(\alpha\beta)_n$ and Hex B a homopolymer $(\beta\beta)_n$. This proposal could also explain deficiency of Hex A and Hex B in variant 0 as a consequence of a mutation in the common subunit β, and the Hex A [and Hex S $(\alpha\alpha)$] deficiency in variant B as a consequence of an an α-subunit mutation (Beutler and Kuhl, 1975; Geiger and Arnon, 1976). This model was supported by hybridization experiments with variant B and variant 0 fibroblasts that gave complementation with the formation of Hex A (Thomas et al., 1974; Galjaard et al., 1974).

V. VARIANT AB AND THE G_{M2}-ACTIVATOR PROTEIN

Hexosaminidases A and B, obtained from the brain of our case with variant AB (C.B.), showed normal kinetic data with synthetic substrates (pNPGlcNAc, pNPGalNAc) and with glycolipid $[^3H]G_{A2}$ in the presence of detergents. The liver extracts of the patient also had some activity against G_{M2}. Thus, the cause of G_{M2} and G_{A2} accumulation remained an enigma. Further studies were stopped when we lost the deep-frozen tissues of the case with variant AB after Christmas of 1973 in a freezer accident. These studies could be resumed only after we got frozen tissue samples of a new case of variant AB in 1974 from Patrick and Ellis (London) (Brett et al., 1973) and fibroblasts of another case (N.S., de Baeque et al., 1975) from Dr. Beratis (New York) in 1976. In our laboratory, E. Conzelmann, who started his thesis work in 1975 with me, checked all possible biochemical parameters of the patient's enzyme and found no abnormalities. I then took samples to B. Geiger and R. Arnon at the Weizmann Institute, where we analyzed the patient's hexosaminidases with immunochemical techniques. No differences between Hex A from normal and variant AB tissue could be detected. In particular, the patient's storage compounds G_{M2} and G_{A2} were split by the patient's Hex A in the presence of detergents at a normal rate, indicating that the defect involved in the disease is not at the genetic level of production of either α or β chains of Hex A (Conzelmann et al., 1978). In the next series of experiments, we searched

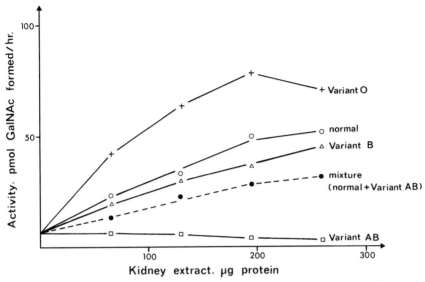

Figure 8.6. Absence of an activating factor from unheated kidney extract of variant AB (Conzelmann and Sandhoff, 1978). Stimulation of the hexosaminidase A-catalyzed hydrolysis of [³H]ganglioside G_{M2} was measured in the presence of extracts (13 mg protein/ml) prepared from human kidneys. Incubation mixtures containing increasing amounts of kidney extracts, 10 μmol of [³H]ganglioside G_{M2}, and 65 ml of purified Hex A were analyzed for [³H]GalNAc formed. Blanks were run for each value without additional Hex A and subtracted. --•--•--•, mixture (1:1, vol/vol) of extracts from normal kidney and variant AB kidney.

in the absence of detergents for possible factors in human tissues stimulating G_{M2} and G_{A2} hydrolysis by purified A and being deficient in variant AB. One such factor was indeed detected in several human tissues and was found to be missing in variant AB tissues (Figure 8.6) (Conzelmann and Sandhoff, 1978).

A partially purified sulfatide activator preparation from Jatzkewitz stimulated G_{M2} and G_{A2} hydrolysis by purified Hex A to some extent; a preparation obtained from Li for the stimulation of ganglioside degradation (Li and Li, 1976) had almost no effect (Sandhoff et al., 1977). So, it was doubtful that the protein factor missing in AB variant (now named G_{M2} activator protein) had a relationship to the activators already described. Therefore we purified the G_{M2} activator protein missing in AB variant tissues from normal human kidney to apparent homogeneity and studied its properties (Conzelmann and Sandhoff, 1979). The human G_{M2} activator protein turned out to be a glycosphingolipid-binding protein. In its presence, clear Michaelis-Menten kinetics were obtained for the G_{M2} and G_{A2} degradation by purified Hex A. The activator complexes the glycolipid molecules and presents them to the enzyme, which otherwise cannot attack the

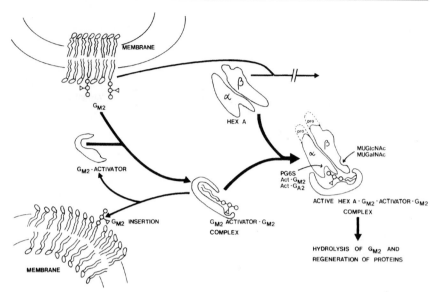

Figure 8.7. Model for the lysosomal catabolism of ganglioside G_{M2} (Sandhoff and Conzelmann, 1984). Hex A cannot attack membrane-bound ganglioside G_{M2}. Instead, the ganglioside is extracted from the membrane by the activator protein and the water-soluble activator–lipid complex is the substrate for the enzymic reaction. Of the two catalytic sites on Hex A, only the one on the α subunit cleaves ganglioside G_{M2}. The hexosaminidase precursor ("prohex A") is also fully active on the activator/G_{M2} complex (Hasilik *et al.* 1982). After the reaction, the product, ganglioside G_{M3}, is reinserted into the membrane and the activator protein is available for another round of catalysis (Conzelmann and Sandhoff, 1979; Conzelmann *et al.*, 1982).

aggregates formed by lipids in aqueous solution. Kinetics obtained in the presence of detergents always suffered from nonlinearities and poor reproducibility. The degradation rate for G_{M2} by Hex A measured in the presence of the G_{M2} activator was high enough to account for the *in vivo* turnover of G_{M2} and exceeded the result obtained eight years before, in the presence of a crude taurocholate preparation by a factor of 160! The activator is highly specific for Hex A: in contrast to detergents, hydrolysis of G_{A2} and G_{M2} by Hex B is almost not enhanced by this protein. The specificity of Hex A and Hex B against G_{M2}, G_{A2}, and globoside in the presence of the G_{M2} activator could account satisfactorily for the lipid storage patterns observed in patients with different variants of infantile G_{M2} gangliosidosis (Figure 8.7).

A water-soluble stoichiometric complex of G_{M2} and activator protein was isolated and identified as the real Michaelis-Menten substrate of Hex A (Conzelmann *et al.*, 1982). Furthermore, the G_{M2} activator was identified *in vitro* as a glycolipid transfer protein. Its species specificity was studied (Burg *et al.*, 1983), and its subcellular and tissue distribution in normal and pathologic tissues was measured in ELISAs (Banerjee *et al.*, 1984). In immunoblots, the mature lysosomal glycoprotein of the G_{M2} activator and its precursor were identified. Both of them were absent in our two cases with AB variant (Burg *et al.*, 1985a; see also Hirabayashi *et al.*, 1983). Metabolic studies with exogenous [³H]G_{M2} revealed a clear block in the G_{M2} catabolism in AB variant fibroblasts, which could be largely released after feeding small amounts of highly purified G_{M2} activator to the mutant cells (Figure 8.8) (Sonderfeld *et al.*, 1985b).

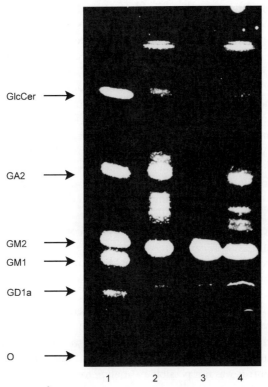

Figure 8.8. Metabolism of [³H]G_{M2} in cells from G_{M2} gangliosidosis, variant AB, as function of exogenously added G_{M2} activator protein (Sonderfeld *et al.*, 1985b). The fibroblasts were fed with [³H]G_{M2} (50 μM) and after 70 h harvested and processed. Total lipid extracts were analyzed by thin-layer chromatography. Lane 1, standards GD1a, G_{M1}, G_{M2}, G_{A2}, GlcCer; lane 2, total lipid extracts of normal cells; lane 3, total lipid extracts of cells from G_{M2} gangliosidosis, variant AB; lane 4, total lipid extracts of cells from G_{M2} gangliosidosis, variant AB, after feeding 30 μg G_{M2} activator protein for 70 h; 0, origin.

The gene of the G_{M2} activator was localized on chromosome 5 (Burg *et al.*, 1985b), and its hydrophobic binding site for G_{M2} was labeled with the help of an N-bromoacetyllyso-G_{M2} (Neuenhofer and Sandhoff, 1985).

VI. VARIANT B1

In 1980, Suzuki and co-workers (Goldman *et al.*, 1980) described two cases with apparent AB variant of G_{M2} gangliosidosis. Though we got fibroblasts of both patients, we analyzed only those of one, in which the G_{M2} activator antigen proved to be absent. Upon moving our laboratory from Munich to Bonn and taking over all the teaching obligations there, we did not analyze the second case and then forgot about it. In the meantime, Li *et al.* (1981) reported that this case (D.N.) had no activator deficiency. They suggested that this patient might have a structural gene mutation of Hex A, such that the enzyme became nonresponsive to stimulation by the G_{M2} activator for the hydrolysis of G_{M2} ganglioside (Hirabayashi *et al.*, 1983). To clarify this point, I asked H. -J. Kytzia, who did his thesis work with me, to study the substrate specificity of the hexosaminidases of two patients with clinical TSD and extensive neuronal G_{M2} storage (D.N. from New York and F.M. from Lyon, France) having apparently both Hex A and G_{M2} activator activity.

Hex A of both patients showed almost normal kinetic constants in cleaving artificial N-acetylglucosaminides and -galactosaminides, but were almost inactive in hydrolyzing a sulfated glucosaminide (a newly described substrate for Hex A, Kresse *et al.*, 1981) and ganglioside G_{M2} in the presence of the G_{M2} activator protein (Table 8.1) (Kytzia *et al.*, 1983). These data immediately suggested two different active sites on normal human Hex A, one on the α subunit and one on the β subunit.

The patients' Hex A apparently has lost one active site whereas the other one was left intact. We soon could prove by suitable competition experiments between different substrates that indeed purified human Hex A from normal tissue carries two distinct active sites with different substrate specificities (see Figure 8.7) (Kytzia and Sandhoff, 1985).

In hybridization experiments, we tried to find out which subunit of Hex A was mutated in variant B1 (Figure 8.9) (Sonderfeld *et al.*, 1985a). In these experiments it was not possible to analyze for the reappearance of any specific protein in the hybrid cells obtained from different mutated cells because all proteins needed for ganglioside G_{M2} degradation were already present in the co-cultivated mutant cells. Therefore, we analyzed the capability of co-cultivated as well as of hybrid cells to degrade exogenously added ganglioside $[^3H]G_{M2}$. This method turned out to be very sensitive because intergenic complementation of mutant cells resulted in a dramatic increase of $[^3H]G_{M2}$ catabolism. Though

Table 8.1. Hexosaminidase Activities Measured in Fibroblast Homogenates Toward Various Substrates[a]

Cells/substrates	4-MU-GlcNAc[b] [nmol/min/mg] [hex A + B]	G_{M2}[c] pmol/h/mg/ [AU[e]]	PG-6S[d] [nmol/min/mg]	Total % Hex A activity
Normal controls:				
1	66.3	59	384	3.34
2	81.2	71	681	7.3
3	55.3	52	425	5.05
Late-infantile variant B	52.2	4	3.3	0.095
Infantile variant AB	80.6	67	460	6.94
Proband F.M.	52.8	50	3.4	0.085
Proband D.N.	115.1	60	25.1	0.071

[a]Determinations were done at least in duplicates. Deviations were always <5%.
[b]4-MU-GlcNAc = 4-methylumbelliferyl-β-D-2-acetamido-2-deoxyglucopyranoside.
[c]Determined in fibroblasts supernatants (100,000g) in the presence of G_{M2} activator and corrected for homogenate values (Conzelmann *et al.*, 1983).
[d]PG-6S = *p*-nitro-phenyl-β-D-glucosaminide-6-sulfate.
[e]AU = activator unit as defined by Conzelmann and Sandhoff (1979).
Source: Kytzia *et al.*, 1983.

presumably only small amounts of active proteins were formed in the hybrid cells, this resulted in a steep increase of turnover rates for G_{M2} degradation (compare also Conzelmann and Sandhoff, 1983/1984). This method to measure turnover rates after cell hybridization as an assay for intergenic complementation may be useful for other problems, too. The hybridization experiments demonstrated variant B1 being allelic to variant B (Sonderfeld *et al.*, 1985a). This clearly indicated that an active site on the α subunit of Hex A is predominantly involved in the hydrolysis of MUGlcNAc-6-S and G_{M2} (as well as G_{A2}) in the presence of the G_{M2} activator and that this active site is inactivated due to a mutation in the α subunit in patients with variant B1. On the other hand, the β subunit of human Hex A carries an active site which is mainly responsible for the hydrolysis of water-soluble N-acetyl-glucosaminides and N-acetyl-galactosaminides such as MUGlcNAc and MUGalNAc, and which is still active in Hex A of patients with variant B1. Kinetic experiments showed furthermore that the free G_{M2} activator competes with the sufated substrate MUGlcNAc-6-S (4-methylumbelliferyl-2-acetamido-2-deoxy-6-sulfoglucopyranoside) for the α site of Hex A and Hex S (Kytzia and Sandhoff, 1985). This indicates that the G_{M2} activator recognizes the α subunit, an observation that is in line with the earlier finding (Conzelmann and Sandhoff, 1979) that Hex A (and not Hex B) almost exclusively degrades glycolipids G_{M2}, G_{A2}, and globoside in the presence of the G_{M2} activator

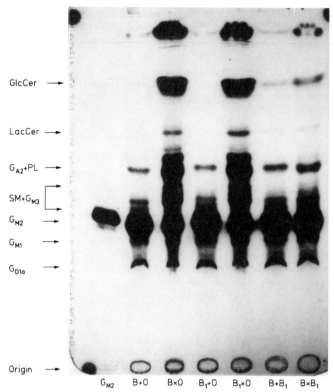

Figure 8.9. Metabolism of exogenously added [³H]-G$_{M2}$ in co-cultivated and fused fibroblasts (Sonderfeld *et al.*, 1985a). Fibroblasts from patients with G$_{M2}$ gangliosidosis, variants B, 0, and B1, were co-cultivated (+) or fused (×) with polyethyleneglycol. Then they were fed with [³H]-G$_{M2}$ for 72 h, harvested, and extracted. The lipids were separated by thin-layer chromatography and the radioactive spots visualized by fluorography. GlcCer, glucosylceramide; G$_{A2}$, GalNAcβ1→4Galβ1→4Glcβ1→1Cer; G$_{M3}$, Gal(3←2αNeuAc)β1→4Glcβ1→1Cer; G$_{M2}$, GalNAcβ1→4Gal (3←2αNeuAc)β1→4Glcβ1→1Cer; G$_{M1}$, Galβ1→3GalNAcβ1→4Gal(3←2αNeuAc)β1→4Glcβ1→1Cer; G$_{D1a}$, Gal (3←2αNeuAc)β1→3GalNAcβ1→4Gal(3←2αNeuAc)β1→4Glcβ1→1Cer; LacCer, lactosylceramide; PL, phospholipid identified as phosphatidylcholine after isolation from the thin-layer plate, alkaline hydrolysis, and chromatographic identification of the labeled fatty acids; SM, sphingomyelin.

protein. Even proHex A already has full activity on G$_{M2}$ in the presence of the G$_{M2}$ activator protein (Hasilik *et al.*, 1982).

A third case with G$_{M2}$ gangliosidosis obtained from Prague was soon diagnosed as variant B1, and a prenatal diagnosis was performed successfully in this family (Conzelmann *et al.*, 1985).

VII. CLINICAL AND BIOCHEMICAL HETEROGENEITY OF G_{M2} GANGLIOSIDOSIS—DEGREE OF ENZYME DEFICIENCY AND DEVELOPMENT OF DIFFERENT CLINICAL SYNDROMES

Deficiencies of Hex A and of both hexosaminidases, A and B, caused by allelic mutations in the α and β subunits, respectively, have been observed in a variety of patients with different clinical forms (infantile, juvenile, adult, and chronic) of G_{M2} gangliosidoses representing a wide variety of clinical symptomatology and in pseudo-deficient patients. Whereas the infantile forms show a very uniform clinical picture, that of juvenile and that of adult forms is quite variable and ranges from spinocerebellar degeneration, muscular atrophy- and amyothrophic lateral sclerosis-like syndromes, to psychiatric symptomatology which hardly resembles that of the infantile form of the disease. On the biochemical level, this heterogeneity is paralleled by a variation of the extent and pattern of the glycolipid accumulation in different regions of the brain. Whereas infantile forms show an excessive and ubiquitous neuronal storage, late-onset forms have a less pronounced accumulation (Jatzkewitz et al., 1965) which is restricted to specific brain regions, mainly the hippocampus, the nuclei of the brainstem, the spinal cord, the granular cells of the cerebellum, and the retina (Jatzkewitz et al., 1965; Escola, 1961; Terry and Weiss, 1963). Obviously, different allelic mutations in one gene locus can lead to extremely different clinical syndromes and neuropathologic forms. A crucial difference observed between different clinical forms, e.g., of variant B of G_{M2} gangliosidosis, are different residual activities of the mutated Hex A of cultured fibroblasts against its natural substrate (Conzelmann et al., 1983/1984). Residual activity assayed in fibroblasts of adult and juvenile patients were small compared with normal levels but significantly higher than that obtained from fibroblasts of infantile patients. Residual activities in the range of 10–20% of the control value already appear to be compatible with normal life. Similar findings in fibroblasts from patients with variant 0 (Kytzia et al., 1984) revealed the importance of small variations of the low residual enzyme activities for the development of different clinical syndromes of a disease.

These observations can be understood qualitatively on the basis of a greatly simplified kinetic model (Conzelmann and Sandhoff, 1983/1984). Assuming a constant influx rate (v_i) of the substrate (e.g., ganglioside) into the lysosomal compartment of an individual cell, the steady-state substrate concentration [S_{eq}] and the turnover rate were simply calculated as a function of the residual enzyme activity (Figure 8.10).

As long as the normal degradative capacity (V_{max}) exceeds the influx rate v_i severalfold (for the G_{M2}-degrading system a factor of about 20-fold was estimated), [S_{eq}] will be way below the K_M value of the degrading enzyme. In this case, substantial reductions of the degrading enzyme's activity (caused by

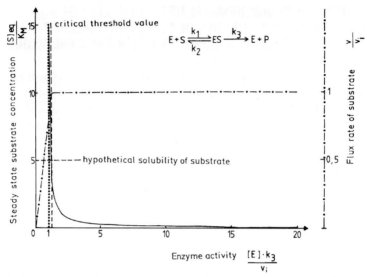

Figure 8.10. Steady-state substrate concentration as a function of enzyme concentration and activity (Conzelmann and Sandhoff, 1983/1984). The model underlying this theoretical calculation assumes influx of the substrate into a compartment at a constant rate (v_i) and its subsequent utilization by the enzyme (for details, see text). ——— $[S]_{eq}$, steady-state substrate concentration; •••• = theoretical threshold of enzyme activity; — — — = critical threshold value, taking limited solubility of substrate into account; •—•—• = turnover rate of substrate (flux rate).

whatever mutation or mechanism) will lead to only a moderate increase of $[S_{eq}]$ and not to an irreversible accumulation, since the remaining enzyme activity is still sufficient to cope with the substrate's influx rate. However, when the residual enzyme activity falls below a critical threshold (and when all available enzyme molecules are saturated with substrate), the turnover of the substrate will be reduced and it will accumulate. The rate of accumulation will be proportional to the difference between threshold activity and the smaller actual residual activity in the lysosome.

Rates of ganglioside biosynthesis (corresponding to influx rates) and degrading enzyme activities (Hex A in the presence of G_{M2} activator) may differ in different organs, cell types, and even between individual neurons. Therefore, the consequence of an incomplete enzyme deficiency may be different for different cells and organs. Cells with a high rate of substrate biosynthesis (corresponding to a high influx rate) and a low level of residual degrading enzyme activity should store first. This means that a certain subset of neurons should be affected preferentially by the storage process, e.g., in the adult form of G_{M2} gangliosidosis. Thus, in contrast to infantile forms, in which all neurons are affected, in late-onset forms function and viability of a subset of neurons should be affected preferentially, causing different clincial symptomatology.

Of course, the oversimplified model, based on plain Michaelis-Menten kinetics without considering any regulation of enzyme activities or intracellular transport of substrates or other relevant parameters, has to be refined in order to contribute to the understanding of the pathogenesis of different diseases. Other factors involved certainly include mechanical distortion of the cells, depletion of precursor pools, and alterations of membrane composition and functions by the storage process. The formation of meganeurites and increase in synaptic spines (Purpura and Suzuki, 1976) may disturb the connectivity of the brain. The formation of lytic and toxic compounds such as glucosylsphingosine in Gaucher disease, galactosylsphingosine in Krabbe disease (Igisu and Suzuki, 1984; Miyatake and Suzuki, 1972; Nilsson et al., 1982; Svennerholm et al., 1980), and lysoganglioside G_{M2} in brains with variants B and 0 of G_{M2} gangliosidosis (Neuenhofer et al., 1986; Rosengren et al., 1987) may also contribute to the pathogenic process in these diseases.

Though the basic defects in G_{M2} gangliosidosis are known and the individual mutations causing them will be soon analyzed in many patients at the level of the genome, the mechanisms of pathogenesis are not yet really understood.

VIII. ADDENDUM

Since this manuscript was prepared, we have continued to investigate the G_{M2}-degrading system on the protein level and on the molecular biological level, as well as the diseases related to this system. Further work on these fields is in progress.

A. Hexosaminidases

Studies were conducted toward understanding the structural requirements for hexosaminidase function and specificity.

Specific photoincorporation of a substrate analog into purified human lysosomal Hex B, tryptic digestion of the covalently labeled enzyme, and isolation and sequencing of the labeled peptides led to the identification of Glu-355 at the substrate-binding site (Liessem et al., 1995).

With the aid of genetically engineered chimeric constructs of the α and β subunits, domains have been identified that confer distinctive substrate specificity to the hexosaminidase isoenzymes (Pennybacker et al., 1996).

B. G$_{M2}$ activator

Studies on the specificity of the G$_{M2}$ activator protein support the model of its mechanism of action as depicted in this chapter (Meier *et al.*, 1991).

Point mutations and deletions within the structural gene of the G$_{M2}$ activator have been identified in four patients with the AB variant of G$_{M2}$ gangliosidosis (Schröder *et al.*, 1991, 1993; Schepers *et al.*, 1996). The mutated proteins are proteolytically labile, so that some of them are degraded in the endoplasmic reticulum and others in later compartments. The resulting loss of the G$_{M2}$ activator causes a block within the G$_{M2}$ degradation in the lysosomes, as demonstrated in metabolic studies in cultivated fibroblasts of the patients. The block can be bypassed by feeding native G$_{M2}$ activator to the culture medium of the mutant fibroblasts (Klima *et al.*, 1993).

C. Pathogenesis

In the meantime, we were able to verify our kinetic model which correlates the clinical course of lysosomal storage diseases with the residual activity of the affected enzyme. We examined this model by measuring flux rates and enzyme activities in cultivated skin fibroblasts from patients with G$_{M2}$ gangliosidoses. The experimentally observed values support our hypotheses, since the flux rate decreases linearly with decreasing enzyme activity only below a threshold value. Patients with adult course of the disease show a significantly higher turnover than patients with juvenile course, and these a higher turnover than patients with infantile course. Smaller differences in the residual enzyme activities correspond to significant differences in the respective flux rates (Leinekugel *et al.*, 1992).

In the meantime, animal models of the G$_{M2}$ gangliosidoses have become available and permit further studies of the pathogenesis of this diseases. In contrast to the human diseases, the mouse models of B variant and 0 variant show very different neurologic phenotypes. The Tay-Sachs model showed no neurologic abnormalities, whereas the Sandhoff model was severely affected. This phenotypic difference between the two mouse models is due to differences in the ganglioside degradation pathway between mice and humans (Sango *et al.*, 1995)

Acknowledgment

I thank Dr. Thomas Kolter for helpful discussions and the Deutsche Forschungsgemeinschaft for supplying the work done in the authors laboratory.

References

Banerjee, A., Burg, J., Conzelmann, E., Carroll, M., and Sandhoff, K. (1984). Enzyme-linked immunosorbent assay for the ganglioside G$_{M2}$-activator protein screening of normal human tissues and bodyfluids, of tissues of G_{M2}-gangliosidosis, and for its subcellular localization. *Hoppe-Seyler's Z. Physiol. Chem.* **365,** 347–356.

Beutler, E., and Kuhl, J. (1975). Subunit structure of human hexosaminidase verified: Interconvertibility of hexosaminidase isoenzymes. *Nature (Lond.)* **258,** 262–264.

Brett, E. M., Ellis, R. B., Haas, L., Ikonne, J. U., Lake, B. D., Patrick, A. D., and Stephens, R. (1973). Late onset G_{M2} gangliosidosis: Clinical, pathological, and biochemical studies on eight patients. *Arch. Dis. Child.* **48,** 775–785.

Burg, J., Banerjee, A., Conzelmann, E., and Sandhoff, K. (1983). Activating proteins for ganglioside G_{M2} degradation by β-hexosaminidase isoenzymes in tissue extracts from different species. *Hoppe-Seyler's Z. Physiol. Chem.* **364,** 821–829.

Burg, J., Banerjee, A., and Sandhoff, K. (1985a). Molecular forms of G_{M2}-activator protein—A study on its biosynthesis in human skin fibroblasts. *Biol. Chem. Hoppe-Seyler.* **366,** 887–891.

Burg, J., Conzelmann, E., Sandhoff, K., Solomon, E., and Swallow, D. M. (1985b). Mapping of the gene coding for the human G_{M2} activator protein to chromosome 5. *Ann. Hum. Genet.* **49,** 41–45.

Burton, R., Handa, S., Howard, R. E., and Vietti, T. (1968). Incorporation of selected isotopes into lipids of humans with cerebral lipidoses: Studies on D-glucosamine-1-14C. *Pathol. Eur.* **3,** 424–430.

Conzelmann, E., and Sandhoff, K. (1978). AB variant of infantile G_{M2}-gangliosidosis: Deficiency of a factor necessary for stimulation of hexosaminidase A-catalyzed degradation of ganglioside G_{M2} and glycolipid G_{A2}. *Proc. Natl. Acad. Sci. (USA)* **75,** 3979–3983.

Conzelmann, E., and Sandhoff, K. (1979). Purification and characterization of an activator protein for the degradation of glycolipids G_{M2} and G_{A2} by hexosaminidase A. *Hoppe-Seyler's Z. Physiol. Chem.* **360,** 1837–1849.

Conzelmann, E., and Sandhoff, K. (1983/1984). Partial enzyme deficiencies: Residual activities and the development of neurological disorders. *Dev. Neurosci.* **6,** 58–71.

Conzelmann, E., Sandhoff, K., Nehrkorn, H., Geiger, B., and Arnon, R. (1978). Purification, biochemical and immunological characterization of hexosaminidase A from variant AB of infantile G_{M2}-gangliosidosis. *Eur. J. Biochem.* **84,** 27–33.

Conzelmann, E., Burg, J., Stephan, G., and Sandhoff, K. (1982). Complexing of glycolipids and their transfer between membranes by the activator protein for lysosomal ganglioside G_{M2} degradation. *Eur. J. Biochem.* **123,** 455–464.

Conzelmann, E., Kytzia, H.-J., Navon, R., and Sandhoff, K. (1983). Ganglioside G_{M2} N-acetyl-β-galactosaminidase activity in cultured fibroblasts of late infantile and adult G_{M2}-gangliosidosis patients and of healthy probands with low hexosaminidase level. *Am. J. Hum. Genet.* **35,** 900–913.

Conzelmann, E., Nehrkorn, H., Kytzia, H.-J., Sandhoff, K., Macek, M., Lehovsky, M., Elleder, M., Jirasek, A., and Kobilkova, J. (1985). Prenatal diagnosis of G_{M2}-gangliosidosis with high residual hexosaminidase A activity (variant B 1; pseudo AB variant). *Pediatr. Res.* **19,** 1220–1224.

de Baeque, C. M., Suzuki, K., Rapin, I., Johnson, A. B., Whethers, D. L., and Suzuki, K. (1975). G_{M2} gangliosidosis, AB variant. Clinico-pathological study of a case. *Acta Neuropathol.* **33,** 207–226.

Escola (1961). Uber die Prozessausbreitung der amaurotischen Idiotie im Zentralnervensystem in verschiedenen Lebensaltern und Besonderheiten der Spätform gegenüber der Pigmentatrophie. *Arch. Psychiat. Nervenkr.* **202,** 95–112.

Frohwein, Y. Z., and Gatt, S. (1967). Isolation of β-N-acetylglucosaminidase and β-N-acetylgalactosaminidase from calf brain. *Biochemistry* **6,** 2775–2783.

Galjaard, H., Hoogeven, A., de Wit-Verbeek, H. A., Reuser, A. J. J., Keijzer, W., Westerveldt, A., and Bootsma, D. (1974). Tay-Sachs and Sandhoff's disease, intergenic complementation after somatic cell hybridization. *Exp. Cell Res.* **87,** 444–448.

Geiger, B., and Arnon, R. (1976). Chemical characterization and subunit structure of human N-acetylhexosaminidases A and B. *Biochemistry* **15,** 3484–3493.

Goldmann, J. E., Yamanaka, T., Rapin, I., Adachi, M., Suzuki, K., and Suzuki, K. (1980). The AB-Variant of G_{M2}-gangliosidosis. *Acta Neuropathol. (Berlin)* Clinical, biochemical and pathological studies of two patients. **52,** 189–202.

Harzer, K., Sandhoff, K., Schall, H., and Kollmann, F. (1971). Enzymatische Untersuchungen im Blut von Überträgern einer Variante der Tay-Sachs'schen Erkrankung (Variante 0). *Klin. Wschr.* **49,** 1187–1191.

Hasilik, A., von Figura, K., Conzelmann, E., Nehrkorn, H., and Sandhoff, K. (1982). Lysosomal enzyme precursors in human fibroblasts. Activation of cathepsin D precursor in vitro and activity of β-hexosaminidase A precursor towards ganglioside G_{M2}. *Eur. J. Biochem.* **125,** 317–321.

Hayashi, M. (1965). Histochemical demonstration of N-acetyl-β-glucosaminidase employing naphtol ASBI N-acetyl-β-glucosaminide as substrate. *J. Histochem. Cytochem.* **13,** 355–360.

Hirabayashi, Y., Li, Y.-T., and Li, S.-C. (1983). The protein activator specific for the enzymic hydrolysis of G_{M2} ganglioside in normal human brain and in brains of three types of G_{M2} gangliosidosis. *J. Neurochem.* **40,** 168–175.

Hultberg, B. (1969). N-Acetylhexosaminidase activities in Tay-Sachs disease. *Lancet* **II,** 1195.

Igisu, H., and Suzuki, K. (1984). Progressive accumulation of toxic metabolite in a genetic leukodystrophy. *Science* **224,** 753–755.

Jatzkewitz, H., and Sandhoff, K. (1963). On a biochemically special form of infantile amaurotic idiocy. *Biochim. Biophys. Acta* **70,** 354–356.

Jatzkewitz, H., Pilz, H., and Sandhoff, K. (1965). Quantitative Bestimmungen von Gangliosiden und ihren neuraminsäurefreien Derivaten bei infantilen, juvenilen und adulten Formen der amaurotischen Idiotie und einer spätinfantilen biochemischen Sonderform. *J. Neurochem.* **12,** 135–144.

Klenk, E. (1937). Die Fettstoffe des Gehirns bei amaurotischer Idiotie und Niemann-Pick'scher Krankheit. *Ber. Ges. Physiol.* **96,** 659–660.

Klenk, E. (1939). Beitrage zur Chemie der Lipidosen. *Hoppe-Seyler's Z. Physiol. Chem.* I. Niemann-Pick'sche Krankheit und amaurotische Idiotie. **262,** 128–143.

Klenk, E. (1940). Beitrage zur Chemie der Lipoidosen. *Hoppe-Seyler's Z. Physiol. Chem.* **267,** 128–144.

Klenk, E. (1942a). Über die Ganglioside, eine neue Gruppe von zuckerhaltigen Gehirnlipoiden. *Hoppe Seyler's Z. Physiol. Chem.* **273,** 76–86.

Klenk, E. (1942b). Über die Ganglioside des Gehirns bei der infantilen amaurotischen Idiotie vom Typ Tay-Sachs. *Ber. Dtsch. Chem. Ges.* **75,** 1632–1636.

Klima, H., Klein, A., van Echten, G., Schwarzmann, G., Suzuki, K., and Sandhoff, K. (1993). Overexpression of a functionally active human GM2 activator protein in *Escherichia Coli*. *Biochem. J.* **292,** 571–576.

Kolodny, E. H., Brady, R. O., and Yolk, B. W. (1969). Demonstration of an alteration of ganglioside metabolism in Tay-Sachs disease. *Biochem. Biophys. Res. Commun.* **37,** 526–531.

Korey, S. R., and Stein, A. (1963). A gangliosidase system. *In* "Brain Lipids and Lipoproteins and the Leucodystrophies, VII. International Congress of Neurology, Rome, 1961" (J. Folch-Pi and H. Bauer, eds.), Elsevier, Amsterdam. pp. 71–82.

Kresse, H., Fuchs, W., Glossel, J., Holtfrerich, D., and Gilberg, W. (1981). Liberation of N-acetylglucosamine-6-sulfate by human β-N-acetylhexosaminidase A. *J. Biol. Chem.* **256,** 12926–12932.

Kuhn, E., and Wiegandt, H. (1963). Die Konstitution der Ganglio-N-tetraose und des Gangliosides GI. *Chem. Ber.* **96,** 866–880.

Kytzia, H.-J., and Sandhoff, K. (1985). Evidence for two different active sites on human hexosaminidase—Interaction of G_{M2} activator protein with hexosaminidase A. *J. Biol. Chem.* **260,** 7568–7572.

Kytzia, H.-J., Hinrichs, U., Maire, I., Suzuki, K., and Sandhoff, K. (1983). Variant of G_{M2}-gangliosidosis with hexosaminidase A having a severely changed substrate specificity. *EMBO J.* **2,** 1201–1205.

Kytzia, H.-J., Hinrichs, U., and Sandhoff, K. (1984). Diagnosis of infantile and juvenile forms of G_{M2} gangliosidosis variant 0—Residual activities toward natural and different synthetic substrates. *Hum. Genet.* **67,** 414–418.

Ledeen, R., and Salsman, K. (1965). Structure of the Tay-Sachs' ganglioside. *Biochemistry* **4,** 2225–2233.

Leinekugel, P., Michel, S., Conzelmann, E., and Sandhoff, K. (1992). Quantitative correlation between the residual activity of β-hexosaminidase A and arylsulfatase A and the severity of the resulting lysosomal storage disease. *Hum. Genet.* **88,** 513–523.

Li, S.-C., and Li, Y.-T. (1976). An activator stimulating the enzymic hydrolysis of sphingoglycolipids. *J. Biol. Chem.* **254,** 1159–1163.

Li, S.-C., Hirabayashi, Y., and Li, Y.-T. (1981). A new variant of type-AB G_{M2}-gangliosidosis. *Biochem. Biophys. Res. Commun.* **101,** 479–485.

Liessem, B., Glombitza, G. J., Knoll, F., Lehmann, J., Kellermann, J., Lottspeich, F., and Sandhoff, K. (1995). Photoaffinity labeling of human lysosomal β-hexosaminidase B. Identification of Glu-355 at the substrate binding site. *J. Biol. Chem.* **270,** 23693–23699.

Makita, A., and Yamakawa, T. (1963). The glycolipids of the brain of Tay-Sachs disease. The chemical structure of globoside and main ganglioside. *Jon. J. Exp. Med.* **33,** 361–368.

Meier, E. M., Schwarzmann, G., Fürst, W., and Sandhoff, K. (1991). The human GM2 activator protein: A substrate specific cofactor of hexosaminidase A. *J. Biol. Chem.* **266,** 1879–1887.

Miyatake, T., and Suzuki, K. (1972). Globoid cell leukodystrophy: Additional deficiency of psychosine galactosidase. *Biochem. Biophys. Res. Commun.* **48,** 538–543.

Neuenhofer, S., and Sandhoff, K. (1985). Affinity labelling of the G_{M2} activator protein. *FEBS Lett.* **185,** 112–114.

Neuenhofer, S., Conzelmann, E., Schwarzmann, G., Egge, H., and Sandhoff, K. (1986). Occurence of lyso-ganglioside G_{M2} (II3-Neu5Ac-gangliotriaosylsphingosine) in G_{M2} gangliosidosis brain. *Biol. Chem. Hoppe Seyler* **367,** 241–244.

Nilsson, O., Mansson, J.-E., Hakansson, G., and Svennerholm, L. (1982). The occurence of psychosine and other glycolipids in spleen and liver from the three major types of Gaucher's disease. *Biochim. Biophys. Acta* **712,** 453–463.

Okada, S., and O'Brien, J. S. (1969). Tay-Sachs disease: Generalized absence of a β-D-N-acetyl-hexosaminidase component. *Science* **165,** 698–700.

Pennybacker, M., Liessem, B., Moczall, H., Tifft, C. J., Sandhoff, K., and Proia, R. L. (1996). Identification of domains in human β-hexosaminidase that determine substrate specificity. *J. Biol. Chem.* **271,** 17377–17382.

Purpura, D. P., and Suzuki, K. (1976). Distortion of neuronal geometry and formation of aberrant synapses in neuronal storage disease. *Brain Res.* **116,** 1–21.

Robinson, D., and Stirling, J. L. (1966). N-acetyl-glucosaminidases in human spleen. *Biochem. J.* **101,** 18P.

Robinson, D., and Stirling, J. L. (1968). N-Acetyl-β-D-glucosaminidases in human spleen. *Biochem. J.* **107,** 321–327.

Robinson, D., Price, R. G., and Dance, N. (1967). Separation and properties of β-galactosidase, β-glucosidase, β-glucuronidase and N-acetyl-β-glucosaminidase from rat kidney. *Biochem. J.* **102,** 525–532.

Rosengren, B., Mansson, J.-E., and Svennerholm, L. (1987). Composition of gangliosides and neutral glycosphingolipids of brain in classical Tay-Sachs and Sandhoff disease: More lyso-G_{M2} in Sandhoff disease? *J. Neurochem.* **49,** 834–840.

Sachs, B. (1887). On arrested cerebral development with special reference to its cortical pathology. *J. Nerv. Ment. Dis.* **14,** 541–553.

Sachs, B. (1896a). A family form of idiocy, generally fatal associated with early blindness. *J. Nerv. Ment. Dis.* **21,** 475–479.

Sachs, B. (1896b). A family form of idiocy, generally fatal, associated with early blindness (amaurotic family idiocy). *NY State J. Med.* **63,** 697–703.

Samuels, S., Korey, S. R., Gonatas, J., Terry, R. D., and Weiss, M. (1963). Studies in Tay-Sachs-Disease, IV Memranous cytoplasmic bodies. *J. Neuropathol. Exp. Neurol.* **22,** 81–97.

Sandhoff, K., and Conzelmann, E. (1984). The biochemical basis of gangliosidoses *Neuropediatrics* **15**(Suppl), 85–92.

Sandhoff, K., Pilz, H., and Jatzkewitz, H. (1964). Über den enzymatischen Abbau von N-acetyl-neuraminsäurefreien Gangliosidresten (Ceramid-oligosacchariden). *Hoppe-Seyler's Z. Physiol. Chem.* **338,** 281–285.

Sandhoff, K. (1965). Disseration: Die Amaurotische Idiotie des Menschen als Storung im Glyko-sphingolipidstoffwechsel. Universitfit Munchen

Sandhoff, K. (1968). Auftrennung der Sauger-N-Acetyl-β-D-hexosaminidase in multiple Formen durch Elektrofokussierung. *Hoppe-Seyler's Z. Physiol. Chem.* **349,** 1095–1098.

Sandhoff, K. (1969). Variation of β-N-acetylhexosaminidase-pattern in Tay-Sachs disease. *FEBS Lett.* **4,** 351–354.

Sandhoff, K. (1970). The hydrolysis of Tay-Sachs ganglioside (TSG) by human N-acetyl-'-D-hexosaminidase A. *FEBS Lett.* **11,** 342–344.

Sandhoff, K., and Wassle, W. (1971). Anreicherung und Charakterisierung˜zweier Formen der menschlichen N-Acetyl-β-D-hexosaminidase. *Hoppe-Seyler's Z. Physiol. Chem.* **352,** 1119–1133.

Sandhoff, K. (1973). Multiple human hexosaminidases. The National Foundation, USA pp. 214–222.

Sandhoff, K., Andreae, U., and Jatzkewitz, H. (1968). Deficient hexosaminidase activity in an exceptional case of Tay-Sachs disease with additional storage of kidney globoside in visceral organs. *Pathol. Eur.* **3,** 278–285.

Sandhoff, K., Harzer, K., Wässle, W., and Jatzkewitz, H. (1971). Enzyme alterations and lipid storage in three variants of Tay-Sachs disease. *J. Neurochem.* **18,** 2469–2489.

Sandhoff, K., Conzelmann, E., and Nehrkorn, H. (1977). Specificity of human liver hexosaminidases A and B against glycosphingolipids G_{M2} and G_{A2}. Purification of the enzymes by affinity chromatography employing specific elution. *Hoppe Seyler's Z. Physiol. Chem.* **358,** 779–787.

Sango, K., Yamanaka, S., Hoffmann, A., Okuda, Y., Grinberg, A., Westphal, H., McDonald, M. P., Crawley, J. N., Sandhoff, K., Suzuki, K., and Proia, R. L. (1995). Mouse models of Tay-Sachs and Sandhoff diseases differ in neurologic phenotype and ganglioside metabolism. *Nat. Genet.* **11,** 170–176.

Schepers, U., Glombitza, G., Lemm, T., Hoffmann, A., Chabas, A., Ozand, P., and Sandhoff, K. (1996). Molecular analysis of a GM2-activator deficiency in two patients with G_{M2}-gangliosidosis AB variant. *Am. J. Hum. Genet.* **59,** 1048–1056.

Schröder, M., Schnabel, D., Suzuki, K., and Sandhoff, K. (1991). A mutation in the gene of a glycolipid binding protein (GM2 activator) that causes GM2 gangliosidosis variant AB. *FEBS Lett.* **290,** 1–3.

Schröder, M., Schnabel, D., Hurwitz, R., Young, E., Suzuki, K., and Sandhoff, K. (1993). Molecular genetics of GM2 gangliosidosis AB variant: A novel mutation and expression in BHK cells. *Hum. Genet.* **92,** 437–440.

Sonderfeld, S., Brendler, S., Sandhoff, K., Galjaard, H., and Hoogeveen, A. T. (1985a). Genetic complementation in somatic cell hybrids of four variants of infantile G_{M2} gangliosidosis. *Hum. Genet.* **71,** 196–200.

Sonderfeld, S., Conzelmann, E., Schwarzmann, G., Burg, J., Hinrichs, U., and Sandhoff, K. (1985b). Incorporation and metabolism of ganglioside G_{M2} in skin fibroblasts from normal and G_{M2} gangliosidosis subjects. *Eur. J. Biochem.* **149,** 247–255.

Srivastava, S. K., and Beutler, E. (1973). Hexosaminidase-A and Hexosaminidase-B: Studies in Tay-Sachs and Sandhoff's disease. *Nature (Lond.)* **241,** 463.

Svennerholm, L. (1962). The chemical structure of normal human brain and Tay-Sachs gangliosides. *Biochem. Biophys. Res. Commun.* **9,** 436–441.

Svennerholm, L., Vanier, M.-T., and Mansson, J.-E. (1980). Krabbe disease: A galactosylsphingosine (psychosine) lipidosis. *J. Lipid Res.* **21,** 53–64.

Tay, W. (1881). Symetrical changes in the region of the yellow spot in each eye of an infant. *Trans. Ophthalmol. Soc. UK* **1,** 55–57.

Terry, R. D., and Korey, S. R. (1960). Membranous cytoplasmic granules in infantile amaurotic idiocy. *Nature (Lond.)* **188,** 1000–1002.

Terry, R. D., and Weiss, M. (1963). Studies in Tay-Sachs disease: II Ultrastructure of the cerebrum. *J. Neuropathol. Exp. Neurol.* **22,** 18–55.

Thomas, G. H., Taylor, H. A. J., Miller, C. S., Axelmann, J., and Migeon, B. R. (1974). Genetic complementation after fusion of Tay-Sachs and Sandhoff cells. *Nature (Lond.)* **250,** 580–582.

9

Subunit Structure of the Hexosaminidase Isozymes

Ernest Beutler
Department of Molecular and Experimental Medicine
The Scripps Research Institute
La Jolla, California 92037

I. Introduction
II. Antibodies Against Hexosaminidase
III. Unravelling the Subunit Structure Immunologically
IV. Converting Hexosaminidase A to Hexosaminidase B
V. Epilogue
References

I. INTRODUCTION

I was a newcomer to the glycolipid storage diseases in the late 1960s and early 1970s. As an active investigator in the field of biochemical genetics for some 15 years, my interests had been centered on the study of red cell abnormalities, first glucose 6-phosphate dehydrogenase deficiency, then sickle-cell disease, and finally galactosemia. Most of my research interests had their beginnings in stimulation provided by a clinical experience, and the glycolipid storage diseases were no exception. In the middle 1960s, Dr. Alfred Knudson, then my colleague in pediatrics at the City of Hope, brought a teenager and her mother to see me. The young woman suffered from fairly severe adult-type Gaucher disease. Over the next few years I felt the same frustration that is experienced by all physicians who are charged with the responsibility for the care of these patients. The discovery in 1965 of the underlying enzyme defect (Brady et al., 1965) provided hope for the first time that something rational and constructive could be done for these patients, and stimulated my interest in this group of diseases. The introduction of the

4-methylumbelliferone derivatives for the assay of lysosomal enzymes (Leaback and Walker, 1961) had made it possible for me to apply my experience in enzymology to this disease without venturing into the unknown, and to me, forbidding, world of lipids. At that time, my son Steven, a college student at the University of California at San Diego, worked as a volunteer in John O'Brien's laboratory. One day John, having read one of my papers on Gaucher disease, called him into his office. "Tell your dad to keep working with those artificial substrates," John said. "That way he'll publish twice as many papers: first the original paper and then the retraction."

Thus, while my first efforts were directed at Gaucher disease, my attention was drawn to this entire group of disorders. The discovery (Okada and O'Brien, 1969) that Tay-Sachs disease was due to a deficiency in one isozyme of hexosaminidase, Hex A, was of particular interest because of the obvious immediate practical implications in heterozygote screening. My earlier experience in attempting to detect carriers for galactosemia (Beutler, 1973), another autosomal recessive disorder, led me to anticipate serious practical difficulties in any program that depended purely on quantitative differences of enzyme activity. The scenario that I foresaw is depicted schematically in Figure 9.1. Biologic variability being what it is, the lower tail of a large, normally distributed distribution is certain to overlap the upper tail of a smaller population with a mean activity one-half that

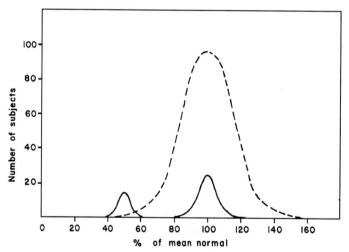

Figure 9.1. A schematic drawing of the distribution of hexosaminidase activity in normal subjects and heterozygotes for Tay-Sachs disease. Although the populations appear to be well separated when small numbers are examined, overlap occurs when large numbers are tested (dashed line). This problem is universal to the detection of heterozygotes based on quantitative enzyme determinations.

of the larger population. This problem could be eliminated by finding a qualitative difference between the mutant and the normal protein. This is what makes a carrier for a disorder such as sickle-cell disease so easy to detect.

It seemed to me that if the mutant gene that caused Tay-Sachs disease directed the formation of cross-reacting material (CRM), a protein that retained antigenicity while losing catalytic potency, the detection of this protein by immunologic means could be very helpful in distinguishing true heterozygotes for Tay-Sachs disease from those who merely had low-normal Hex A activity. A similar approach had been taken at about this time in heterozygote detection for hemophilia (Zimmerman et al., 1971), a disease that presented the same kind of diagnostic dilemma. This, then, was the reasoning that brought me into the area of Tay-Sachs disease. Soon it became apparent to me that there were many other interesting questions about the biochemistry of the hexosaminidase system, and our studies were broadened to try to answer some of these.

II. ANTIBODIES AGAINST HEXOSAMINIDASE

Satish K. Srivastava joined my laboratory at the City of Hope in about 1966 as a postdoctoral fellow and remained as a member of the junior faculty. Trained in Lucknow, India, Satish was a careful, honest, and tireless worker. I asked him to collaborate with me on this project, and he played a key role in many of the studies that were to follow. Our initial approach was to purify both Hex A and Hex B to homogeneity, to raise antibodies against the purified enzymes, and then to look for CRM in fibroblasts and serum from patients with the disease.

Our initial results (Srivastava and Beutler, 1972) showed us that using antibodies to detect Tay-Sachs carriers would not be easy. First of all, we discovered that the antibody against Hex A reacted with Hex B and vice versa, an observation also made independently by Carroll and Robinson (Carroll and Robinson, 1973). This was going to make it more difficult to find CRM in the blood of patients with Tay-Sachs disease. Worse yet, titration curves and immunoelectrophoresis showed that there was no CRM in fibroblasts from most Tay-Sachs patients.

It was enough to dampen my enthusiasm. What I had been looking for was an operationally simple means for the detection of heterozygotes. What we had found made the prospects for the development of such a procedure seem very dim indeed. But in reaching this conclusion we learned how to purify the hexosaminidases, a task that had not been accomplished previously, and we were stimulated to think about the relationship between Hex A and Hex B. We now attempted to produce an antibody that was specific against the two hexosaminidases. We absorbed the anti-Hex A with Hex B and the anti-Hex B with Hex A. We succeeded in making an anti-Hex A antiserum that no longer reacted with Hex B;

the reverse was not true. Try as we would, when we absorbed anti-Hex B serum with Hex A, it lost its activity not only against Hex A but also against Hex B.

III. UNRAVELLING THE SUBUNIT STRUCTURE IMMUNOLOGICALLY

What had been particularly puzzling to me was the question of how one simple Mendelian disorder, Tay-Sachs disease, could abolish the formation of Hex A without producing a deficiency of Hex B, while another simple Mendelian disorder, Sandhoff disease, could produce absence of both isozymes. We were not the only ones, and not even the first, to ponder this enigma, and several models had been proposed to account for it. The most popular of these is shown in Figure 9.2. It proposed that Hex B was the precursor of Hex A. The lesion in Sandhoff disease would, according to this model, be in the formation of Hex B. This would, of course, also abolish the formation of Hex A. In contrast, the lesion in Tay-Sachs disease would be in the subsequent reaction, *viz.*, the conversion of Hex B to Hex A. Our data showing that the antibodies produced against either pure Hex A or pure Hex B reacted also with the other isozyme was certainly consistent with this model.

But we also found that livers from some patients with Sandhoff disease contained Hex A antigen but no Hex B, a finding difficult to reconcile with the proposal that Hex B was the precursor and Hex A the product. Moreover, our studies with the absorption of the anti-A and anti-B serum had suggested that Hex A contained an antigen not represented in Hex B, but that the reverse was not true. This finding was also not easy to reconcile with the idea that Hex B was converted to Hex A.

Our findings led us to conceive a model that might aptly be named the "common subunit model," in which the two isozymes share a subunit, Hex A possessing a subunit not present in Hex B. Specifically, we proposed that Hex A was a heteropolymer containing α and β subunits, whereas Hex B was a homopolymer of β subunits. I found this model to be a satisfying one because it made it relatively simple to account for all of the known data. Since Hex B contained only the β subunits, all antibodies raised against it would be removed by Hex A, which also

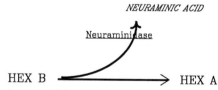

Figure 9.2. A common theory once believed to account for the relationship between hexosaminidase B (Hex B) and hexosaminidase (Hex A). It was believed that Hex A might be the asialo derivative of Hex B.

contains β subunits. However, since Hex A contained a unique subunit, namely, the α subunit, antibody directed against the α subunit would not be absorbed completely by treatment with Hex B. It made it simple, too, to account for the findings in both Tay-Sachs disease and Sandhoff disease. In Tay-Sachs disease the mutation would be one that limited formation of the α subunit, so that Hex A could not be made. The model would even predict that the amount of Hex B should be increased in quantity, and this had been reported to be the case. A mutation affecting the β subunit, on the other hand, would prevent synthesis of both Hex A and Hex B, since both isozymes contained β subunits. In the case of Sandhoff disease, what happened to the α subunits that were formed? Here, too, the model was satisfying, for a more highly charged production designated Hex S had been described (Ropers and Schwantes, 1973), and this, we believed, should consist of a homopolymer of subunits.

IV. CONVERTING HEXOSAMINIDASE A TO HEXOSAMINIDASE B

As we were performing structural studies to verify this model, a very interesting abstract was published by Carmody and Rattazzi in (1973). In their studies they had tested the first model, as shown in Figure 9.2, by treating Hex A with neuraminidase to determine whether it could, indeed, be converted to Hex B. Conversion appeared to occur, but careful investigators that they are, Carmody and Rattazzi used heat- and EDTA-inactivated preparations of neuraminidase as a control. To their surprise, conversion still occurred. I was eager, of course, to repeat these results. I was relieved to find that the *Cholera vibrio* neuraminidase that Carmody and Rattazzi had used did not have to be grown up and purified in our laboratory. We could buy it from Sigma. As soon as this reagent arrived, I outlined an experiment to Doris Villacorte, a capable and industrious technician who had been working with us on this problem for some time. I suggested to Doris that she treat some of our homogeneous Hex A with the neuraminidase preparation, with heat-inactivated neuraminidase, and that she incubate a control preparation of homogeneous Hex A in the same buffer. After incubation, electrophoretic analysis should be carried out. Fifteen minutes later, Doris reappeared in my office. "Should I put merthiolate in the control?," she asked. I queried why she would want to consider doing that. "The neuraminidase has merthiolate in it as a preservative," she explained. The care with which Doris approached her work really paid off. It was the merthiolate control that gave us the answer. Merthiolate converted Hex A to Hex B very efficiently in a concentration- and time-dependent manner. Sulfhydryl agents had long been used to separate hemoglobin chains. Merthiolate dissociated the α and β chains of Hex A and allowed the β chains to reassociate. We quickly published an abstract of these results (Beutler *et al.*, 1974) and then attempted to publish them in a widely read international journal.

Frankly, I thought they were quite significant. We sent the paper to *Nature*, and after a delay of several months, received the following critique:

> (1) The authors have converted hexosaminidase A to an aryl-alkyl mercurimercaptide which now has some of the chromatographic and electrophoretic properties of hexosaminidase B. This finding is interesting, but the title is misleading and biochemically incorrect. Actually this deception is recognized by the authors themselves where they say the "hex-B–like enzyme" (on page 3, pare 2 line 1). The product of their reaction of hexosaminidase A with the organic mercurials certainly is not *natural* hexosaminidase B and publication of this contribution would absolutely confuse the unsophisticated reader.
>
> (2) The author's results are irreconcilable with the findings of Sandhoff (*Birth Defects, Orig. Art. Series IX*, 1973, p. 214–221) who was able to convert hex A to hex B in good yield supply by incubating hex A in dilute buffer at pH 5.0 for 30 min. in the absence of mercurials or other preservatives. Neither did he employ neuraminidase in these incubations.

Enclosed was a paper that described the experiments alluded to in the comments. Needless to say, I did not appreciate the implication that I was trying to mislead the reader. Indeed, further events showed that our formulation had been entirely correct, and that we had not merely formed a mercaptide from Hex A, giving it a mobility that simulated the mobility of Hex B, but had actually formed Hex B. I felt quite confident that the common subunit theory was correct, whatever the subunit size might be, but in informal conversations with some of my colleagues, sensed that they were less convinced than I. If Hex B were comprised of β subunits and Hex S of α subunits, could we produce authentic Hex's from these two homopolymers? We tried a variety of dissociation and reassociation of subunits, altering pH, salt strength, etc., and succeeded in finding conditions under which this could be accomplished (Beutler and Kuhl, 1975).

Subsequently, we performed structural studies on the enzyme in an effort to test the common subunit theory. The data fit very well, but we were unable to dissociate the α subunit into the size that we *thought* it should have. Based on the classical models of proteins that are composed of two different subunits, e.g., hemoglobin and lactate dehydrogenase, I nurtured the preconceived idea that the α and β subunits should be approximately the same size. Thus we initially assigned a structure of α_3/β_3, and subsequently with additional studies of enzyme and subunit molecular weight a size of α_2/β_2. Working together with Akira Yoshida, we appeared at long last to be able to make Hex A into subunits of this single size (Beutler *et al.*, 1976).

Having resolved to my own satisfaction the question of the relationship of Hex A and Hex B, I abandoned this area of research and returned to that of

Gaucher disease, which has continued to interest and challenge my colleagues and myself to the present day. With the advantage of 10 years of hindsight, there was still more to learn about the basic structure of the hexosaminidases than I had realized, and others made substantial further contributions to the understanding of this complex series of isozymes.

V. EPILOGUE

If we were to do it over, would we have done it differently? In some respects, the answer is yes. We initiated our investigations by purifying Hex A and Hex B to homogeneity. Straightforward though this strategy was, it occurred to me only later that purification of the enzyme would not really have been required. We could merely have raised antibody against crude preparations of Hex A and Hex B. By using their ability to remove enzyme activity from a solution, we could have measured the slope of the titration curve and determined whether or not CRM was present. Why didn't we? I think it simply didn't occur to me to do so. There were, of course, advantages to having in hand pure antigen, and purification of the hexosaminidase proved to be a rather straightforward and not too difficult task. At least it seems so now, more than 15 years later. One aspect of our work proved to be inaccurate. The studies of Mahuran et al. (1982) of the structure of hexosaminidase A revealed that what we had considered to be the β chain actually consisted of two slightly different β chains. In addition, their studies clarified our initial inability to cleave the α chains into subunits equal in size to the β chains (Mahuran et al. 1982). It appears that the α chain could not be cleaved because it is actually a single subunit twice as large as the β chain (Mahuran and Lowden, 1980). What is not entirely clear is how we finally succeeded in convincing ourselves that each hexosaminidase molecule contained two α chains. I have recently reviewed this question with Dr. Yoshida, who was my collaborator on those studies. He suggested that the relatively drastic treatment used could actually have broken a peptide bond; a precedent for this exists (Ui and Sorimachi, 1976), and he has personally encountered this problem in other proteins, including human α-mannosidase. Alternatively, the S-carboxymethylated subunit may have become insoluble and was inadvertently discarded.

On the whole, I found our excursion into the biochemical genetics of the hexosaminidases to be very gratifying. We approached an interesting problem in what I considered to be a logical manner and were able to arrive at an intellectually satisfying solution which has in large measure stood the test of time.

References

Beutler, E. (1973). Screening for galactosemia. Studies of the gene frequencies for galactosemia and the Duarte variant. Isr. J. Med. Sci. 9, 1323–1329.

Beutler, E., and Kuhl, W. (1975). Subunit structure of human hexosaminidase verified interconvertibility of hexosaminidase isozymes. *Nature* **258**, 262–264.

Beutler, E., Villacorte, D., and Srivastava, S. K. (1974). Non enzymatic conversion of hexosaminidase-A to hexosaminidase-B by merthiolate. *IRCS* **2**, 1090.

Beutler, E., Yoshida, A., Kuhl, W., and Lee, J. E. S. (1976). The subunits of human hexosaminidase A. *Biochem. J.* **159**, 541–543.

Brady, R. O., Kanfer, J. N., and Shapiro, D. (1965). Metabolism of glucocerebrosides. II. Evidence of an enzymatic deficiency in Gaucher's disease. *Biochem. Biophys. Res. Commun.* **18**, 221–225.

Carmody, P. J., and Rattazzi, M. C. (1973). Is neuraminidase responsible for the in vitro conversion of human hexosaminidase A to hexosaminidase B? *Am. J. Hum. Genet.* **25**, 19A (abstr.).

Carroll, M., and Robinson, D. (1973). Immunological properties of N-acetyl-β-D-glucosaminidase of normal human liver and of G_{M2}-gangliosidosis liver. *Biochem. J.* **131**, 91–96.

Leaback, D. H., and Walker, P. G. (1961). Studies on glucosaminidase. IV. The fluorimetric assay of N-acetyl-β-glucosaminidase. *Biochem. J.* **78**, 151–156.

Mahuran, D., and Lowden, J. A. (1980). The subunit and polypeptide structure of hexosaminidases from human placenta. *Can. J. Biochem.* **58**, 287–294.

Mahuran, D. J., Tsui, F., Gravel, R. A., and Lowden, J. A. (1982). Evidence for two dissimilar polypeptide chains in the β2 subunit of hexosaminidase. *Proc. Natl. Acad. Sci. (USA)* **79**, 1602–1605.

Okada, S., and O'Brien, J. S. (1969). Tay-Sachs disease: Generalized absence of a β-D-N-acetylhexosaminidase component. *Science* **165**, 698–700.

Ropers, H. H., and Schwantes, U. (1973). On the molecular basis of Sandhoff's disease. *Humangenetik* **20**, 167–170.

Srivastava, S. K., and Beutler, E. (1972). Antibody against purified human hexosaminidase B cross-reacting with human hexosaminidase A. *Biochem. Biophys. Res. Commun.* **47**, 753–759.

Ui, N., and Sorimachi, K. (1976). The polypeptide chain structure of thyroglobulin. *In* "Thyroid Research" (J. Robbins, ed.), pp. 177–179. Exper. Pta. Medical, Amsterdam.

Zimmerman, T. S., Ratnoff, O. D., Littell, A. S., Zimmerman, T. S., Ratnoff, O. D., and Littell, A. S. (1971). Detection of carriers of classic hemophilia using an immunologic assay for antihemophilic factor (factor VIII). *J. Clin. Invest.* **50**, 255–258.

10

Molecular Genetics of the β-Hexosaminidase Isoenzymes: An Introduction

Edwin H. Kolodny
Department of Neurology
New York University School of Medicine
New York, New York

    I. Personal Recollections
   II. Biochemical Genetics of the Hexosaminidases
  III. Evolution of Molecular Biology
      A. Composition of DNA
      B. Restriction Enzymes and Cloning
  IV. Analysis of DNA
   V. Classification of Mutations
  VI. Detection of Known Mutations
      A. Polymerase Chain Reaction
      B. Dot (Slot) Blotting
      C. Restriction Digestion
      D. ARMS Test
  VII. Screening for New Mutations
      A. Heteroduplex Analysis
      B. Single-Strand Conformational Polymorphism Analysis (SSCP)
      C. Chemical Cleavage of Mismatch Analysis (CCM)
      D. Denaturing Gradient Gel Electrophoresis (DGGE)
 VIII. From Enzyme to Gene Structure
      A. Isolation of the α-Chain Gene, HEXA
      B. Isolation of the β-Chain Gene, HEXB
      C. Homology between HEXA and HEXB
      D. Cloning of the G_{M2} Activator Protein Gene—GM2A

Advances in Genetics, Vol. 44

    IX. Mutations in the G_{M2} Gangliosidoses
       A. Classic Infantile Tay-Sachs Disease
       B. Adult G_{M2} Gangliosidosis
       C. B_1 Variant
       D. Insoluble α Mutation
       E. β-Chain Mutations
    X. Genetically Engineered Animal Models
    XI. Conclusions
       References

I. PERSONAL RECOLLECTIONS

The discovery by Robinson and Stirling in 1967 that lysosomal hexosaminidase could be separated into two distinct electrophoretic species provided Okada and O'Brien (1969) with the clue that led to their description two years later of hexosaminidase A deficiency in Tay-Sachs disease. At that time, I was working with Dr. Roscoe Brady at the National Institutes of Health (NIH) as a Special Fellow. We were investigating the enzymatic hydrolysis of G_{M2} ganglioside by various tissues using a radioactively labeled substrate prepared by *in vivo* labeling. Dr. Brady has described how we discovered that skeletal muscle tissue from Tay-Sachs patients were deficient in G_{M2} ganglioside hexosaminidase activity (Kolodny *et al.*, 1969). How I came to be in Dr. Brady's lab and the effect of this experience on my future work is an interesting story.

    As a volunteer at a state hospital in Waltham, Massachusetts, during my college days, I had noticed a child with Tay-Sachs disease who was a relative of a childhood friend. Years later, when I was looking for a fellowship in neurochemistry at the NIH, Dr. Brady mentioned his interest in Tay-Sachs disease. I recalled the dispair I had felt years earlier on my first exposure to Tay-Sachs disease at the state hospital in Waltham and so I jumped at the chance to do something to solve the problem. Curiously, one of the first families I met with Tay-Sachs disease after beginning my work with Dr. Brady was the Gershowitz family, who lived in Maryland not far from the NIH. Harold and Bayla Gershowitz had a son, Steven, with Tay-Sachs disease, who provided us with the opportunity to follow the disease from its earliest stages. An exploration of their family tree (Figure 10.1) revealed Boston relatives who had had children with Tay-Sachs disease. One of these was actually the child (IV 4) I had seen in the state hospital in Waltham. Another (III 1), I learned, while a patient at the Children's Hospital in Boston during the early 1930s, had been examined by none other than Bernard Sachs, who had confirmed the diagnosis. That is as close as I have come to Barney Sachs!

    The Gershowitz family also provided us with our first opportunity to do genetic counseling based on hexosaminidase assays of individual members

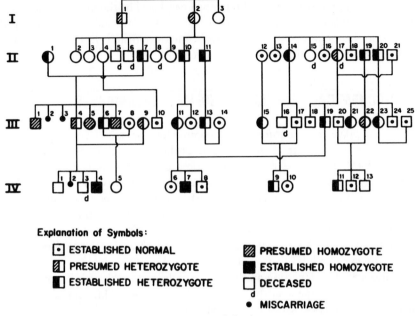

Figure 10.1. G family.

(Figure 10.1). Note that subject IV 9 in Figure 10.1 inherited his Tay-Sachs gene not from his father (III 15), who was the offspring of a carrier (II 14), but from his mother (III 17). This was a surprise to this Ashkenazi-Jewish family, since III 17 was a non-Jew. Dr. Yoshi Suzuki in Dr. Kunihiko Suzuki's lab at the University of Pennsylvania had taught me the cellulose acetate technique for the separation of hexosaminidases A and B, and I was using it to back up the heat denaturation assay of O'Brien *et al.* (1970). In May 1970, I received a call from Dr. Mike Kaback, then a member of the Pediatrics Department staff at the Johns Hopkins Hospital, to help in the diagnosis of a new child born into a family with a Tay-Sachs sibling. The youngster proved not to have Tay-Sachs, and her grateful parents, Karen and Dr. Bob Zeiger, went on to help Mike in setting up the Baltimore Tay-Sachs Prevention Program. Then, in 1973, after they moved to Boston, they were instrumental in helping us to establish the Tay-Sachs Prevention Program in Boston.

My work in carrier detection of Tay-Sachs disease led to the study of several families with the Sandhoff variant of the disease. I found that to identify Sandhoff disease carriers, I could not use the formula that O'Brien *et al.* (1970) had described for Tay-Sachs heterozygotes. Instead, what I observed was a reduced amount of hexosaminidase activity that was predominantly in the A form. This was reported at the symposium on sphinogolipidoses held at the Kingsbrook Jewish Medical Center in 1971 (Kolodny, 1972).

For the next two years, I remained perplexed by the marked differences in hexosaminidase isoenzyme activity in Tay-Sachs and Sandhoff carriers, until reports on the subunit structure, chemical composition, and chromosome assignment of the hexosaminidases began to appear.

II. BIOCHEMICAL GENETICS OF THE HEXOSAMINIDASES

More than 50 years passed after Sachs reported his first case of Tay-Sachs disease (Sachs, 1887) before the chemical nature of the stored lipid was elucidated by Klenk (1942). The cause of the ganglioside accumulation, a deficiency of hexosaminidase A, was not found for another quarter-century (Okada and O'Brien, 1969). As the previous chapters have shown, progress has been rapid since then. Accomplishments of the next decade included descriptions of the subunit structure (Beutler et al., 1976), chemical composition (Geiger and Arnon, 1976), substrate specificity (Sandhoff et al., 1977), and chromosomal localization of the hexosaminidases (Gilbert et al., 1975). In the 1980s, the steps involved in their intracellular processing and targeting to lysosomes became known, a sulfated substrate became available that could more readily differentiate the two major hexosminidases (Kresse et al., 1981), and, before the end of the decade, the genes encoding both subunits and the G_{M2}-activator protein had been isolated and sequenced. The last 10 years have been perhaps the most fruitful, with the elucidation of the many molecular defects that cause Tay-Sachs disease and its variants.

III. EVOLUTION OF MOLECULAR BIOLOGY

The foundation of our present understanding of the molecular genetics of the G_{M2} gangliosidoses is based on a series of landmark discoveries in molecular biology that occurred contemporaneously with the growth of our knowledge of the hexosaminidases.

A. Composition of DNA

DNA, the fundamental genetic material in chromosomes (Avery et al., 1944; Hershey et al., 1953) is a polymer of nucleotides, each consisting of a purine or pyrimidine base, a phosphate, and deoxyribose joined into polynucleotide chains by $5'-3'$-phosphodiester bonds. In 1953, Watson and Crick proposed a double-helix model of DNA in which two polynucleotide chains running in opposite directions are joined by complementary base pairs, one base pair consisting of a small pyrimidine, cytosine or thymine, and a large purine base, either adenine or guanine. The sequence of these base pairs determines the order of the 20 different amino acids that comprise individual proteins. Nirenberg, Crick, and Khorana showed in the early 1960s that this genetic code was based on 61 codons, each consisting of a triplet of three successive bases that codes for one particular amino acid and three codons that signal the termination of peptide synthesis (Nirenberg et al., 1964; Khorana, 1968). In many cases, more than one triplet codes for the same amino acid.

The information contained in the nuclear DNA, known as genomic DNA, is transferred to specific messenger RNA (mRNA) molecules by complementary base-pair matching as in DNA, except that uracil replaces thymine. Each messenger RNA is single-stranded and specifies a single gene product. The process of making RNA from DNA is called transcription and takes place in the nucleus. Messenger RNA is, in turn, translated into polypeptides, which are assembled on ribosomes in the cytoplasm.

The coding sequences of genes, known as exons, are interrupted by non-coding sequences, called introns. Mature mRNA molecules are formed by splicing together the exons and excising the introns. The splice junctions of introns are determined by consensus sequences composed of the dinucleotide GT at the 5' (donor) end and AG at the 3' end. Genes also contain flanking regions important in regulating the initiation and termination of protein synthesis. At the 5' end upstream from coding region are promotor sequences that signal to the enzyme RNA polymerase II the presence of a DNA sequence appropriate for transcription. These usually include a TATA box and a CCAAT box as well as additional sequences that enhance gene transcription. At the 3' end of the gene, the flanking sequence AATAAA is the signal for the addition of a poly-A tail to the end of the mRNA strands. The poly-A tail aids in maintaining stability of the mRNA molecule in the cytoplasm.

B. Restriction enzymes and cloning

The discovery by Smith and Wilcox, in 1970, of restriction endonucleases provided the foundation for recombinant DNA technology. These enzymes recognize short sequences of specific nucleotide base pairs (restriction sites), producing fragments of DNA with ends of known base composition. Utilizing these enzymes, Cohen, Boyer, and their colleagues, in 1973, demonstrated that DNA molecules could be cloned by ligation into a vector that replicates autonomously within a host cell and, in this way, many copies of a single fragment of DNA could be produced. Bacterial plasmids specially constructed for this purpose are used. They are circular bits of double-stranded DNA that contain restriction sites that can be opened up by specific restriction enzymes to allow foreign DNA cut with the same enzyme to be spliced into them with DNA ligase. The recombinant plasmid is then introduced by transfection into bacteria, where the foreign DNA is amplified more than 10^6-fold (Okayama and Berg, 1982). A gene for resistance to an antibiotic is usually included in the hybrid plasmid DNA. The antibiotic resistance gene allows only those cells that have incorporated the plasmid vector to survive in media containing the antibiotic. The same restriction enzyme used originally to form the recombinant DNA is used to cut out the fragments of cloned DNA. Phage (virus that infect bacteria), cosmids (plasmids in which lambda phage "cos" sites have been inserted), and YACs (yeast artificial chromosomes) are also used for cloning. These are specialized vectors that allow much larger insert sizes than the maximum 10-kb insert of a plasmid vector.

Cloning involves the preparation of multiple idential copies of a DNA fragment representing either a whole gene, a part of a gene, or other autonomous DNA. For this purpose, libraries are constructed with the objective of including within them at least one representative copy of all of the DNA sequences (Maniatis et al., 1978). A genomic DNA library is created by fragmenting nuclear DNA by digestion with restriction enzymes (often rare cutters that cleave at infrequent recognition sites) or by mechanical shearing and then cloning of the fragments in a host bacterial cell. Another type of library, containing cDNA (the base sequence of the coding regions of genes), is prepared from the total mRNA of a particular tissue and may therefore be enriched in particular species of mRNA. The mRNA is treated with the viral enzyme reverse transcriptase to form first strand cDNA. A double-stranded cDNA is then constructed by the action of DNA polymerase, inserted into an appropriate vector, and allowed to proliferate within the host cell.

To screen a genomic or cDNA library for a DNA fragment or gene of interest, the library is grown in the appropriate host on agar plates at a density that will permit individual colonies to be isolated at a later time. The colonies are then transferred to a nitrocellulose or nylon membrane so that a replica of the original plate is obtained. The filter is treated with alkali to denature the DNA and then baked in an oven. A radioactive probe is then applied to the filter, the filter is washed, and X-ray film is placed on the filter. The probe anneals (hybridizes) to DNA fragments that have a homologous (i.e., complementary) sequence of nucleotide bases. The dark signals that appear on the X-ray film by autoradiography may be used as a guide to pick the corresponding clones from the original agar plate for further study (Grunstein and Hogness, 1975).

The probe used in screening a library may be an oligonucleotide, a single-stranded cDNA, an mRNA, or an antibody to a specific protein. If partial amino acid sequence of the encoded protein is known, then oligonucleotides (17–25 bases in length) can be prepared that are complementary to the predicted nucleotide base sequence for that particular region of the protein. More than a single oligonucleotide would normally be used, because of the degeneracy in the genetic code in that most amino acids are determined by more than a single codon (nucleotide triplet). For this reason, peptides containing the fewest number of possible codon combinations are selected for assembly of the probe. The 5′ amino-terminal region is often utilized, as it contains the methionine start codon, which has only a single possible triplet combination of nucleotides. The oligonucleotide set is radiolabeled with ^{32}P (Rigby et al., 1977) and hybridized with the library. Often only a single member of the set actually anneals to the DNA.

If the protein sequence is unknown, the filter can be hybridized with total mRNA. The mRNA that binds to its corresponding DNA is eluted from the filter, in vitro translated, and its product identified by immunoprecipitation with the appropriate antibody. This technique is useful when the mRNA is enriched for the gene of interest and requires that an antibody to the protein is available.

An expression vector system such as λgt 11 will cause the cell to synthesize a part of the normal gene product. The protein from positive clones is detected using radiolabeled antibody and the corresponding clones are selected from the original ager plate, propagated, and their DNA isolated and characterized. Another cloning technique that is possible when antibody to the protein of interest is available utilizes polyribosomes to immunoselect both nascent protein and the mRNA directing its synthesis. The mRNA is removed from the precipitated polyribosomes and used to prepare a cDNA libray.

IV. ANALYSIS OF DNA

Fragments of DNA generated by restriction endonuclease cleavage can be separated according to their size by agarose gel electrophorosis. DNA fragments of known size are run alongside the DNA of interest. The negatively charged DNA migrates toward the anode, with the larger fragments moving more slowly than the smaller ones. The individual bands are revealed by staining the gel with ethidium bromide, which inserts itself between the bases and fluoresces under ultraviolet light.

As many as 1 million fragments may be generated by cutting genomic DNA with one or more restriction endonucleases. To detect a single fragment or group of stucturally related fragments, the DNA on a gel is denatured to separate the two strands and then transferred to a nylon or nitrocellulose filter by the capillary action of buffer flow through the gel. The filter is made ready for hybridization by baking or ultraviolet crosslinking. This technique of Southern blotting, named after its inventor, E. M. Southern (1975), is particularly useful for studying gene structure and for the detection of large insertions and deletions and major rearrangements in the gene sequence.

The RNA equivalent of Southern blotting is Northern blotting (Alwine, 1977). Total RNA is separated on a gel, transferred to a filter by blotting, and probed with radiolabeled DNA. This method allows both the size and abundance of specific mRNA molecules to be determined.

Detection of mutations requires ultimately the ability to determine the exact nucleotide sequence of genes. This became possible with the development of the base-specific chemical cleavage method by Maxam and Gilbert (1977) and dideoxy sequencing described by Sanger *et al.* (1975). Both methods depend on the power of polyacrylamide gels to resolve nucleic acid fragments that differ in length by a single base, and both require an amplification step.

The Maxam-Gilbert method utilizes four partial but base-specific cleavages of a DNA fragment that is end-labeled with ^{32}P. The labeled fragments are separated by size on polyacrylamide urea gels at high voltage and examined by autoradiography. Starting from the bottom of the gel, the base sequence is read from the autoradiograph showing the ladder of each of the base-specific tracks.

Dideoxy sequencing is now the more commonly used method. The DNA fragment is cloned in M13, a single-stranded filamentous virus. A short oligonucleotide primer that is complementary to the region adjacent to the sequence to be analyzed is annealed to the cloned single-stranded DNA template. A DNA polymerase (Klenow fragment or Sequenase®) catalyzes the addition of the complementary bases. The reaction is carried out in four separate tubes, each of which contains a mixture of nucleotides (one of which is radioactive) and one of four different dideoxynucleotides. When the dideoxytriphosphates of adenine, guanine, cytosine, and thymine are incorporated into the elongating DNA, they block further extension. A series of fragments is generated, their lengths corresponding to the position of the normal nucleotide in the sequence. Electrophoretic separation of the four sets of reactants in adjacent slots on a long acrylamide gel followed by radioautography reveals a ladderlike arrangement of bands from which the DNA sequence can be deduced.

V. CLASSIFICATION OF MUTATIONS

Heritable changes in DNA may involve a single DNA sequence (simple mutations), an exchange between two allelic or nonallelic sequences, or larger-scale chromosome abnormalities (see Table 10.1). The most common mutations are base substitutions, i.e., replacement of one pyrimidine (C or T), by another or one purine (A or G) by another (transition), or substitution of a pyrimidine by a purine or vice versa (transversion). Transitions are more common than transversions. In certain cases of single-base substitution, the altered codon does not change the amino acid specified. This is due to the degeneracy of the genetic code, which minimizes the effects of mutations, especially those in the third base of the codon (third-base wobble). These synonymous substitutions are neutral mutations but can activate a cryptic splice site, causing genetic disease.

Missense mutations are base substitutions that result in alteration of the amino acid residue. It may be conservative, in which case the new amino acid has properties similar to the one it replaces, and may have little effect on the properties of the protein. When the missense mutation is nonconservative, the substituted amino acid has different properties from the original amino acid and therefore the

Table 10.1. Classes of Mutations

Base substitutions	
Transitions	Substitution of a pyrimidine (C to T) by a pyrimidine, or of a purine (A or G) by a purine.
Transversions	Substitution of a pyrimidine by a purine, or of a purine by a pyrimidine.
Synonymous	Silent mutations which do not change the sequence of the gene product (neutral mutation).
Nonsynonymous	The sequence of a polypeptide or RNA is altered.
Missense	Altered codon specifies a different amino acid.
Nonsense	A codon specifying an amino acid is replaced by a stop codon.
Gene conversion	Substitution of multiple bases.
Insertions	
May produce frameshift unless multiple of three bases inserted.	
Deletions	
May produce frameshift unless multiple of three bases deleted.	

effects on the protein may be more severe. A nonconservative substitution occurs more often with base substitutions at the first and second codon positions, whereas a conservative substitution is more likely with third-codon position changes.

Nonsense (chain termination) mutations are particularly disruptive because they convert a codon specifying an amino acid to one causing termination of polypeptide synthesis. If this occurs near the 5′ end of the coding sequence, the protein is severely truncated and is likely to be without function and rapidly degraded by the cell. In the case of a mutation in a stop codon that instead specifies an amino acid, the polypeptide chain is extended and may function poorly or be unstable.

Alterations in nucleotide sequences at intron–exon boundaries can also dramatically alter RNA splicing and gene expression. Single-base changes at the donor site (the last three bases in the preceding exon and the first six residues of the intron) or acceptor site (the last 10 residues of the intron and the first residue of the next exon) can inactivate existing splice sites or create new ones. This produces a large insertion in the processed mRNA or creates alternative donor or acceptor splice sites that compete with the normal splice sites during RNA processing.

One or more nucleotides added or deleted from the reading frame of an mRNA product will alter the amino acid sequence from that point to the carboxy terminus. The new reading frame could include a stop codon, causing premature chain termination. It may also affect exon/intron splicing. If the mutated sequence involves a loss or gain of a multiple of the three bases, a frameshift does not occur.

Other classes of mutations may result from unequal crossing-over during meiosis. These include rare deletions, partial gene duplication, fusion genes, and the trinucleotide repeat expansions. Yet another class of mutations affects transcription through alterations upstream in the promotor regions.

Table 10.2. Mutation Analysis

I. Known mutations
   A. Dot/slot blotting by ASOH
   B. Restriction endonuclease digestion
   C. ARMS test (multiplex PCR)

II. New mutations
   A. Heteroduplex analysis
   B. Single-strand conformational polymorphism (SSCP)
   C. Chemical cleavage of mismatches (CCM)
   D. Denaturing gradient gel electrophoresis (DGGE)

VI. DETECTION OF KNOWN MUTATIONS

The more abundant the gene of interest, the easier it should be technically to demonstrate the presence of a mutation (see Table 10.2). Otherwise, as in the case of an autoradiograph, for example, a long exposure time may be required to reveal the signals from DNA bonds to which the probe has hybridized. Under these circumstances, the background activity may be so substantial that the DNA bands being sought are obscured.

A. Polymerase chain reaction

The polymerase chain reaction (Mullis *et al.*, 1987; Saiki *et al.*, 1988) has dramatically improved the ability to screen for mutations, to perform family studies, and to generate amounts of individual DNA fragments that are sufficient for direct sequencing. Two primers, approximately 15–20 base pairs in length, are prepared. One is complementary to the end of the target sequence on the coding or sense strand, the other, to sequence at the opposite end at the antisense strand. The region between these two primers is amplified in a series of three-step cycles that are repeated 20–30 times, producing 10^6–10^8 copies in a few hours. In the first step, the two complementary strands of DNA are separated by a short heat treatment. In the second step, the primers anneal to the respective, now separated, strands. The third step involves extension of the primers by addition of nucleotide bases using the existing DNA as a template. A thermostable DNA polymerase is used that remains catalytically active in spite of the high temperature used in the heat denaturation step. The method now permits amplification of nucleic acid fragments that are several kilobases in length with high fidelity.

B. Dot (slot) blotting

For detecting a known mutation, the PCR-amplified material can be analyzed directly by hybridization to allele-specific oligonucleotides (ASO) (Studencki *et al.*, 1985). The DNA is spotted on a solid support as a series of dots or slots and hybridized with a synthetic oligonucleotide of 19–20 bases in length. Under appropriately stringent conditions, only perfectly matched sequences will hybridize. A pair of ASO probes is complementary to the wild-type sequence and the other containing a single mismatch complementary to a point mutation. The former anneals only with the normal DNA sequence and the latter, only with the mutated sequence. If the DNA is from a heterozygote with one normal allele and one mutated allele, both labeled probes produce a signal in the autoradiograph.

C. Restriction digestion

The need for a labeled probe can be avoided if the mutation to be surveyed alters a restriction enzyme site. Following restriction enzyme digestion, DNA fragment sizes are examined by agarose gel electropheresis and ethidium bromide staining. If the mutation creates a new restriction site, then use of the appropriate enzyme will generate an additional fragment, smaller than that produced by cleavage of the corresponding wild-type nucleic acid. Similarly, if the mutation eliminates a restriction enzyme cleavage site, then the amplified fragment will not cut; i.e., the cut fragments will be larger than expected.

D. ARMS test

PCR amplification can be combined with allele-specific oligonucleotide primers to detect known mutations in a procedure known as the ARMS test. One common primer and two slightly different versions of a second primer are used. One version contains the normal sequence, while the other is specific for the mutant sequence. A multiplex reaction system can be created using the common primer together with additional primers so that several mutations can be screened simultaneously. Size differences in the expected products distinguish among the various muations.

VII. SCREENING FOR NEW MUTATIONS

Detection of an unknown sequence change is more difficult. Sequencing the entire coding region as well as intron–exon junctions of a gene is most likely to reveal all single-base substitutions as well as small deletions and insertions. However, far less laborious screening methods are available. These include heteroduplex analysis, single-strand conformational polymorphism (SSCP), chemical cleavage of mismatches (CCM), and denaturing gradient gel electrophoresis (DGGE).

A. Heteroduplex analysis

Allowing single-mutant DNA strands to base-pair with complementary strands from the wild-type allele results in a heteroduplex that moves in a polyacrylamide gel more slowly than a homoduplex. Insertions, deletions, and most single-base substitutions within PCR products of under 200 base pairs can be detected with this technique (Keen *et al.*, 1991). Mismatches in heteroduplexes formed between the test DNA and a wild-type RNA sequence can be revealed by RNAase A cleavage (RNAase protection assay).

B. Single-strand conformational polymorphism analysis (SSCP)

PCR-amplified DNA samples are denatured and electrophoresed in a nondenaturing polyacrylamide gel. The single-stranded DNA tends to fold up on itself, forming weak intramolecular bonds. Therefore, its mobility in a nondenaturing gel depends on both chain length and conformation. Comparison is made with the wild-type pattern. SSCP is best suited to fragments less than 200 base pairs in length (Sheffield *et al.*, 1993).

C. Chemical cleavage of mismatch analysis (CCM)

A heteroduplex is formed of normal and mutant DNA that differ by a single base. The normal DNA is radioactively labeled. A mismatched cytosine is chemically modified by hydroxylamine, and a mismatched thymidine, by osmium tetroxide. Piperidine is then used to cleave the chemically modified DNA strand and the fragments are separated by denaturing polyacrylamide gel electrophoresis. After autoradiography, the approximate position of the mutation can be estimated by fragment size.

D. Denaturing gradient gel electrophoresis (DGGE)

DGGE utilizes a gradient of increasing amounts of a chemical denaturant or a temperature gradient in an electrophoretic gel through which DNA duplexes migrate (Cariello and Skopek, 1993). At some position in the gel the strands melt and separate, and further migration is halted. Primer design is crucial in order to achieve the right melting profile. The addition of a poly-GC tail (a GC clamp) to each primer improves the sensitivity.

VIII. FROM ENZYME TO GENE STRUCTURE

The discovery of hexosaminidase A deficiency as described by Brady (Chapter 6) and Okada and O'Brien (Chapter 7) was the key that led to a more complete understanding of the molecular events causing Tay-Sachs disease. Initially, the focus was on the characterization of hexosaminidases A and B. Geiger and Arnon (1976) showed that the chemical composition of the two enzymes was similar.

Immunologic studies also pointed to a structural relationship between hexos-aminidases A and B, prompting Srivastava and Beutler (1973) to propose the existence of two subunits, α and β, with hexosaminidase A being a heteropolymer of αβ and hexosaminidase B a homopolymer of β subunits. The two-subunit model was substantiated by genetic linkage studies that showed dependence of human hexosaminidase A on the formation of hexosaminidase B (Lalley *et al.*, 1974), and subsequently, mapping of hexosaminidase A expression to chromosome 15 and of hexosaminidase B expression, to chromosome 5 (Gilbert *et al.*, 1975). Additional details of the structural relationship between these two enzyme components are provided by Sandhoff in Chapter 8 and Beutler in Chapter 9.

The structural similarities between the two hexosaminidases account for the fact that they can hydrolyze many of the same neutral water-soluble natural and artificial substrates containing terminal β-linked GlcNAc or GalNAc. However, Kresse *et al.* (1981) were able to show that the α subunit was specific for negatively charged substrates. This finding led to the recognition of two different active sites on hexosaminidase A (Kytzia and Sandhoff, 1985) and the development of an ar-tificial substrate, 4-methylumbelliferyl-GlcNAc-6-SO_4 (4-MUGS) that is specific for the α chain of hexosmainidase A. For the enzymatic hydrolysis of G_{M2} ganglio-side, a small helper protein is needed. Sandhoff's work first hinted at the existence of such an activator factor in 1969 (Sandhoff, 1969), which he and Conzelmann were able to purify and characterize 10 years later (Conzelmann *et al.*, 1979).

The next advance in our understanding of the molecular causes of Tay-Sachs disease and its variants was the elucidation of the cellular events occurring in the biosynthesis of the hexosaminidases. The availability of antibodies to hex-osaminidases A and B and to isolated α chain enabled Hasilik and Neufeld (1980a, 1980b) to show how precursor forms of the α and β chains are transformed to mature enzyme. They demonstrated the posttranslational modification of the β-hexosaminidase subunits by glycosylation, oligosaccharide phosphorylation, α–β chain association, and limited proteolysis. These studies, which Neufeld reviews in Chapter 14, provide a framework for understanding the gene mutations reponsible for the various forms of G_{M2} gangliosidosis.

A. Isolation of the α-chain gene, *HEXA*

Cloning of the genes for the α and β chains of hexosaminidase presented a formidable challenge because of their low abundance, representing no more than 0.01% of the total translatable mRNA of human placenta. The effort to clone each of these genes occurred in parallel, revealing marked similarities in their structure. This homology has prompted the suggestion that both genes arose from a common ancestral gene.

Myerowitz and Proia (1984) isolated the first cDNA clone, a 119-base pair fragment complementary to the 3' end of the α-chain message. They used antisera prepared to hexosaminidase A to immunoselect from polyribosomes of human lung fibroblasts mRNA that was capable of translation to the α and β chains of hexosaminidase. This resulted in a 1000-fold enrichment from the starting mRNA. They then used ^{32}P labeled cDNA prepared from both β-hexosaminidase-enriched and -depleted polyribosomal mRNA to screen plasmids transformed with the immunoprecipitated mRNA for inserts to sequences complementary to the hexosaminidase mRNA. This effort resulted in the discovery of one clone that hybridized to a greater extent with the enriched probe than to the deleted probe. It hybridized with a message from placental (poly-A$^+$) RNA that programmed the translation of a polypeptide that Myerowitz and Proia were able to identify immunologically as α chain.

With this cDNA clone as a probe, Myerowitz et al. (1985) next obtained from an adult human liver library a 393-base pair cDNA whose sequence at the 3' terminus was identical to that of the probe. This, in turn, was used to screen a human fibroblast cDNA library from which an 891-base pair α-chain cDNA insert was obtained. Finally, they reprobed the adult human liver library with a 300-base pair fragment from the 5' terminus of the new clone, and this led them to a clone containing the entire coding sequence for the α-chain cDNA of human β-hexosaminidase.

The new probe hybridized to two mRNA species of 2.1 kb and 2.6 kb in length, respectively, both of which were deficient from Tay-Sachs cells. Using a panel of mouse–human hybrids, they confirmed that their cDNA mapped to chromosome 15. They further showed that its deduced amino acid structure was homologous to an 11-amino acid peptide found in the β chain by O'Dowd et al. (1985).

Korneluk et al. (1986) isolated cDNA clones coding for the α subunit by an alternative approach. They digested purified hexosaminidase A with trypsin to obtain peptides which were then isolated and sequenced. From one 7-amino acid peptide they derived two sets of oligonucleotides, which they used to screen an SV40-transformed human fibroblast cDNA library. Three cDNA clones were obtained which contained the deduced sequences of five α-chain peptides. As in the case of the cDNA isolated by Myerowitz et al. (1985), their cDNA also mapped to human chromosome 15 and specified two different-sized mRNA species. Both species of mRNA were either absent or reduced in fibroblast cell lines of infantile and juvenile Tay-Sachs patients.

To clone the α-chain gene and examine its structure, Proia and Soravia (1987) utilized human genomic libraries constructed with cosmid and bacteriophage vectors. With the 119-base pair cDNA that had been isolated by Myerowitz and Proia (1984), they isolated two overlapping cosmid clones that contained most of the exonic sequences but not the 5' end of the gene. The 5' end of the gene was found by utilizing as a probe a 525-base pair cDNA fragment previously isolated by Myerowitz et al. (1985). The location of exonic sequences was determined with α-chain cDNA probes and the position of the exon–intron borders identified by comparison of the genomic and cDNA sequences. The introns sequences conformed to the GT/AG rule at their donor and acceptor sites, respectively.

Other characteristics of eukaryotic genes were also confirmed including the presence in the 5' untranslated region of an AT-rich sequence, a sequence similar to the CCAAT box motif, and, between them, a GC-rich region. The amino terminus of the coding sequence contained a 22-amino acid hydrophobic stretch that served as a signal sequence for cleavage, so that a leucine at position 23 became the amino terminus of the mature peptide. The α-chain gene itself was approximately 35 kb in length and contained 14 exons.

B. Isolation of the β-chain gene, *HEXB*

From purified hexosaminidase B, O'Dowd *et al.* (1985) obtained peptides by cyanogen bromide cleavage from which they derived the sequence for an oligonucleotide probe. With this probe, they screened a cDNA library of SV40-transformed human fibroblasts, resulting in the isolation of a cDNA clone containing the deduced sequence of one 31-amino acid peptide fragment and another 11-amino acid fragment, which included the probe sequence. They further showed that the cDNA hybridized to a series of somatic cell hybrids, all of which contained human chromosome 5, and that it detected a 2.2-kb mRNA that coded for the pre-β-chain of hexosaminidase. This mRNA was absent from one of three cases of Sandhoff disease which they examined.

Proia (1988) used a λgt11 liver cDNA library to clone the β-chain gene. The oligonucleotide probe he employed was derived from a portion of the β-chain peptide sequence published earlier by O'Dowd *et al.* (1985). The 1.7-kb insert that they isolated included a 64-base pair fragment from the 3' untranslated region and a further 12 base pairs upstream from the first ATG start site.

Restriction fragments of the β-chain cDNA were then prepared and used to probe three human genomic libraries constructed in bacteriophage λ vectors. From the overlapping genomic clones obtained, a restriction map of the genomic DNA was prepared and the precise intron–exon junctions identified. Exonic sequences were identified by hybridizing oligonucleotides that corresponded to β-chain coding sequence with genomic DNA. The gene itself was approximately 40 kb in length and, as in the case of the α-chain gene, contained a coding region of 14 exons. The intron–exon junctions also conformed to the AG/GT rule for the donor/acceptor site consensus sequences.

C. Homology between *HEXA* and *HEXB*

Already by 1985, the remarkable homology in coding sequence between the α and β chains was appreciated (Myerowitz *et al.*, 1985). Korneluk *et al.* (1986) showed that alignment of the pre-α and pre-β polypeptides revealed 55% homology in nucleotides and 57% homology in amino acids. Proia's work (1988) indicated that the homology extended to the number of exons and introns with 12 of the 13 introns interrupting the genes at corresponding positions.

This homology has facilitated studies of conserved sequences in the two genes. For example, *in vitro* mutagenesis of the Arg^{211} residue of the β-chain cDNA corresponding to Arg^{178} of the α-chain cDNA has revealed that Arg^{211} is the essential residue in the active site of hexosaminidase B. This explains why mutations in the codon for Arg^{178}, the analogous site in the α-chain gene, eliminate the protein's catalytic activity without affecting dimer formation or intracellular targeting (Brown *et al.*, 1989).

D. Cloning of the G_{M2} activator protein gene—*GM2A*

The G_{M2} activator protein gene, GM2A, was cloned shortly after the cloning of HEXA and HEXB. Schröder *et al.* (1989) prepared a mixture of oligonucleotide probes corresponding to four different areas of amino acid sequence and used these to screen an SV40-transformed human cDNA library. They eventually obtained an 821-base pair fragment which hybridized to all four oligonucleotide probes.

The fragment isolated by Schröder *et al.* was then used to screen a second cDNA library which had been prepared from the cultured fibroblasts of a patient with juvenile Sandhoff disease. This yielded clones that contained the 3′ terminus but none that included the initiation codon or 5′ untranslated sequence. Additional sequence from the 5′ end of the cDNA was obtained using anchored PCR methodology.

Genomic clones were isolated from a human brain tissue library of genomic DNA. From this library, Shröeder *et al.* (1989) isolated clones which together represented a gene of approximately 16 kb and contained 4 exons and 3 introns. The 5′ end, however, remained incomplete. The exon–intron junctions followed the AG/GT rule, but they did not find an ATG initation codon or obvious polyadenylation signal.

Nagarajan *et al.* (1992) subsequently reported the presence of two different G_{M2} activator mRNAs from both fibroblasts and placenta. They found that these two species had a common 5′ end but diverged at the 3′ end. Correspondingly, they observed that highly purified G_{M2} activator isolated from human tissues always appeared as a doublet during electrophoresis.

A processed pseudo-gene for GM2A has also been identified. Xie *et al.* (1992) PCR-amplified both genes and hybridized the products to a human/hamster somatic hybrid cell panel. The pseudo-gene mapped to chromosome 3 and the functional gene to chromosome 5.

IX. MUTATIONS IN THE G$_{M2}$ GANGLIOSIDOSES

The G$_{M2}$ gangliosidoses can arise due to inherited defects in *HEXA*, *HEXB*, or *GM2A*. The remaining sections in this chapter are devoted to descriptions of the molecular abnormalities in these genes. The comments that follow are meant to provide the reader with a general background of the early developments in this field.

A. Classic infantile Tay-Sachs disease

The high frequency of the Tay-Sachs gene among Ashkenazi Jews had spawned the belief that its presence in this population was due to a founder effect and that a single mutation would therefore be found. This concept was momentarily reinforced when both Myerowitz and Gravel reported the same splice mutation at a national symposium on Tay-Sachs disease held in November 1987. Each independently found a single nucleotide altered at the 5' boundary of intron 12, from a guanosine to a cytosine, creating a new Dde I restriction site in the mutant (Myerowitz, 1988; Arpaia *et al.*, 1988). However, both of their studies was based on their use of the same cell line, GM 2986. Therefore, it was acknowledged that further studies would be needed to test the hypothesis of genetic homogeneity for a single mutation as the cause of Tay-Sachs disease in Ashkenazi Jews.

That more than a single mutation was responsible for Tay-Sachs disease was already known from the work of Myerowitz and Hogikyan (1986, 1987), who had shown a 7.6-kb deletion at the 5' end of the α-chain gene from genomic DNA of a French-Canadian Tay-Sachs patient from Eastern Quebec. The presence of similarly oriented Alu sequences at the 5' and 3' boundaries of the deletion suggested to these workers that the deletion may have arisen during homologous recombination from unequal crossing-over between Alu sequences. They did not find this deletion in DNA hybridization analyses of genomic DNA from Ashkenazi patients (Myerowitz and Hogikyan, 1986).

In Tay-Sachs disease of both Ashkenazic Jews and non-Jewish French-Canadians, the absence of detectable mRNA or enzyme protein prompted Myerowitz and Gravel and their colleagues to examine genomic DNA for mutations. Myerowitz (1988) found that, after amplifying the region encompassing the splice mutation by PCR and subsequently hybridizing with an allele-specific oligonucleotide probe, only 30% of Ashkenazi obligate heterozygotes carried this mutation. Myerowitz and Costigan (1988) subsequently reported a second α-chain mutation in Ashkenazi Jewish patients, a 4-base pair insertion in exon 11 that produces a premature termination signal. By PCR amplification and dot-blot analysis with sequence-specific probes, they found this mutation to be present in approximately 70% of Tay-Sachs carriers from the Ashkenazi Jewish population.

B. Adult G_{M2} gangliosidosis

A third α-chain mutation in Ashkenazi Jews was found to be responsible for the later-onset form of G_{M2} gangliosidosis (Navon and Proia, 1989; Paw et al., 1989). In this variant, a β-chain precursor is synthesized but fails to associate with α-chains to form mature hexosaminidase A (Frisch et al., 1984; d'Asso et al., 1984). From the fibroblasts of a patient with the adult-onset disorder, Paw et al. (1989) obtained a segment of mRNA with the RNAase protection assay that contained a mutation. The cDNA prepared from this mRNA and the corresponding segment of genomic DNA were amplified by PCR and sequenced revealing a G-to-A transition at the 3' end of exon 7, causing a substitution of serine for glycine and eliminating a Scr F1 restriction site. Paw et al. (1989) found the same mutation in fetal fibroblasts with an α-β association defective phenotype and in cells from five patients with chronic G_{M2} gangliosidosis.

Navon and Proia (1989) reported this point mutation in eight adult Ashkenazi patients from five unrelated families. Two patients also carried the splice junctional mutation in exon 12, and the remaining six carried the 4-base pair insertion error in exon 11. Their findings confirmed the prediction that patients with this form of G_{M2} gangliosidosis are compound heterozygotes with reduced α-chain mRNA due to the presence of one allele for classic infantile Tay-Sachs disease (d'Asso et al., 1984).

C. B_1 variant

In a cDNA clone of a patient with the B_1 variant of G_{M2} gangliosidosis, Ohno and Suzuki discovered a guanosine-to-adenine transversion in base #533 of exon 5 that produced an arginine-to-histidine amino acid substitution (Ohno and Suzuki, 1988). The mutation affects the substrate specificity of the resultant protein so that it retains the abilility to hydrolyze 4-methylumbelliferyl-β-N-acetylglucosaminide (4-MUGlcNAc) but fails to cleave 4-MU-GlcNAc-β-sulfate and G_{M2} ganglioside (Kytzia et al., 1983). In a subsequent study, Tanaka et al. (1988) showed that, unlike classic Tay-Sachs disease, the B_1 mutation has a wide geographic and ethnic distribution. In this report, they identified with allele-specific oligonucleotide probes four additional patients with the same G-to-A change in base #533. Further investigation (Tanaka et al., 1990) revealed an additional patient with the B_1 variant who had a C-to-T change in base 532, causing a substitution in the same amino acid involved in the more common B1 mutation.

D. Insoluble α mutation

Two non-Jewish Italian patients with Tay-Sachs disease were described who had an unusual form of α-subunit processing defect. In both patients, a normal amount of α chain is synthesized, but it fails to phosphorylate, cannot move out of the endoplasmic reticulum, and does not associate with β chain to form native hexosaminidase A (Proia and Neufeld, 1982; Zokaeem et al., 1987). In the case of Proia and Neufeld (1982), Nakano et al. (1988) showed that there was a G-to-A substitution in exon 13, changing a normal glutamic acid to lysine. These authors hypothesized that the alteration in electric charge created by this amino acid substitution interfered with the normal transport of the α chain through the endoplasmic reticulum and on to the Golgi apparatus.

The patient of Zokeeem et al. (1987) contained a shortened α subunit. Lau and Neufeld (1989) found that the fibroblasts from this patient had a homozygous deletion of a cytosine residue at base #1510 of the coding sequence. This deletion caused a frameshift producing an alteration in the next four amino acids and then premature termination. One of the missing amino acids is a cysteine residue that may be important in disulfide bond formation and correct folding of the protein.

E. β-Chain mutations

Molecular heterogeneity also exists in Sandhoff disease. With a cDNA coding for the β chain of β-hexosaminidase, O'Dowd et al. (1986) examined mRNA and genomic DNA from the cell lines of 16 patients with Sandhoff disease. Among 11 cases of infantile Sandhoff disease, they found 3 with normal size and amount of mRNA, 4 with reduced levels of mRNA, and 4 with no detectable mRNA. Two of the latter had partial gene deletions. Each of the 5 juvenile variant cell lines had detectable mRNA, but in one the amount was reduced. On the basis of these studies, one could anticipate that, as in the case of the α-chain gene, a variety of different mutations will be found that are responsible for the deficiency in β-chain synthesis and activity that characterized the Sandhoff variant of G_{M2} gangliosidosis.

Indeed, in cultured fibroblasts from a patient with a juvenile form of Sandhoff disease (MacLeod et al., 1977), Nakano and Suzuki (1989) identified a single nucleotide transversion from a guanosine to an adenine at 26 bases from the 3' terminus of intron 12. This alteration created a consensus sequence for the 3' splice site of an intron CAG/G, and therefore produced a 24-base insertion from the 3' terminus of intron 12. The insertion was in frame and appeared between exons 12 and 13. It added eight additional amino acids to the enzyme protein between amino acids #491 and #492 of the primary sequence.

In summary, more than 100 mutations have now been described collectively in the three genes. These include missense mutations causing an amino acid substitution, nonsense mutations leading to chain termination, deletions and insertions that result in a frameshift, and splice-site alterations that produce abnormal mRNAs. Mutations in the region of codons 170–211 of the HEXA gene are particularly likely to interfere with the catalytic process, as occurs in the B_1 variant, because of their proximity to the Arg^{178} residue. In cases where a mutation occurs in the donor splice site, such as +1 IVS, the previous exon is removed and the resultant mRNA is unstable. Also, many HEXA mutations are at CpG dinucleotides, considered to be mutational "hot spots."

In general, the phenotype for any particular mutation is best delineated in the genotypically homozygous state. In compound heterozygotes, the biochemically more severe allele determines the phenotype. The absence of both immunologically detectable α subunits and residual enzyme activity is associated with the most severe forms of G_{M2} gangliosidosis.

X. GENETICALLY ENGINEERED ANIMAL MODELS

Animal models of inherited metabolic diseases are useful for studying their developmental pathology and pathogenesis and for testing novel treatments such as enzyme or gene therapy. Spontaneously occurring animal models of the G_{M2} gangliosidoses have been described in dogs (Cummings et al., 1985; Singer and Cork, 1989), cats (Cork et al., 1977; Neuwelt et al., 1985), and Yorkshire swine (Pierce et al., 1976; Kosanke et al., 1978). The majority resemble Sandhoff disease in humans. In the case of the Japanese Spaniel dog, the G_{M2} activator protein is defective (Ishikawa et al., 1987).

Animal models can also be generated artificially. Mice have many advantages. Much is known about the genetics of the laboratory mouse, with considerable synteny existing between the genetic material of mice and humans. Mice are easily bred, and they can be bred relatively cheaply. They have a short life span (~2–3 years), a short generation time (~3 months), and an average female can produce 4–8 litters with an average litter size of 6–8 pups. For these reasons, mice have been favored for the production of transgenic animals, including "knockout" models of human disease.

Transgenic animals are generated by inserting a foreign gene, a transgene, either into the host chromosomes randomly or by gene targeting into a preselected endogenous gene. If the random integration event alters endogenous gene expression by creating an insertional mutation, a phenotypic alteration may result. The exogenous DNA is transmitted to daughter cells during cell division and passed on to subsequent generations in a Mendelian fashion. This is equivalent to whole animal expression cloning.

Embryonic stem (ES) cells can be readily grown in culture and manipulated for this purpose. In one scenario, the normal mouse gene is replaced and inactivated through homologous recombination with a preselected cloned mouse gene that, while altered, still retains much of its original sequence. ES cells are obtained from the inner cell mass of isolated blastocysts ($3\frac{1}{2}$-day-old embryos), cultured *in vitro*, genetically modified by insertion of a transgene or by gene targeting, then injected back into the host blastocyst and reimplanted in a pseudopregnant mouse. A chimera is likely to form, composed of cells from the blastocyst and those of the implanted ES cells. Backcrosses between chimeras and subsequent inbreeding will produce mice that are homozygous for the genetic modification.

To select for cells which have properly integrated into the targeting vector, a gene for resistance to an antibiotic such as neomycin may be included. Additionally, to thwart the survival of cells that have undergone nonhomologous recombination, a suicide gene such as thymidine kinase (TK) can be included within the vector, positioned alongside the gene of interest. Addition of neomycin to the cell culture media ensures that only those cells that integrate the vector will survive. The inclusion of gancyclovir in the media as well will have no effect on the cells if the recombination event excludes the flanking sequence containing the TK gene, but will cause the cell to be destroyed if the TK gene is also integrated in the host genome. The actual transfer of the targeting construct into mouse ES cells is done by electroporation, a process in which pulses of high voltage are delivered to cells, causing a temporary relaxation in the permeability properties of the plasma membranes.

The murine α- and β-chain genes, designated *Hexa* and *Hexb*, respectively, have been cloned and found to have significant homology with the cognate human sequences (Yamanaka *et al.*, 1994a, 1994b). These accomplishments and the development of knockout mouse models of Tay-Sachs disease and Sandhoff disease (Taniike *et al.*, 1995; Sange *et al.*, 1995; Phaneuf *et al.*, 1996) are described by Proia in Chapter 18 of this volume.

XI. CONCLUSIONS

Scarcely a quarter-century has elapsed from recognition of the enzyme deficiency in Tay-Sachs disease to the engineering of an animal knockout for this disease. Progress in our understanding of the molecular causes of Tay-Sachs disease and other variants of G_{M2} gangliosidosis has closely paralleled the development of techniques for cloning genes and detecting mutations. Direct benefits continue to accrue to patients and their families in more accurate diagnoses, carrier detection, and prenatal diagnoses. Numerous challenges lie ahead, especially in the application of this knowledge to the prevention and treatment of these diseases.

References

Alwine, J. C., Kemp, D. J., and Stark, G. R. (1977). Method for detection of specific RNAs in agarose gels by transfer to diazobenzylox-methyl paper and hybridization with DNA probes. *Proc. Natl. Acad. Sci. (USA)* **74,** 5350–5354.

Arpaia, E., Dumbrille-Ross, A., Maler, T., Neote, K., Tropak, M., Troxel, C., Stirling, J. S., Pitts, J. S., Bapat, B., Lamhonwah, A. M., Mahuran, D. J., Schuster, S. M., Clarke, J. T. R., Lowden, J. A., and Gravel, R. A. (1988). Identification of an altered splice site in Ashkenazi Tay-Sachs disease. *Nature* **333,** 85–86.

Avery, O. T., MacLeod, C. M., and McCarty, M. (1944). Studies of the chemical nature of the substance inducing transformation of pneumococcal types; Induction of transformation by desoxyribonucleic acid fraction isolated from pneumococcus type III. *J. Exp. Med.* **79,** 137–158.

Beutler, E., Yoshida, A., Kuhl, W., and Lee, J. E. S. (1976). The subunits of human hexosaminidase A. *Biochem. J.* **159,** 541–543.

Brown, C. A., Neote, K., Leung, A., Gravel, R. A., and Mahuran, D. J. (1989). Introduction of the alpha subunit mutation associated with the B1 variant of Tay-Sachs disease into the beta subunit produces a beta-hexosaminidase B without catalytic activity. *J. Biol. Chem.* **264,** 21705–21710.

Cariello, N. F., and Skopek, T. R. (1993). Mutational analysis using denaturing gradient gel electrophoresis and PCR. *Mutat. Res.* **288,** 103–112.

Cohen, S. N., Chang, A. C., Boyer, H. W., and Helline, R. B. (1973). Construction of biologically functional bacterial plasmids in vitro. *Proc. Natl. Acad. Sci. (USA)* **70,** 3240–3244.

Conzelmann, E., and Sandhoff, K. (1979). Purification and characterization of an activator protein for the degradation of glycolipids G_{M2} and G_{A2} by hexosaminidase A. *Hoppe-Seylers Z. Physiol. Chem.* **360,** 1837–1849.

Cork, L. C., Munnel, J. F., Lorenz, M. D., Murphy, J. Y., Baker, H. J., and Rattazzi, M. C. (1977). G_{M2} ganglioside lysosomal storage disease in cats with beta-hexosaminidase deficiency. *Science* **196,** 1014–1017.

Cummings, J. F., Wood, P. A., Walkley, S. U., de Lahunta, A., and DeForest, M. E. (1985). GM_2 gangliosidosis in a Japanese Spaniel. *Acta Neuropathol. (Berl.)* **67,** 247–253.

d'Azzo, A., Proia, R. L., Kolodny, E. H., Kaback, M. M., and Neufeld, E. F. (1984). Faulty association of α- and β-subunits in some forms of β-hexosaminidase A deficiency. *J. Biol. Chem.* **259,** 11070–11074.

Frisch, A., Baram, D., and Navon, R. (1984). Hexosaminidase A deficient adults: Presence of α-chain precursor in cultured skin fibroblasts. *Biochem. Biophys. Res. Commun.* **119,** 101–107.

Geiger, B., and Arnon, R. (1976). Chemical characterization and subunit structure of human β acetylhexosaminidases A and B. *Biochemistry* **15,** 3484–3493.

Gilbert, F., Kucherlapati, R., Creagan, R. P., Murnane, M. J., Darlington, G. J., and Ruddle, F. H. (1975). Tay Sachs' and Sandhoff's diseases: The assignment of genes for hexosaminidase A and B to individual human chromosomes. *Proc. Natl. Acad. Sci. (USA)* **72,** 263–267.

Grunstein, M., and Hogness, D. S. (1975). Colony hybridization: A method for the isolation of cloned DNAs that contain a specific gene. *Proc. Natl. Acad. Sci. (USA)* **72,** 3961–3965.

Hasilik, A., and Neufeld, E. F. (1980a). Biosynthesis of lysosomal enzymes in fibroblasts. Synthesis as precursors of higher molecular weight. *J. Biol. Chem.* **255,** 4937–4945.

Hasilik, A., and Neufeld, E. F. (1980b). Biosynthesis of lysosomal enzymes in fibroblasts. Phosphorylation of mannose residues. *J. Biol. Chem.* **255,** 4946–4950.

Hershey, A. D., Dixon, J., and Chase, M. (1953). Nucleic acid economy in bacteria infected with bacteriophage T2. I. Purine and pyrimidine composition. *J. Gen. Physiol.* **36,** 777–789.

Ishikawa, Y., Li, S.-C., Wood, P. A., and Li, Y.-T. (1987). Biochemical basis of type AB G_{M2} gangliosidosis in a Japanese Spaniel. *J. Neurochem.* **48,** 860–864.

Keen, J., Lester, D., Inglehearn, C., Curtis, A., and Bhattacharya, S. (1991). Rapid detection of single base mismatches as heteroduplexes on Hydrolink gels. *Trends Genet.* **7,** 5.

Khorana, H. G. (1968). Nucleic acid synthesis in the study of the genetic code. *In* "Nobel Lectures: Physiology or Medicine (1963–1970)," Elsevier, Amsterdam, 1973.

Klenk, E. (1942). Uber die Ganglioside des Gehirns bei der infantilen amaurotischen Idiotie vom Typ Tay-Sachs. *Ber. Dtsch. Chem. Ges.* **75,** 1632–1636.

Kolodny, E. H. (1972). Sandhoff's disease: Studies on the enzyme defect in homozygotes and detection of heterozygotes. *In* "Sphingolipids, Sphingolipidoses and Allied Disorders" (B. W. Volk and S. M. Aronson, eds.), pp. 321–341. Plenum, New York.

Kolodny, E. H., Brady, R. O., and Volk, B. W. (1969). Demonstration of an alteration of ganglioside metabolism in Tay-Sachs disease. *Biochem. Biophys. Res. Commun.* **37**, 526–531.

Korneluk, R. G., Mahuran, D., Neote, K., Klavins, M. H., O'Dowd, B. F., Tropak, M., Willard, H. F., Anderson, M. J., Lowden, J. A., and Gravel, R. A. (1986). Isolation of cDNA clones coding for the α-subunit of β-hexosaminidase: Extensive homology between the α- and β-subunits and studies on Tay-Sachs disease. *J. Biol. Chem.* **261**, 8407–8413.

Kosanke, S. D., Pierce, K. R., and Bay, W. W. (1978). Clinical and biochemical abnormalities in porcine G$_{M2}$-gangliosidosis. *Vet. Pathol.* **15**, 685–699.

Kresse, H., Fuchs, W., Glössl, J., Holtfrerich, D., and Gilberg, W. (1978). Liberation of N-acetylglucosamine-6-sulfate by human beta-N-acetylhexosaminidase. *J. Biol. Chem.* **256**, 12926–12932.

Kytzia, H. J., and Sandhoff, K. (1985). Evidence for two different active sites on human beta-hexosaminidase A. Interaction of G$_{M2}$ activator protein with beta-hexosaminidase A. *J. Biol. Chem.* **260**, 7568–7572.

Kytzia, H.-J., Hinrichs, U., Maire, I., Suzuki, K., and Sandhoff, K. (1983). Variant of G$_{M2}$-gangliosidosis with hexosaminidase A having a severely changed substrate specificity. *EMBO. J.* **2**, 1201–1205.

Lalley, P. A., Rattazzi, M. C., and Shows, T. B. (1974). Human β-D-N-acetylhexosaminidase A and B: Expression and linkage relationship in somatic cell hybrids. *Proc. Natl. Acad. Sci. (USA)* **71**, 1569–1573.

Lau, M. M., and Neufeld, E. F. (1989). A frameshift mutation in a patient with Tay-Sachs disease causes premature termination and defective intracellular transport of the alpha-subunit of beta-hexosaminidase. *J. Biol. Chem* **264**, 21376–21380.

MacLeod, P. M., Wood, S., Jan, J. E., Applegarth, D. A., and Dolman, C. L. (1977). Progressive cerebellar ataxia, spasticity, psychomotor retardation, and hexosaminidase deficiency in a 10-year-old child: Juvenile Sandhoff disease. *Neurology* **27**, 571–573.

Maniatis, T., Hardison, R. C., Lacy, E., Lauer, J., O'Connell, C., Quon, D., Sim, G. K., and Efstratiadis, A. (1978). The isolation of structural genes from libraries of eucaryotic DNA. *Cell* **15**, 687–701.

Maxam, A. M., and Gilbert, W. (1977). A new method for sequencing DNA. *Proc. Natl. Acad. Sci. (USA)* **74**, 560–564.

Mullis, K. B., and Faloona, F. (1987). Specific synthesis of DNA in vitro via a polymerase catalyzed chain reaction. *Meth. Enzymol.* **155**, 335–350.

Myerowitz, R. (1988). Splice junction mutation in some Ashkenazi Jews with Tay-Sachs disease: Evidence against a single defect within this ethnic group. *Proc. Natl. Acad. Sci. (USA)* **85**, 3955–3959.

Myerowitz, R., and Costigan, F. C. (1989). The major defect in Ashkenazi Jews with Tay-Sachs disease is an insertion in the gene for the α-chain of β-hexosaminidase. *J. Biol. Chem.* **262**, 18587–18589.

Myerowitz, R., and Hogikyan, N. D. (1986). Different mutations in Ashkenazi Jewish and non-Jewish French Canadians with Tay-Sachs disease. *Science* **232**, 1646–1648.

Myerowitz, R., and Hogikyan, N. D. (1987). A deletion involving Alu sequences in the β-hexosaminidase α chain gene of French Canadians with Tay-Sachs disease. *J. Biol. Chem.* **262**, 15396–15399.

Myerowitz, R., and Proia, R. L. (1984). cDNA clone for the alpha-chain of human beta-hexosaminidase: Deficiency of alpha-chain mRNA in Ashkenazi Tay-Sachs fibroblasts. *Proc. Natl. Acad. Sci. (USA)* **81**, 5394–5398.

Myerowitz, R., Piekarz, R., Neufeld, E. F., Shows, T. B., and Suzuki, K. (1985). Human β-hexosaminidase α-chain: Coding sequence and homology with the β-chain. *Proc. Natl. Acad. Sci. (USA)* **81**, 7830–7834.

Nagarajan, S., Chen, H., Li, S., Li, Y., and Lockyer, J. M. (1992). Evidence for two cDNA clones encoding human G_{M2}-activator protein. *Biochem. J.* **282**, 807–813.

Nakano, T., and Suzuki, K. (1989). Genetic cause of a juvenile form of Sandhoff disease: Abnormal splicing of β-hexosaminidase β chain gene transcript due to a point mutation within intron 12. *J. Biol. Chem.* **264**, 5155–5158.

Nakano, T., Muscillo, M., Ohno, K., Hoffman, A. J., and Suzuki, K. (1988). A point mutation in the coding sequence of the β-hexosaminidase α-gene results in defective processing of the enzyme protein in an unusual G_{M2}-gangliosidosis variant. *J. Neurochem.* **51**, 984–987.

Navon, R., and Proia, R. L. (1989). The mutations in Ashkenazi Jews with adult G_{M2} gangliosidosis, the adult form of Tay-Sachs disease. *Science* **243**, 1471–1474.

Neuwelt, E. A., Johnson, W. G., Blank, N. K., Pagel, M. A., Maslen-McClure, C., McClure, M. J., and Wu, P. M. (1985). Characterization of a new model of G_{M2}-gangliosidosis (Sandhoff disease) in Korat cats. *J. Clin. Invest.* **76**, 482–490.

Nirenberg, M. W., and Leder, P. (1964). RNA codewords and protein synthesis. *Science* **145**, 1399–1407.

O'Brien, J. S., Okada, S., Chen, A., and Fillerup, D. L. (1970). Tay-Sachs disease. Detection of heterozygotes and homozygotes by serum hexosaminidase assay. *N. Engl. J. Med.* **283**, 15–20.

O'Dowd, B. F., Quan, F., Willard, H. F., Lamhonwah, A.-M., Korneluk, R. G., Lowden, J. A., Gravel, R. A., and Mahuran, D. J. (1985). Isolation of cDNA clones coding for the β subunit of human β-hexosaminidase. *Proc. Natl. Acad. Sci. (USA)* **82**, 1184–1188.

O'Dowd, B. F., Klavins, M. H., Willard, H. F., Gravel, R., Lowden, J. A., and Mahuran, D. J. (1986). Molecular heterogeneity in the infantile and juvenile forms of Sandhoff disease (O-variant G_{M2}-gangliosidosis). *J. Biol. Chem.* **261**, 12680–12685.

Ohno, K., and Suzuki, K. (1988). Mutation in G_{M2}-gangliosidosis B, variant. *J. Neurochem.* **50**, 316–318.

Okada, S., and O'Brien, J. (1969). Tay-Sachs disease: General absence of a β-D-N-acetylhexosaminidase component. *Science* **164**, 698–700.

Okayama, H., and Berg, P. (1982). High-efficiency cloning of full-length cDNA. *Mol. Cell Biol.* **2**, 161–170.

Paw, B. H., Kaback, M. M., and Neufeld, E. F. (1989). Molecular basis of adult-onset and chronic G_{M2}-gangliosidosis in patients of Ashkenazi-Jewish origin: Substitution of serine for glycine at position 269 of the α-subunit of β-hexosaminidase. *Proc. Natl. Acad. Sci. (USA)* **86**, 2413–2417. Erratum (1989). *Proc. Natl. Acad. Sci. (USA)* **86**, 5025.

Phaneuf, D., Wakamatsu, N., Huang, J.-Q., Borowski, A., Peterson, A. C., Fortunato, S. R., Ritter, G., Igdoura, S. A., Morales, C. R., Benoit, G., Akerman, B. R., Leclerc, D., Hanai, N., Marth, J. D., Trasler, J. M., and Gravel, R. A. (1996). Dramatically different phenotypes in mouse models of human Tay-Sachs and Sandhoff diseases. *Hum. Mol. Genet.* **5**, 1–14.

Pierce, K. R., Kosanke, S. D., Bay, W. W., and Bridges, C. H. (1976). Animal model: Porcine cerebrospinal lipodystrophy (G_{M2} gangliosidosis). *Am. J. Pathol.* **83**, 419–422.

Proia, R. L. (1988). Gene encoding the human β-hexosaminidase α-chain: Extensive homology of intron placement in the α- and β-chain genes. *Proc. Natl. Acad. Sci. (USA)* **85**, 1883–1887.

Proia, R. L., and Neufeld, E. F. (1982). Synthesis of β-hexosaminidase in cell-free translation and in intact fibroblasts. An insoluble precursor α chain in a rare form of Tay-Sachs disease. *Proc. Natl. Acad. Sci. (USA)* **79**, 6360–6364.

Proia, R. L., and Soravia, E. (1987). Organization of the gene encoding human β-hexosaminidase α-chain. *J. Biol. Chem.* **262**, 5677–5681.

Rigby, T. W. J., Dieckmann, M., Rhodes, C., and Berg, P. (1977). Labeling deoxyribonucleic acid to high specific activity in vitro by nick translation with DNA polymerase I. *J. Mol. Biol.* **113**, 237–251.

Robinson, D., and Stirling, J. (1967). N-acetyl-β-glucosaminidases in human spleen. *Biochem. J.* **107,** 321–327.

Sachs, B. (1887). On arrested cerebral development, with special reference to its cortical pathology. *J. Nerv. Ment. Dis.* **14,** 541–553.

Saiki, R. K., Gelfand, D. H., Stoffel, S., Scharf, S. J., Higuchi, R., Horn, G. T., Mullis, K. B., and Erlich, H. A. (1988). Primer-directed enzymatic amplification of DNA with a thermostable DNA polymerase. *Science* **239,** 487–494.

Sandhoff, K. (1969). Variation of beta-N-acetylhexosaminidase-pattern in Tay-Sachs disease. *FEBS Lett.* **4,** 351–354.

Sandhoff, K., Conzelmann, E., and Nehrkorn, H. (1977). Specificity of human liver hexosaminidases A and B against glycosphingolipids G_{M2} and G_{A2}. Purification of the enzymes by affinity chromatography employing specific elution. *Hoppe Seylers. Z. Physiol. Chem.* **358,** 779–787.

Sanger, F., Nicklen, S., and Coulson, A. R. (1977). DNA sequencing with chain-terminating inhibitors. *Proc. Natl. Acad. Sci. (USA)* **74,** 5463–5467.

Sango, K., Yamanaka, S., Hoffman, A., Okuda, Y., Grinberg, A., Westphal, H., McDonald, M. P., Crawley, J. N., Sandhoff, K., Suzuki, K., and Proia, R. L. (1995). Mouse models of Tay-Sachs and Sandhoff diseases differ in neurologic phenotype and ganglioside metabolism. *Nat. Genet.* **11,** 170–176.

Schröder, M., Klima, H., Nakano, T., Kwon, H., Quintern, L. E., Gärtner, S., Suzuki, K., and Sandhoff, K. (1989). Isolation of a cDNA encoding the human G_{M2} activator protein. *FEBS Lett.* **251,** 197–200.

Sheffield, V. C., Beck, J. S., Kwitek, A. E., Sandstrom, D. W., and Stone, E. M. (1993). The sensitivity of single-strand conformation polymorphism analysis for the detection of single base substitutions. *Genomics* **16,** 325–332.

Singer, H. S., and Cork, L. C. (1989). Canine GM_2 gangliosidosis: Morphological and biochemical analysis. *Vet. Pathol.* **26,** 114–120.

Smith, H. O., and Wilcox, K. W. (1970). A restriction enzyme from *Hemophilus influenzae*: I. Purification and general properties. *J. Mol. Biol.* **51,** 379–391.

Southern, E. M. (1975). Detection of specific sequences among DNA fragments separated by gel electrophoresis. *J. Mol. Biol.* **98,** 503–517.

Srivastava, S. K., and Beutler, E. (1973). Hexosaminidase-A and hexosaminidase-B: Studies on Tay-Sachs and Sandhoff's disease. *Nature* **241,** 463.

Studencki, A. B., Conner, B. J., Impraim, C. C., Teplitz, R. L., and Wallace, R. B. (1985). Discrimination among the human $β^A$, $β^S$, and $β^C$-globin genes using allele-specific oligonucleotide hybridization probes. *Am. J. Hum. Genet.* **37,** 42–51.

Tanaka, A., Ohno, K., and Suzuki, K. (1988). G_{M2}-gangliosidosis B_1 variant: A wide geographic and ethnic distribution of the specific β-hexosaminidase β chain mutation originally identified in a Puerto Rican patient. *Biochem. Biophys. Res. Commun.* **156,** 1015–1019.

Tanaka, A., Ohno, K., Sandhoff, K., Maire, I., Kolodny, E. H., Brown, A., and Suzuki, K. (1990). G_{M2}-gangliosidosis B_1 variant: Analysis of beta hexosaminidase alpha gene abnormalities in seven patients. *Am. J. Hum. Genet.* **46,** 329–339. Erratum (1991). *Am. J. Hum. Genet.* **48,** 176.

Taniike, M., Yamanaka, S., Proia, R. L., Langaman, C., Bone-Turrentine, T., and Suzuki, K. (1995). Neuropathology of mice with targeted disruption of *Hexa* gene, a model of Tay-Sachs disease. *Acta Neuropathol.* **89,** 296–304.

Watson, J. D., and Crick, F. H. C. (1953). Genetical implications of the structure of deoxyribonucleic acid. *Nature* **171,** 964–967.

Xie, B., Kennedy, J. L., McInnes, B., Auger, D., and Mahuran, D. (1992). Identification of a processed pseudogene related to the functional gene encoding the G_{M2} activator protein: Localization of the pseudogene to human chromosome 3 and the functional gene to human chromosome 5. *Genomics* **14,** 796–798.

Yamanaka, S., Johnson, O. N., Norflus, F., Boles, D. J., and Proia, R. L. (1994a). Structure and expression of the mouse β-hexosaminidase genes, *Hexa* and *Hexb*. *Genomics* **21**, 588–596.

Yamanaka, S., Johnson, M. D., Grinberg, A., Westphal, H., Crawley, J. N., Taniike, M., Suzuki, K., and Proia, R. L. (1994b). Targeted disruption of the *Hexa* gene results in mice with biochemical and pathologic features of Tay-Sachs disease. *Proc. Natl. Acad. Sci. (USA)* **91**, 9975–9979.

Zokeeem, G., Bayleran, J., Kaplan, P., Hechtman, P., and Neufeld, E. F. (1987). A shortened β-hexosaminidase α-chain in an Italian patient with infantile Tay-Sachs disease. *Am. J. Hum. Genet.* **40**, 537–547.

11

Cloning the β-Hexosaminidase Genes

Richard L. Proia

Genetics of Development and Disease Branch
National Institute of Diabetes and Digestive and Kidney Diseases
National Institutes of Health
Bethesda, Maryland 20892

When I entered Liz Neufeld's lab at the National Institutes of Health (NIH) in 1981 as a postdoctoral fellow, she presented me with a challenging project—the cloning of the β-hexosaminidase genes. It was clear that this step was required if significant progress in the field was to be made. Only by determining the gene structures would it be possible to identify the underlying defects in Tay-Sachs and Sandhoff diseases. The cloning of the genes could also yield critical information about the structure, function, and evolution of the β-hexosaminidase system.

This was in the early days of the recombinant DNA revolution, and the technological advances that have made the task of gene cloning now straightforward had not yet taken hold. A few years earlier, Andre Hasilik and Liz Neufeld had developed a panel of antibodies to β-hexosaminidase that they had used to characterize the biosynthetic pathway of the enzyme in human fibroblasts (Hasilik and Neufeld, 1980a, 1980b). Armed with these potent antibodies, we decided to clone the β-hexosaminidase genes based on a polysome immunopurification approach. In this procedure, metabolically active cells are cracked open and polysomes— the cell's protein synthesis machinery—are isolated. The polysome contains an attached polypeptide chain in the process of translation that can be used as a tag for isolation of the β-hexosaminidase messenger RNA (mRNA) using specific

Copyright © 2001 by Academic Press
0065-2660/01 $35.00

antibodies. The isolated polysomes can be dissociated and the recovered mRNA reverse transcribed into cDNA for cloning. A drawback in our approach was that the antibodies were made against the human enzyme. This meant that we needed to isolate polysomes from human cells or tissues. Metabolically active human tissues suitable for polysome isolation were nearly impossible to obtain, so we were forced to grow cultured human fibroblasts. Although well suited for polysome isolation, human fibroblasts yielded only small amounts of material. As a result, we had to grow up well over 1000 large flasks of cells to carry out this experiment. After developing a cell-free translation assay to monitor the purification of the mRNA (Proia and Neufeld, 1982), I succeeded in isolating a tiny amount of highly purified β-hexosaminidase mRNA from the human fibroblasts. I handed the mRNA off to Rachel Myerowitz, who expertly transformed it into a collection of 44 cDNA clones. Now we faced the problem of figuring out which, *if any*, of these clones corresponded to β-hexosaminidase. We were severely disadvantaged because we had no protein sequence information to match with the sequence of the cDNA. My approach was to prepare two hybridization probes— one was a β-hexosaminidase- "enriched" probe made from the highly purified polysomal mRNA preparation, and the other was a β-hexosaminidase-"depleted" probe made from the polysomal mRNA preparation with the β-hexosaminidase mRNA removed.

Figure 11.1 shows the experiment where I hybridized these probes to duplicate sets of DNA from the 44 potential β-hexosaminidase clones. Only a single clone reacted strongly with the enriched probe (arrow) but weakly with the depleted probe, providing evidence that it was a β-hexosaminidase cDNA. We went on to show that it was an α-subunit cDNA. This represented the first cloning of a β-hexosaminidase gene sequence (Myerowitz and Proia, 1984). Since we did not have protein sequence information, our evidence showing that this clone was the α-subunit cDNA was, by necessity, indirect and provoked many anxious moments for me. In subsequent work it was shown that the sequence of the original clone did indeed encode a very small portion of the subunit—only 119 base pairs out of the approximately 2000-bp subunit cDNA (Myerowitz et al., 1985)! From this tiny segment of the α-subunit, Rachel Myerowitz and Kuni Suzuki went on to isolate and sequence the entire α-subunit cDNA (Myerowitz et al., 1985). I used this tiny probe to isolate and characterize the 35-kb β-hexosaminidase A (*Hex A*) gene (Proia and Soravia, 1987) (Figure 11.2). After completing the *Hex A* structure, I continued on and isolated and characterized the structurally and evolutionary related β-hexosaminidase B (*Hex B*) gene (Proia, 1988) (Figure 11.3).

The β-hexosaminidases are built from two structurally similar polypeptide subunits, α and β. The subunits assemble into dimers giving rise to three isozymes, S(αα) A (αβ) and B (ββ). Due in large part to a functionally different active site carried by each subunit (Kytzia and Sandhoff, 1985), this multiple

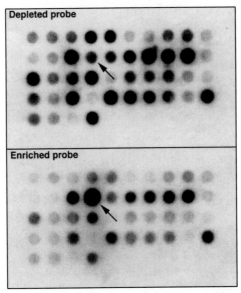

Figure 11.1. The first cloning of a β-hexosaminidase gene. The exposed film shows the differential hybridization of 44 cDNA clones prepared from immunopurified β-hexosaminidase mRNA with a β-hexosaminidase-"enriched" probe and a β-hexosaminidase-"depleted" probe. Note that one clone (arrow) hybridizes more strongly with the enriched probe than with the depleted probe. This clone contained a portion of the α-subunit cDNA sequence (Myerowitz and Proia, 1984).

isozyme system is able to degrade a wide range of glycoconjugates containing β-linked N-acetylhexosaminyl residues.

The organization of the α- and β-subunit genes is very similar (Fig. 11.4). Considering the enormous variation in length of eukaryotic genes, the α gene (35 kilobases) and B gene (40 kilobases) are close in size. Both genes have 13 introns dividing their protein coding regions into 14 exons. The overall spacing of the exons in the two genes is also similar; the exons are widely spaced at the 5′ ends and tightly packed at the 3′ ends.

The protein sequences of the α- and β-subunits are identical at over 50% of their amino acids (Rorneluk et al., 1986; Proia, 1988). When the α and β coding sequences are aligned for maximal identity as shown in Figure 11.5, 12 of the 13 exons interrupt at precisely the same positions. Only the 5′ most intron in each gene is noncoincident. Within genes derived from a common ancestor, as with the globin (Maniatis et al., 1980) serine protease (Rodgers, 1985; Craik et al., 1984), vitellogen in (Wahli, 1980), and ACTH proenkephalin (Node, 1982) families, there is conservation of intron location with respect to coding regions. The striking conservation of intron positioning in the β-hexosaminidase genes is clear evidence that the duplication of a progenitor gene was a step in the

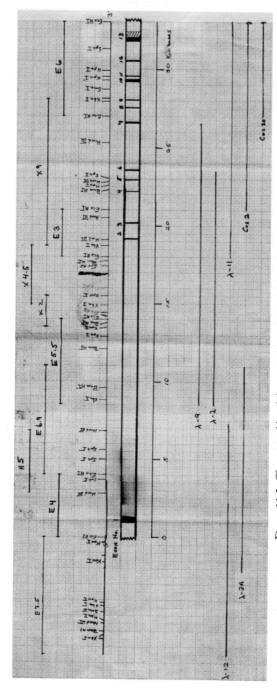

Figure 11.2. The original hand-drawn working scheme of the β-hexosaminidase A gene.

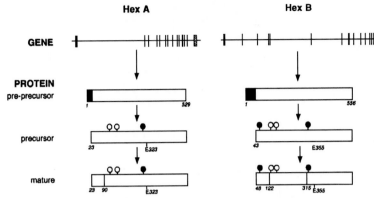

Figure 11.3. The molecular biology of the β-hexosaminidase system. The intron–exon architecture of the β-hexosaminidase A (Hex A) and β-hexosaminidase B (Hex B) genes are shown at the top of the figure (Proia and Soravia, 1987; Proia, 1988). The vertical lines represent exons and the horizontal lines represent introns. Below the gene structures, the β-hexosaminidase polypeptides are represented by the rectangles. The numbers refer to amino acid positions in the sequence. Pre-precursors are the forms containing signal sequences obtained by cell-free translation (black boxes). The precursors are the forms with the signal sequences removed but before lysosomal processing (Little et al., 1988; Quon et al., 1989). The mature forms are proteolytically processed in lysosomes. The vertical lines in the rectangles represent the sites of proteolytic processing. The circles represent N-linked oligosaccharides (Sonderfeld-Fresko and Proia, 1989; Weitz and Proia, 1992). The filled circles represent the highly phosphorylated N-linked oligosaccharides. E323 and E355 are the glutamic acid residues that function as proton donors in the catalytic action of β-hexosaminidase (Fernandes et al., 1997; Pennybacker et al., 1997).

evolution of the α- and β-subunits. The similarity of the genes in their size and their exon spacing is likely a reflection of the primordial gene structure prior to its duplication.

Although the organization of the genes is quite similar, significant differences exist especially in the 5′ regions: (1) intron 1 in each gene is

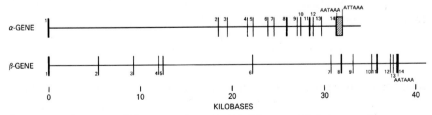

Figure 11.4. Comparison of the structural organization of the α- and β-chain genes. The exons are represented by the numbered black boxes.

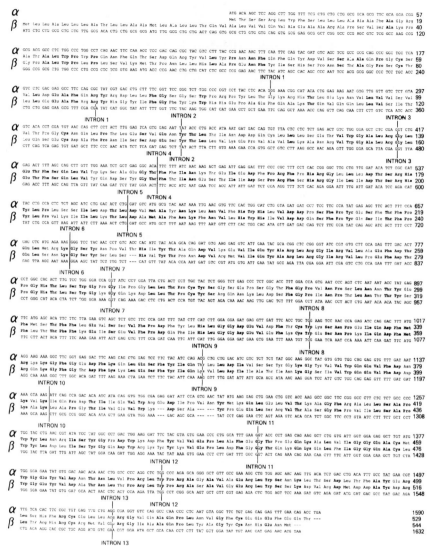

Figure 11.5. Relative position of itrons in the α and β protein coding regions. Identical amino acids at the same position are in bold type. The α-chain sequence is from Myerowitz *et al.*, 1985; the β-chain sequence is from Proia, 1988.

nonhomologously placed; (2) the amino acid sequences carried by the first and second exons in the two genes are much less similar (26% identity) than those specified by the remaining exons (63% identity); and (3) the amino-terminus of the β-polypeptide is considerably longer than the amino-terminus of the α-polypeptide (Figure 11.5). This local diversity in length and sequence as well as the discordance in intron-exon structure may have been introduced by a mechanism called "intron sliding." It has been proposed that mutations causing splice junctional slippage result in length and sequence variability by the extension or contraction of exons (Craik *et al.*, 1983). Such a hypothesis is attractive because, after the duplication event, the sliding of intron 1 in one of the β-hexosaminidase genes would account for the positional noncoincidence of the first introns, and for the sequence divergence and length variation at the amino-terminal end of the subunits. Intron deletion or insertion during the evolution of the β-hexosaminidase genes may also be a plausible explanation for the difference in intron-exon structures; however, additional mechanisms would have to be invoked to explain the variation of size and sequence at the amino-termini of the subunits. "Exon shuffling," i.e., the recruitment of exonic units from unrelated genes (Gilbert, 1985), is unlikely to have occurred due to the diminished but distinct sequence identity between the 5' exons of the two genes. The ultimate resolution of the mechanisms involved in the evolution of the α and β genes will require the isolation and examination, from a more primitive organism, of a β-hexosaminidase gene that has not been duplicated.

Gene duplication during evolution, as exemplified by the globin family, is a mechanism for generating phenotypic diversity without interfering with the function of preexisting gene products. Through duplication and diversification of a single β-hexosaminidase gene, a multiple isozyme system has evolved that is able to accommodate complex range of substrates.

The establishment of these β-hexosaminidase gene structures enabled other laboratories to go on and identify the molecular defects causing Tay-Sachs and Sandhoff diseases (Gravel *et al.*, 1995; Myerowitz, 1997). Mutations in the α-subunit gene result in Tay-Sachs disease, whereas the clinically related syndrome, Sandhoff disease, is caused by mutations in the β-subunit gene. The elucidation of the β-hexosaminidase α- and β-subunit gene structures (Proia and Soravia, 1987; Proia, 1988) was necessary to enable the eventual identification of the genetic lesions that underlie these disorders. As described in this volume, significant advances have already been made toward understanding these mutations.

In addition to the identification of mutations, the cloned β-hexosaminidase sequences were fundamental for the elucidation of important structural and functional features of the enzyme. We now understand in fine detail the biosynthesis, glycosylation, phosphorylation, processing, and catalytic mechanism of the enzyme (summarized in Figure 11.3). The cloning of the β-hexosaminidase genes has stimulated great progress in understanding the defects underlying Tay-Sachs disease and the molecular biology of the β-hexosaminidase system. The path of research on Tay-Sachs disease which began over 100 years ago continues to lead to a comprehensive understanding of the disease and, hopefully, an effective treatment.

Acknowledgments

This project evolved during my postdoctoral tenure with Elizabeth Neufeld. I am grateful to her for allowing me the freedom to pursue this work. I also gratefully acknowledge the efforts of my co-workers on the work I have described. I thank Jen Reed for graphics.

References

Craik, C. S., Qui-Lim, C., Swift, G. H., Quinto, C., MacDonald, R. J., and Rutter, W. J. (1984). Structure of related rat pancreatic trypsin genes. *J. Biol. Chem.* **259**, 14255–14264.

Craik, C. S., Rutter, W. J., and Fletteric, R. (1983). Slice junctions: Association with variation in protein structure. *Science* **220**, 1125–1129.

Fernandes, M. J., Yew, S., Leclerc, D., Henrissat, B., Vorgias, C. E., Gravel, R. A., Hechtman, P., and Kaplan, F. (1997). Identification of candidate active site residues in lysosomal β-hexosaminidase A. *J. Biol. Chem.* **272**(2), 814–820.

Gilbert, W. (1985). Genes-in-pieces revisited. *Science*. **228**, 823–824.

Gravel, R. A., Clarke, J. T. R., Kaback, M. M., Mahuran, D., Sandhoff, K., and Suzuki, K. (1995). The G_{M2} gangliosidoses. *In* "The Metabolic and Molecular Basis of Inherited Disease α" (C. R. Scriver, A. L. Beaudet, W. S. Sly, and D. Valle, eds.), pp. 2839–2879. McGraw-Hill, New York.

Hasilik, A., and Neufeld, E. F. (1980a). Biosynthesis of lysosomal enzymes in fibroblasts. Phosphorylation of mannose residues. *J. Biol. Chem.* **255**(10), 4946–4950.

Hasilik, A., and Neufeld, E. F. (1980b). Biosynthesis of lysosomal enzymes in fibroblasts. Synthesis as precursors of higher molecular weight. *J. Biol. Chem.* **255**(10), 4937–4945.

Korneluk, R. G., Mahuran, D. J., Noete, K., Kavins, M. H., O'Dowd, B. F., Tropak, M., Willard, H. F., Anderson, M., Lowden, J. A., and Gravel, R. A. (1986). Isolation of cDNA clones coding for the α-subunit of human β-hexosaminidase. *J. Biol. Chem.* **261**, 8407–8413.

Kytzia, H. J., and Sandhoff, K. (1985). Evidence for two different active sites on human β-hexosaminidase. A. *J. Biol. Chem.* **260**, 7568–7572.

Little, L. E., Lau, M. M., Quon, D. V., Fowler, A. V., and Neufeld, E. F. (1988). Proteolytic processing of the α-chain of the lysosomal enzyme, β-hexosaminidase, in normal human fibroblasts. *J. Biol. Chem.* **263**(9), 4288–4292.

Maniatis, T., Fritsch, E. F., Laver, J., and Lawn, R. M. (1980). The molecular genetics of human hemoglobins. *Ann. Rev. Gene.* **14**, 145–178.

Myerowitz, R. (1997). Tay-Sachs disease-causing mutations and neutral polymorphisms in the Hex A gene. *Hum. Mutat.* **9**(3), 195–208.

Myerowitz, R., Piekarz, R., Neufeld, E. F., Shows, T. B., and Suzuki, K. (1985). Human β-hexosaminidase α chain: Coding sequence and homology with the β chain. *Proc. Natl. Acad. Sci. U.S.A.* **82**(23), 7830–7834.

Myerowitz, R., and Proia, R. L. (1984). cDNA clone for the α-chain of human β-hexosaminidase: Deficiency of α-chain mRNA in Ashkenazi Tay-Sachs fibroblasts. *Proc. Natl. Acad. Sci. U.S.A.* **81**(17), 5394–5398.

Noda, M., Teranish, Y., Takahashi, H., Toyosato, M., Notake, M., Nakanish, S., and Numa, S. (1982). Isolation and structural organization of the human preproenkephalin gene. *Nature* **297**, 431–434.

Pennybacker, M., Schuette, C. G., Liessem, B., Hepbildikler, S. T., Kopetka, J. A., Ellis, M. R., Myerowitz, R., Sandhoff, K., and Proia, R. L. (1997). Evidence for the involvement of Glu-355 in the catalytic action of human β-hexosaminidase B. *J. Biol. Chem.* **272**(12), 8002–8006.

Proia, R. L. (1988). Gene encoding the human β-hexosaminidase β chain: Extensive homology of intron placement in the α- and β-chain genes. *Proc. Natl. Acad. Sci. U.S.A.* **85**(6), 1883–1887.

Proia, R. L., and Neufeld, E. F. (1982). Synthesis of β-hexosaminidase in cell-free translation and in intact fibroblasts: An insoluble precursor α chain in a rare form of Tay-Sachs disease. *Proc. Natl. Acad. Sci. U.S.A.* **79**(20), 6360–6364.

Proia, R. L., and Soravia, E. (1987). Organization of the gene encoding the human β-hexosaminidase α-chain. *J. Biol. Chem.* **262**(12), 5677–5681.

Quon, D. V., Proia, R. L., Fowler, A. V., Bleibaum, J., and Neufeld, E. F. (1989). Proteolytic processing of the β-subunit of the lysosomal enzyme, β-hexosaminidase, in normal human fibroblasts. *J. Biol. Chem.* **264**(6), 3380–3384.

Rodgers, J. (1985). Exon shuffling and intron insertion in serine protease genes. *Nature.* **315,** 458–459.

Sonderfeld-Fresko, S., and Proia, R. L. (1989). Analysis of the glycosylation and phosphorylation of the lysosomal enzyme, β-hexosaminidase B, by site-directed mutagenesis. *J. Biol. Chem.* **264**(13), 7692–7697.

Wahli, W., Dawid, I. B., Wyler, T., Weber, R., and Ryffle, G. U. (1980). Comparitive analysis of the structural organization of two closely related vitellogenin genes in X. *laevis. Cell* **20,** 107–117.

Weitz, G., and Proia, R. L. (1992). Analysis of the glycosylation and phosphorylation of the α-subunit of the lysosomal enzyme, β-hexosaminidase A, by site-directed mutagenesis. *J. Biol. Chem.* **267**(14), 10039–10044.

The Search for the Genetic Lesion in Ashkenazi Jews with Classic Tay-Sachs Disease

Rachel Myerowitz
St. Mary's College of Maryland
St. Mary's City, Maryland 20686

This manuscript is dedicated to the memory of my mother, Eva Steinberg Myerowitz, a woman of true inner and outer beauty, elegance, brilliance, and warmth, who will forever hold my admiration and love.

In 1981, while a postdoctoral fellow in the laboratory of Elizabeth F. Neufeld at the National Institutes of Health, I decided to apply the emerging techniques of recombinant DNA to the study of lysosomal enzymes and the diseases resulting from deficiencies therein. This was a project laden with risk, as I was naive in the ways of molecular biology, as was Dr. Neufeld's laboratory. Moreover, the low abundance of lysosomal enzymes made them difficult to clone. Nonetheless, I began, intending to clone α-L-iduronidase. Two and one-half years later, Liz Neufeld, Rick Proia, and I celebrated the isolation of a cDNA clone (Myerowitz and Proia, 1984) encoding a fragment of the polypeptide defective in Tay-Sachs disease, the α-chain of human β-hexosaminidase. Isolation of a full-length cDNA encoding the entire α-chain polypeptide soon followed (Myerowitz *et al.*, 1985), as did the isolation and characterization of the α-chain gene (Proia and Soravia, 1987). Equipped with these probes and pictures of the gene and moved by an irrational desire to discover the mutation resulting in classic Tay-Sachs disease in the Ashkenazi Jewish population, I began the hunt. An unexpected bonus of the search was uncovering the α-chain mutation in the non-Jewish French Canadian population from eastern Quebec. This group, like the Ashkenazi Jews, carries the gene for the classic form of Tay-Sachs disease 10 times more frequently than the general population (Anderman *et al.*, 1977). Because patients with classic Tay-Sachs disease from either the French Canadian or Ashkenazi Jewish group are clinically and biochemically indistinguishable, the possibility had been raised

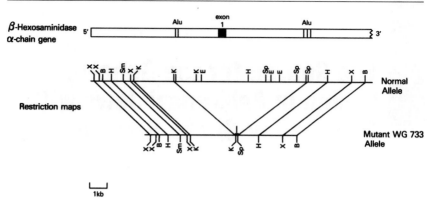

Figure 12.1. Comparison of the restriction maps of the 5' end of the β-hexosaminidase α-chain normal gene and the mutant gene of a non-Jewish French Canadian patient with classic Tay-Sachs disease. The top part of the figure is a schematic of the 5' end of the normal α-chain gene. Exon 1 is designated by a solid box, followed by an open box for intron 1. Alu repetitive elements in the vicinity of the breakage points are indicated. The restriction map of the normal α-chain gene is from Proia and Soravia (1987). The map of the mutant gene below it (WG733) is from Myerowitz and Hogikyan (1987). Vertical arrows indicate the breakage and reunion points involved in the deletion of the French Canadian patient. Restriction endonuclease abbreviations are B, BamHI; E, EcoRI; H, HindIII; K, KpnI; Sm, SmaI; Sp, SphI; X, XbaI.

that both populations harbor the same α-chain mutation (O'Brien, 1983). Southern analysis of the α-chain gene from French Canadian and Ashkenazi Jewish patients disproved that notion; the α-chain gene from the former displayed an obvious deletion at the 5' end, whereas that from the latter group harbored a more subtle genetic lesion not detectable by Southern blots (Myerowitz and Hogikyan, 1986). To fully characterize the French Canadian deletion in terms of size, precise location, and environment, we isolated a clone encompassing the deletion from a genomic library constructed with DNA from the fibroblasts of a French Canadian patient (WG733) with classic Tay-Sachs disease. Comparison of the restriction map of the mutant with that of the normal gene (Figure 12.1) showed that the deletion was 7.6 kb long, included part of intron 1, all of exon 1, and extended 2000 base pairs upstream past the putative promotor region. Sequence analysis of the deletion junction in the mutant and corresponding regions of the normal gene demonstrated the presence of similarly oriented Alu sequences at the 5' and 3' deletion boundaries (Myerowitz and Hogikyan, 1987). These data are consistent with the possibility that the deletion may have arisen during homologous recombination from unequal crossing-over between Alu sequences.

Elucidation of the mutation underlying classic Tay-Sachs disease in the Ashkenazi Jewish population posed more of a challenge. Restriction enzyme digestion patterns of the mutant α-chain gene were identical to those of the normal gene, indicating that the mutation involved a small deletion or single base change. Theoretically the lesion could reside anywhere along the 40 kb gene. But, in light of the absence of α-chain mRNA in Ashkenazi Jewish patients cells (Myerowitz and Proia, 1984), I reasoned that the mutation probably disturbed transcription or RNA processing and focused my attention on regions of the gene governing these events. Toward that end I isolated the entire α-chain gene (save 1500 base pairs encompassing exon 8) from an Ashkenazi Jewish patient, GM 2968, with classic Tay-Sachs disease, and compared its nucleotide sequences with those of the normal gene in the promoter region, exon and splice junction regions, and polyadenylation area. The only difference observed between these sequences was at the 5′ boundary of intron 12, where a guanosine in the conserved splice junction dinucleotide G–T had been altered to a cytidine (Figure 12.2). Because changes in these invariant bases generally result in mRNA splicing errors, I assumed that the alteration was functionally significant and developed a method to assay for it in patients and carriers. The assay is based on the fact that a new Dde1 restriction enzyme site is created by the guanosine-to-cytosine change. Amplification of the region encompassing the mutation by polymerase chain reaction followed by digestion of the amplified product with Dde1, then Southern analysis of the sample, reveals a 120-base-pair DNA fragment from the normal allele in contrast to an 85-base pair piece from the mutant.

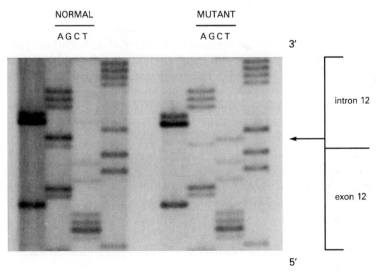

Figure 12.2. Nucleotide sequence analysis of the α-chain genes from a normal subject and from an Ashkenazi Jewish patient with classic Tay-Sachs disease in the region surrounding the 5′ boundary of exon 12. The change of a guanosine to a cytidine at the 5′ border of intron 12 in GM 2968 is indicated by an arrow.

It has generally been assumed that only one mutation gives rise to the classic form of Tay-Sachs disease in the Ashkenazi Jewish population (O'Brien, 1983). Since my experimental strategy rested on this assumption, one can imagine how surprised I was to see that GM 2968, the cell strain originally used to clone the α-chain gene, had only one allele with the splice junction mutation (Figure. 12.3). Analysis of the parental DNA showed that this allele was inherited from the father. In a second family, the affected child also had one allele with the splice-junction mutation, this time inherited from the mother. In a third family, the affected child did not exhibit this mutation in either allele, and neither parent carried it. The simplest explanation for these results was that at least two mutations, rather than the assumed one, casued the classic form of Tay-Sachs disease in the Ashkenazi Jewish population (Myerowitz, 1988). To gain some idea of the frequency of this mutant allele in this ethnic group, we tested obligate heterozygotes for this mutation and came up with a preliminary number of 30%. Did this mutation have functional significance, or was it merely a neutral polymorphism? The fact that 20 Ashkenazi Jews designated noncarriers by enzymatic assay were negative for this alteration supported but did not prove the assumption that the splice junction mutation was functionally significant. Later studies demonstrated the existence of abnormal α-chain mRNAs in Ashkenazi patients (Ohno and Suzuki, 1988).

How many mutations then give rise to classic Tay-Sachs disease in the Ashkenazi Jewish population? I started the search believing it was one. Theoretically, my results suggested that a multiplicity of defects could underlie classic Tay-Sachs in the Ashkenazi Jewish population. Intuitively, I felt that this was not the case and continued my research under the working hypothesis that only one other lesion was present in this ethnic group. Since the α-chain gene from Ashkenazi Jewish patient GM 515 had assayed negative for the splice junction mutation (Figure 12.3) and was therefore presumably homozygous for the second unknown defect, I began my search for the "second" mutation with eagerness and excitement by isolating the entire α-chain gene from a genomic library constructed from the DNA of patient GM 515. Comparison of its nucleotide sequence with that of the normal gene in the promoter region, exon and splice junction regions, and polyadenylation area revealed only one difference, a 4-base pair insertion (5-TATC-3') in exon 11 (Figure 12.4) identical in sequence to the four bases preceding it (Myerowitz and Costigan, 1988). The mutation causes a shift in the reading frame that results in a nonsense mutation 9 base pairs from the insertion (Figure 12.4). As has been observed in other systems, a mutation within an exon frequently results in a deficiency of mRNA rather than production of a truncated protein.

I developed a PCR-based assay for the insertion mutation which involved amplification of a segment of exon 11 inclusive of the region of insertion and detection of the lesion by probing duplicate samples of the amplified product with

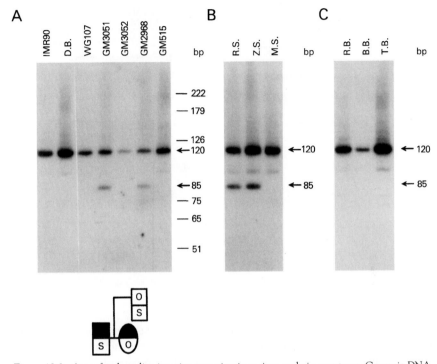

Figure 12.3. Assay for the splice junction mutation in various α-chain genotypes. Genomic DNA samples were assayed for the replacement of a guanosine with a cytidine at the 5′ boundary of exon 12 of the α-chain of β-hexosaminidase (**A**) Lanes: IMR90 (normal), D.B. (Ashkenazi normal), WG107 (non-Jewish French Canadian with classic Tay-Sachs), GM3051 (Ashkenazi obligate heterozygote; father of GM2968), GM3052 (Ashkenazi obligate heterozygote; mother of GM2968), GM2968 (Ashkenazi with classic Tay-Sachs), GM515 (Ashkenazi with classic Tay-Sachs). (**B**) Lanes: R.S. (Ashkenazi with classic Tay-Sachs), Z.S. (Ashkenazi obligate heterozygote; mother of R.S.), M.S. (Ashkenazi obligate heterozygote; father of R.S.). (**C**) Lanes: R.B. (Ashkenazi with classic Tay-Sachs), B.B. (Ashkenazi obligate heterozygote; father of R.B.), T.B. (Ashkenazi obligate heterozygote; mother of R.B.). The schematic below **A** shows the genotypes of GM2968 (affected child), GM3052 (mother), and GM3051 (father). The shaded areas signify the normal α-chain allele. S, splice junction mutation; O, another mutation.

oligonucleotides specific for either the normal or mutant sequence (Myerowitz, 1988). Again, I was in for another surprise as I, with a rapidly beating heart, turned on the lights in the dark room and peered at the X-ray film only to see that cell strain GM 515 contained only one allele with the insertion mutation (Figure 12.5). The other allele could not contain the splice junction mutation, as we had already shown it to be lacking in this patient. Was my asssumption of two mutations in the Ashkenazi Jewish population wrong? In fact, my intuition had

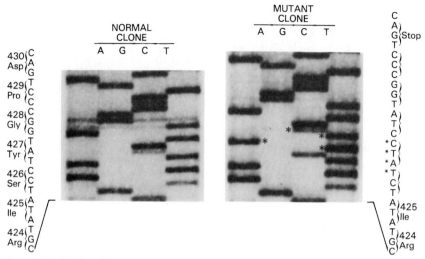

Figure 12.4. Nucleotide sequence analysis of normal and mutant α-chain genes in region of exon 11 containing the insertion. The insertion is marked with asterisks.

served me well, as subsequent analysis of the frequency of the insertion mutation in the Ashkenazi Jewish population by testing for its presence in heterozygote carriers yielded a number of 70%. The second mutant allele in patient GM 515 turned out to be a private family mutation identified as a single base change in exon 11 (Shore *et al.*, 1992). The etiology of the high gene frequency of classic Tay-Sachs

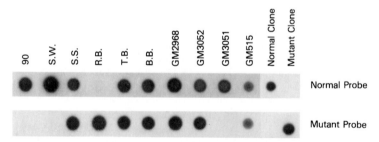

Figure 12.5. Assay for exon 11 insertion defect in various α-chain genotypes. Genomic DNA samples were assayed for the insertion defect in exon 11 of the α-chain of β-hexoseaminidase. IMR 90 (normal), S.W. (Ashkenazi normal), S.S. (Ashkenazi obligate heterozygote carrier), R.B. (Ashkenazi with classic Tay-Sachs), B.B. (Ashkenazi obligate heterozygote, fater of R.B.), T.B. (Ashkenazi obligate heterozygote, mother of R.B.), GM 2968 (Ashkenazi with classic Tay-Sachs), GM 3052 (Ashkenazi obligate heterozygote, mother of GM 2968), GM 3051 (Ashkenazi obligate heterozygote, father of GM 2968), GM 515 (Ashkenazi with classic Tay-Sachs).

disease had been the subject of controversy. The discovery of two, rather than one, Tay-Sachs–causing lesions in the Ashkenazi Jewish population supports the idea that the high frequency in this population is the result of selective advantage rather than the founder effect.

The hunt was over. But the search for the Tay-Sachs mutation has afforded me the chance to experience the incomparable elation associated with the moment of discovery that rarely if ever occurs in a research career. I had indeed been lucky.

References

Anderman, E., Scriver, C. R., Wolfe, L. S., Dansky, L., and Andermann, F. (1977). Genetic variants of Tay-Sachs disease: Tay-Sachs disease and Sandhoff's disease in French Canadians, juvenile Tay-Sachs disease in Lebanese Canadians, and a Tay-Sachs screening program in the French-Canadian population. *Prog. Clin. Biol. Res.* **18,** 161–188.

Myerowitz, R. (1988). Splice junction mutation in some Ashkenazi Jews with Tay-Sachs disease: Evidence against a single defect within the ethnic group. *Proc. Natl. Acad. Sci. (USA)* **85,** 3955–3959.

Myerowitz, R., and Costigan, F. C. (1988). The major defect in Ashkenazi Jews with Tay-Sachs disease is an insertion in the gene for the α-chain of β-hexosaminidase. *J. Biol. Chem.* **263,** 18587–18589.

Myerowitz, R., and Hogikyan, N. D. (1986). Different mutations in Ashkenazi Jews and non-Jewish French Canadians with Tay-Sachs disease. *Science* **232,** 1646–1648.

Myerowitz, R., and Hogikyan, N. D. (1987). A deletion involving Ala sequences in the β-hexosaminidase α-chain gene of French Canadians with Tay-Sachs disease. *J. Biol. Chem.* **262,** 15396–15399.

Myerowitz, R., and Proia, R. (1984). cDNA clone for the α-chain of human β-hexosaminidase: Deficiency of α-chain mRNA in Ashkenazi Tay-Sachs fibroblusts. *Proc. Natl. Acad. Sci. (USA)* **81,** 5394–5398.

Myerowitz, R., Piekarz, R., Neufeld, E. F., Shows, T. B., and Suzuki, K. (1985). Human β-hexosaminidase α-chain: Coding sequence and homology with the β-chain. *Proc. Natl. Acad. Sci. (USA)* **82,** 7830–7834.

O'Brien, J. S. (1983). The gangliosidoses. In "The Metabolic Basis of Inherited Disease" (J. B. Stanbury, J. B. Wyngaarden, D. S. Fredrickson, J. L. Goldstein, and M. S. Brown, eds.), pp. 945–969. McGraw-Hill, New York.

Ohno, K., and Suzuki, K. (1988). Multiple abnormal β-hexosaminidase α-chain mRNAs in a compound heterozygous Ashkenazi Jewish patient with Tay-Sach disease. *J. Biol. Chem.* **263,** 18563–18567.

Proia, R. L., and Soravia, E. (1987). Organization of the gene encoding the human β-hexosaminidase α-chain. *J. Biol. Chem.* **262,** 5677–5681.

Shore, S., Tomczak, J., Grebner, E. E., and Myerowitz, R. (1992). An unusual genotype in an Ashkenazi Jewish patient with Tay-Sachs disease. *Hum. Mutat.* **1,** 486–490.

13

The β-Hexosaminidase Story in Toronto: From Enzyme Structure to Gene Mutation

Don J. Mahuran*
The Research Institute, The Hospital for Sick Children
and
Department of Laboratory Medicine and Pathobiology
University of Toronto
Toronto, Ontario, Canada M5G 1X8

Roy A. Gravel
Departments of Cell Biology and Anatomy,
and
Biochemistry and Molecular Biology
University of Calgary
Calgary, Alberta, Canada T2N 1N4

  I. Introduction
 II. Structures of Hexosaminidase A and Hexosaminidase B
III. Isolation of cDNA Clones Coding for the α and β Chains
 IV. Extensive Homology Between the Deduced α and β
     Primary Structures
  V. Posttranslational Processing of the Pre-pro-α
     and Pre-pro-β Chains
     A. Proteolytic Processing
     B. Glycolytic Processing
 VI. Structure–Function Relationships
     A. Catalytic Site
     B. Substrate-Binding Sites
VII. Molecular Heterogeneity in Tay-Sachs and Sandhoff Diseases
     References

*To whom correspondence should be addressed. E-mail: hex@sickkids.on.ca. Fax: (416) 813-8700.
Telephone: (416) 813-6161.

I. INTRODUCTION

Our involvement with β-hexosaminidase (Hex) and its disorders began through our association with Dr. J. Alexander (Sandy) Lowden. He was instrumental in setting up the Canadian arm of the North American Tay-Sachs screening program in the early 1970s at The Hospital for Sick Children (Lowden *et al.*, 1974) and established a Medical Research Council of Canada Program on the sphingolipidoses in 1976. One of us, Mahuran, began as a postdoctoral fellow with Lowden in 1976 to work on the structural characterization of Hex. Gravel also joined the group at that time as a co-investigator to provide a genetic input to the project. In 1983, Lowden left the group to found the Research Development Corporation of the Hospital for Sick Children, while Mahuran was appointed to staff and replaced him as a project director. We worked together until 1989 to provide a biochemical and genetic understanding of the Hex system in man. In 1989, Gravel left Toronto to take up a position as scientific director at the McGill University—Montreal Children's Hospital Research Institute. The following describes our experiences working together on the genetics and biochemistry of Hex until 1989 and continues with Mahuran carrying the banner in Toronto until the present time. This was a very active period of research on Hex, involving the efforts of many laboratories. Particularly relevant to this chapter were the work of Elizabeth Neufeld, Andrej Hasilik, Rachel Myerowitz, and Rick Proia, whose work paralleled our own, but who utilized different technologies. This combination of approaches clarified the structures and functions of the complex isoenzyme system that makes up Hex in man.

II. STRUCTURES OF HEXOSAMINIDASE A AND HEXOSAMINIDASE B

Our early years emphasized the structural characterization of Hex. Mahuran's initial project was to duplicate the organic synthesis of the highly specific Hex affinity ligand developed by Geiger *et al.* (1974). It was a difficult synthesis, and one not to be repeated unnecessarily. That first column, made in 1976, was used until 1991. These early studies produced the unexpected result that Hex A is a trimer of the structure $\alpha_1\beta_2$, in contrast to the then-accepted structure of $\alpha_2\beta_2$ (reviewed by Mahuran *et al.*, 1985). It was found that while the β subunits of Hex A and Hex B did contain two polypeptides of 25–30 kDa, the α subunit of Hex A was composed of a single chain of 50–60 kDa. These results were not so easily accepted by our colleagues, and it was over 2 years before they finally reached publication in the Canadian literature (Mahuran and Lowden, 1980).

At about this time, Hasilik and Neufeld (1980) published their landmark paper showing that the mature subunits were formed posttranslationally from larger pro-α and pro-β polypeptide precursors (pro-α = 67 kDa, pro-β = 63 kDa). They also described the mature α subunit as being composed of a single α polypeptide chain (54 kDa), in agreement with our structure, but they viewed the mature β subunit as composed of a single 29-kDa chain and other "smaller fragments" (24, 22, and 19 kDa), which did not correlate with our data. Despite this difference, their information helped to explain some puzzling results of our own. We had separated reduced and alkylated Hex A and Hex B on isoelectric focusing slab gels (IEF-PAGE), run in urea and detergent, and had found that the polypeptide chains forming the β subunit in both isozymes could be separated into an acidic group and a basic group of heterogeneous bands. Using two-dimensional IEF/SDS-PAGE, we found that the basic group had a slightly lower apparent M_r (26 kDa) than the acidic group (29 kDa). We proposed, in keeping with the biosynthetic processing data, that the "single" 25- to 30-kDa band observed on sodium dodecyl sulfate (SDS) gels in fact represented distinctly different polypeptides formed through the specific internal hydrolysis of the larger pro-β precursor chain once it had entered the lysosome (Mahuran *et al.*, 1982) .

Proof that the β band from SDS gels indeed corresponded to two distinct protein species was obtained by Florence Tsui, working in Gravel's laboratory. She came up with the idea of using a modification of a Cleveland digest (Bordier and Crettol-Jarvinen, 1979) to examine the peptide composition of the β proteins. The method consisted of using a two-dimensional electrophoretic system in which

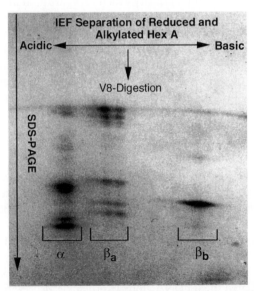

Figure 13.1. Two-dimensional peptide maps of reduced and alkylated Hex A. The polypeptides were initially separated by IEF-PAGE. An unstained IEF-gel sample lane was then turned horizontally (acidic end to the left) and embedded on top of a SDS-PAGE stacking gel. Digestion with V8-protease occurred in the stacking gel overnight and the resulting peptides separated by M_r in the running gel.

the reduced and alkylated polypeptides comprising the β subunit were resolved by SDS- or IEF-PAGE in the first dimension, digested *in situ* by V8-protease or papain, and the peptides separated by SDS-PAGE in the second dimension. With either M_r or charge separation in the first dimension, the peptide patterns revealed two distinct components to the β subunits of Hex B and Hex A, whether purified from placenta (Mahuran *et al.*, 1982) or immunoprecipitated from fibroblast(Tsui *et al.*, 1983) (Figure 13.1). She went on to show that the peptide pattern of the β species from fibroblasts was nearly identical to that of the pro-β, while that of the mature α chain was highly homologous to the pro-α polypeptide pattern. These data confirmed, biochemically, Hasilik and Neufeld's cell labeling results (above). Thus, we defined the structure of mature Hex A as $\alpha(\beta_a\beta_b)$ and mature Hex B as 2 $(\beta_a\beta_b)$ (Mahuran *et al.*, 1982), where β_a and β_b denote the acidic (β_a) and basic (β_b) halves of the pro-β precursor, formed through an internal cleavage of the ~63-kDa precursor chain in the lysosome. Formal proof of the model would have to await determination of the primary structure of α and β subunits and their precursors.

III. ISOLATION OF cDNA CLONES CODING FOR THE α AND β CHAINS

With the advent of recombinant DNA technology it became possible to determine, unequivocally, the total primary structure of rare proteins. Gravel committed his laboratory to establishing this technology, while Mahuran concentrated on developing high-performance liquid chromatography (HPLC) techniques for the isolation of proteins and peptides at a sufficient purity to allow amino acid sequence analysis. We began with the isolation of cDNA clones that code for the β subunit. First, we wished to determine if the β_a and β_b primary structures were contained within the pro-β chain and establish the proteolytic mechanism directing their formation. Second, unlike the α subunit in Hex A, β subunits could be isolated in pure form directly from Hex B. Thus, no method had to be developed immediately for the preparative separation of the α and β subunits. The approach proved successful when Brian O'Dowd, a Ph.D. student with Mahuran, used a Hex B peptide sequence to define an oligonucleotide probe. He used the probe to screen a cDNA library in Gravel's laboratory and isolated a cDNA clone coding for the pre-pro-β polypeptide chain (O'Dowd *et al.*, 1985).

A similar approach was used to isolate a cDNA coding for the pre-pro-α chain. However, in this case the reduced and alkylated α chains were first separated from the β_a and β_b components of placental Hex A by HPLC M_r-sieve chromatograph in 6 M guanidine-HCl. Robert Korneluk, a postdoctoral fellow with Gravel, isolated the α cDNA clone (Korneluk *et al.*, 1986). Myerowitz and Proia had also isolated a pre-pro-α cDNA clone (Myerowitz *et al.*, 1985; Myerowitz and Proia, 1984). Their approach differed in that they used an antiserum to screen an expression library.

The successful isolation of the cDNA clones formed the beginning of a new phase in Hex research. It would allow determination of the primary structure and biosynthetic processing of the enzyme and bring research on Tay-Sachs and Sandhoff diseases to the direct analysis of the mutant genes.

IV. EXTENSIVE HOMOLOGY BETWEEN THE DEDUCED α AND β PRIMARY STRUCTURES

The first surprising result came when we began sequencing the α cDNA. The deduced primary structure of the α chain was strikingly homologous to the β cDNA sequence (Figure 13.2). They were found to share 57% amino acid identity (Korneluk et al., 1986). Such extensive homology was not anticipated, because the two polypeptides are encoded on different chromosomes. However, it should have been suspected because of their biochemical similarities; e.g., they were known to contain related active sites, any combination of the two subunits produced an active Hex isozyme (Hex S, α-α; Hex A, α-β; and Hex B, β-β), and antiserum made against one subunit often weakly cross-reacted with the other. Thus, the proteins appear to have been derived from a common ancestral gene, a suggestion later reinforced by gene structure studies (see later). Furthermore, we felt that regions in the aligned sequences that have been well conserved would likely be important for proper folding, substrate binding, or catalysis. This hypothesis is now generally accepted for Hex and other glycosyl hydrolases, which have been grouped into families based on sequence homology; Hex has been placed in family 20 (Henrissat, 1991; Henrissat and Bairoch, 1993).

The coding sequence homology was underscored by the related structure of the two genes. Proia and Soravia (1987) cloned the *HEXA* gene and showed that it is about 35 kb long and contained 14 exons. Proia (1988) and Neote *et al.* (1988) cloned the *HEXB* gene and found that it too contains 14 exons and is about 45 kb long. Examination of the intron–exon junctions showed that all but the first occur at identical locations. It therefore appears that the two genes were derived by duplication of a primordial gene and eventually found their way to different chromosomes, *HEXA* on chromosome 15 and *HEXB* on chromosome 5.

V. POSTTRANSLATIONAL PROCESSING OF THE PRE-PRO-α AND PRE-PRO-β CHAINS

A. Proteolytic processing

Once we had deduced the primary structure of the pre-pro-α and pre-pro-β poly-peptides, it became possible to localize the internal site of hydrolysis generating

```
Chain                                                                          aa#

alpha                                    MTSSRLWFSLLLAAAFAGRATA                  22
                                            *    *      *    * *
beta                               MELCGLGLPRPPMLLALLLATLLAAMLALLTQVALVVQVAEA    42

alpha    -------------[LWPWPQNFQTSDQRYVLYPNNFQFQYDVSSAAQPGCSVLDEAFQRY            68
                     ***  *            *  *  **       *   *   *   * *** **
beta     ARAPSVS[AKPGPALWPLPLSVKMTPNLLHLAPENFYISHSPNSTAGPSCTLLEEAFRRY           10

alpha    RDLLFG]SGSWPRPYLTGKRH{TLEKNVLVVSVVTPGCNQLPTLESVENYTLTINDDQCL          126
             **      *         *       **        *    *     * * ***
beta     HGYIFG]FYKWHHEPAEFQAK{TQVQQLLVSITLQSECDAFPNISSDESYTLLVKEPVAV         159
         Fusion   #1
alpha    LLSET|VWGALRGLETFSQLVWKSAEGTFFINKTEIEDFPRFPHRGLLLDTSRHYLPLSS         185
          *    |*************** ***  **   *  * ***  ***  *  ********
beta     LKANR|VWGALRGLETFSQLVYQDSYGTFTINESTIIDSPRFSHRGILIDTSRHYLPVKI         218

alpha    ILDTLDVMAXNKLNVFHWHLVDDPSFPXFSFTFPFLMRKGSYNPVTHIYTAQDVKEVIEY         245
         **  ***  **  **  **  ***  ***  ****  *  *****  ****    *  **   **  ****
beta     ILKTLDAMAFNKFNVLHWHLVDDQSFPXQSITFPFLSNKGSYSLS-HVYTPNDVRMVIEY         277
                                                         Fusion   #2
alpha    ARLRGIRVLAEFDTPGHTLSWGPGIPGLLTPCYS GSEP|SGTFGPVNPSLNNTYEFMS         302
         *********  ************  *   *******     |    ***  **  **  **  *
beta     ARLRGIRVLPEFDTPGHTLSWGKGQKDLLTPCYS}RQNK|(LDSFGPINPTLNTTYSFLT         334

alpha    TFFLEVSSVFPDFYLHLGGDEVDFTCWKSNPEIQDFMRKKGFGEDFKQLESFYIQTLLDI         362
         ***  *  *  ****   *******  *  **  ***  ***  ******  ****  ***  *******     ***
beta     TFFKEISEVFPDQFIHLGGDEVEFKCWESNPKIQDFMRQKGFGTDFKKLESFYIQKVLDI         394

alpha    VSSYGKGYVVWQEVFDNKVKIQPDTIIQVWREDIPVNYMKELELVTKAGFRALLSAPWYL         422
            **   *******  ***   *  **   **   *       *   **   **   **      *******
beta     IATINKGSIVWQEVFDDKVKLAPGTIVEVWK-DSA--YPEELSRVTASGFPVILSAPWYL         451

alpha    NRISYGPDWKDFYVVEPLAFEGTPEQKALVIGGEACMWGEYVDNTNLVPRLWPRAGAVAE         482
         ****  **     *  ****  *  **      **  *  ******  ******  ***  *******  **  *
beta     DLISYGQDWRKYYKVEPLDFGGTQKQKQLFIGGEACLWGEYVDATNLTPRLWPRASAVGE         511

alpha    RLWSNKLTSDLTFAYERLSHFRCELLRRGVQAQPLNVGFCEQEFEQT}                     529
         ****  *    *    **  **    **      **  ****   *    *
beta     RLWSSKDVRDMDDAYDRLTRHRCRMVERGIAAQPLYAGYCNHENM)                       556
```

Figure 13.2. Alignment of the deduced primary structures of the pre-pro-α and pre-pro-β chains of human β-hexosaminidase A. The signal peptide is in italics and underlined. The mature α_p and β_p chains are enclosed in [ ]. The mature α_m and β_b chains are enclosed in { }. The mature β_a chain is enclosed in ( ). Residues between the end of one set of brackets and the beginning of another are believed to be removed during maturation. Residues believed to be involved in the active sites (Brown and Mahuran, 1991; Fernandes *et al.*, 1996; Liessem *et al.*, 1995; Tse *et al.*, 1996a) are double-underlined, as are glycosylations sites known to contain mannose 6-phosphate residue(s) before final processing in the lysosome. Other sites containing oligosaccharides are underlined. The most highly conserved domain within 15 aligned Hex-related sequences is dotted-underlined (Tse *et al.*, 1996a). The α/β junctions of two chimeric Hex proteins, Fusion 1 and Fusion 2 (Tse *et al.*, 1996b), are indicated by vertical lines. The Fusion 2 junction has been previously shown to be located in a hydrophilic loop structure (Sagherian *et al.*, 1993).

the β_a and β_b chains as well as other processing sites in the pro-isozymes. We accomplished this by examining the partial N-terminal sequences of each of the HPLC-purified mature α, β_a, and β_b polypeptides purified from placenta (Mahuran et al., 1988) and of the pro-β (uncleaved) form of Hex B secreted by Tay-Sachs fibroblasts, the latter an experiment of John Stirling(Stirling et al., 1988) done while visiting Toronto. Purified α chain was obtained from Hex A by HPLC M_r-sieve chromatography. In order to isolate the constituent β_a and β_b chains from Hex B, an HPLC–ion-exchange technique was developed by which the reduced and alkylated chains were separated in 7 M urea using a linear salt gradient.

N-terminal sequencing identified the processing sites of the pre-pro-α and pre-pro-β polypeptides (Figure 13.2). Little et al. (1988) confirmed the site of cleavage generating the pro-α chain from fibroblast protein excised from an SDS gel.

The pro-β N-terminus we determined from the secreted Hex B (Stirling et al., 1988) differed from the one we originally predicted using rules established for signal peptides by von Heijne (1986), assuming that protein synthesis began at the first in-frame ATG (Met) (Korneluk et al., 1986). The actual site we found would conform to von Heijne's rules if the second or third Met were to initiate protein synthesis. However, Kuldeep Neote, a Ph.D. student with Gravel, confirmed through cellular expression studies using deletion and frame-shift constructs that, although either the second or third ATG can function as the site for translation initiation, producing functional signal peptides, when the first ATG is present it is used 100% of the time (Neote et al., 1990a).

The N-terminal sequences of the mature polypeptides initially suggested that 65 amino acids are removed from the pro-α sequence, while 79 amino acids are removed from the pro-β sequence in the lysosome (Figure 13.2), with β_b corresponding to the N-terminal half and β_a at the C-terminal half of the β sequence (Figure 13.2). Interestingly, the N-termini of the mature α and β_b polypeptides were in identical positions in the aligned sequences, despite the absence of significant homology in this region. However, the N-terminus of the α chain was "ragged" [about half starting at T (Thr) and half at the next amino acid, L (Leu)], implying that the initial N-terminal cleavage took place somewhere upstream from this site and was followed by cycles of exopeptidase digestion (Mahuran and Gravel, 1988). In more recent work by Martin Hubbes (an M.Sc. student of a colleague of Mahuran's, John Callahan), it was discovered that the N-terminal segments of the pro chains are not entirely lost during maturation (Hubbes et al., 1989; Sonderfeld-Fresko and Proia, 1989). They are instead retained in the mature subunit by disulfide bonds formed by Cys residues that are aligned within their primary structures, α-Cys^{58} or β-Cys^{91} (Figure 13.2). We have suggested that these peptides be referred to as α_p and β_p (Hubbes et al., 1989). From N-terminal sequencing and amino acid analysis of the HPLC-isolated peptides it was estimated that α_p contains residues Leu^{23}-Gly^{74}, and β_p contains residues

Ala^{50}-Gly^{107} (Figure 13.2) (Hubbes *et al.*, 1989). The initial identification of β_p was made by Proia's group while investigating the potential N-linked oligosaccharide attachment sites in the β subunit (Sonderfeld-Fresko and Proia, 1989).

Both the β_a and β_b chains were found in the β subunits of Hex A and Hex B, but appeared to contain one interesting difference. The N-terminus of β_a was displaced by one amino acid residue, depending on its isozyme of origin. Thus, in Hex B, the N-terminus occurred uniformly at K (Lys), while in Hex A, the N-terminal amino acid was at the next residue, L (Mahuran *et al.*, 1988). An explanation for this observation was obtained during a study by Mahuran on the structure of intracellular Hex I (Hex with an intermediate p*I*, i.e., between Hex B and Hex A). He found that Hex I from human placenta is made up of processing intermediates of both subunits of mature Hex A (Mahuran, 1990). Interesting, intermediates with the same N-termini were identified by pulse-chase and microsequencing experiments in human fibroblasts conducted by Little *et al.* (1988) and Quon *et al.* (1989). Taken together, these studies demonstrate the specificity of the maturation process within the lysosome. The tissue Hex Is have more basic p*I*s than mature Hex A, due to the removal of a majority of basic amino acids during maturation. Since we now know that the K group is removed sometime after the cleavage of the β_a and β_b chains (Mahuran, 1990; Sagherian *et al.*, 1993), the ion-exchange separation of Hex B from Hex A used by Mahuran favored the purification of forms of Hex A lacking this basic group, and forms of Hex B containing it in their β subunits.

The internal cleavage site generating the β polypeptides was determined from the sequence of a tryptic peptide corresponding to the C-terminus of the β_b chain and was later confirmed by an M.Sc. student in Mahuran's laboratory, Carmelina Sagherian (Sagherian *et al.*, 1993). The sequence of this peptide should have terminated with an R^{312} (Arg) but stopped abruptly with S^{311} (Ser) (Figure 13.2). Thus, the sequence unaccounted for between this peptide and the N-terminus of the β_a chain must have been removed during processing. These are RQN $\pm$ K (Arg-Gln-Asn $\pm$ Lys). This sequence would carry a strong positive charge in the lysosome, making the domain strongly hydrophilic. We surmise that in the β subunit of both be pro-isozymes, this sequence must be extended at the surface of the protein and subject to proteolytic attack. It is interesting to note that on either side of the cleavage site in the β subunit, there is extensive homology (indicated by * in Fig. 13.2) with the α subunit. However, at the cleavage site itself, there is no homology at all between the deduced sequences. Studies of this cleavage site by Sagherian *et al.* using *in vitro* mutagenesis and cellular expression showed that cleavage occurs at either of the two basic residues, but on their N-terminal sides. This results in the immediate loss of the sequence RQN and a slower loss of the final K due to another step in the maturation of the Hex A or Hex B β subunit (Mahuran, 1990). Replacement of the β sequence (RQNK) with the α sequence (GSEP) produced a normally functional and stable Hex B in which no β_a or β_b chains were generated (Sagherian *et al.*, 1993). This study also revealed that the cleavage site is located within a disulfide loop structure between Cys^{309} and Cys^{360} (Figure 13.2).

B. Glycolytic processing

To characterize the structure of the mature isozymes fully, and for a better understanding of the mechanism directing Hex to the lysosome, it was necessary to define the location and structure of the attached oligosaccharides. This could ultimately assist in the identification of the recognition markers for the phosphotransferase responsible for the phosphorylation of mannose residues on the enzyme. To accomplish this, O'Dowd developed HPLC techniques for the preparative isolation of the major glycopeptides of Hex. Two methods were used to isolate the glycopeptides. In the simpler approach, the terminal mannose-containing glycopeptides were purified by Concanavalin A-Sepharose chromatography. The peptide portion of each isolated peak was sequenced to localize it within the deduced amino acid sequence and the oligosaccharide structure determined by ^{1}H-NMR. Three glycopeptides were identified: (1) one, containing a Man_3-GlcNAc$_2$ oligosaccharide, at the first of three possible N-linked glycosylation sites [Asn-X-(Ser or Thr)] of the α subunit; (2) another Man_3-GlcNAc$_2$ structure at the first of four possible sites in the mature β subunit (a fifth site resides in the pro-sequence); and (3) a mixture of Man_{5-7}-GlcNAc$_2$ structures found associated at the second possible attachment site in the β sequence (Figure 13.2). Thus, both of the latter oligosaccharides reside on the mature β_b polypeptide (O'Dowd *et al.*, 1988).

In the second approach, the glycopeptides (including those that would not bind Con A) were detected by separating the total tryptic/chymotryptic digest on reverse-phase HPLC and testing each peak for glucosamine. One additional glycopeptide was identified by this method. It was localized to the third possible site in the mature β sequence. This placed it near the amino terminus of the β_a chain (Figure 13.2). It retained a single GlcNAc reside as its oligosaccharide (O'Dowd *et al.*, 1988).

The fourth site in the mature β sequence is located 60 amino acids from the C-terminus of the β_a chain. A tryptic peptide containing this site had previously been isolated and sequenced to reveal an underivitized, i.e., unglycosylated, Asn residue (Korneluk *et al.*, 1986). Thus, it is not glycosylated.

The identification of the β_p chain (above) was precipitated by the analysis of the glycosylation and phosphorylation sites of Hex B, by site-directed mutagenesis and cellular expression (Sonderfeld-Fresko and Proia, 1989). The deduced sequence of the β_p chain contains a potential Asn-linked oligosaccharide attachment site. This study demonstrated that, in addition to the sites identified by O'Dowd *et al.*, the β_p site was not only used, but often contained a phosphorylated mannose residue. Furthermore, the β_p peptide was retained in the mature β subunit after its protyolitic cleavage from the pro-β chain. However, the major site of phosphorylation was found to be the single site on the $β_a$ chain that O'Dowd *et al.* (1988) had shown was rapidly processed in the lysosome to a single Gluc-NAc residue. This extensive carbohydrate "processing" also explains the several bands described by Hasilik and Neufeld in 1980 (above) for the $β_b$ pattern from fibroblast extracts (Sonderfeld-Fresko and Proia, 1989). A similar methodology was used to identify the occupied sites in the α subunit (Weitz and Proia, 1992). In this study, oligosaccharides were documented to be attached to all three sites in the α subunit, with the primary site of phosphorylation aligning with the single glycosylation site on the $β_a$ chain (the third α site, Figure 13.2). Unlike the $β_p$ chain, no possible attachment site is present on the $α_p$ peptide. The $α_p$ peptide was not detected in this study, but was found in purified, reduced, and denatured Hex A from placenta (Hubbes *et al.*, 1989).

Interestingly, in this study the $α_p$ peptide could only be detected by OD_{280} as it eluted from an HPLC sieve column, not by SDS-PAGE analysis (Hubbes *et al.*, 1989). Thus the primary targets for the phosphotransferase are oligosaccharides attached to aligned Asn residues in the α and β subunits.

VI. STRUCTURE–FUNCTION RELATIONSHIPS

It is now well established that whereas the active sites of both the α and β subunits (in dimeric forms) are able to hydrolyze many of the same neutral artificial, e.g., MUG (Hou *et al.*, 1996), and several natural substrates [reviewed by Mahuran *et al.* (1985)], only the catalytic site in the α subunit can hydrolyze negatively charged substrates such as β-linked glucosamine 6-sulfate containing glycosaminoglycans (Hou *et al.*, 1996; Kresse *et al.*, 1981; Kytzia and Sandhoff, 1985) and artificial substrates (Hou *et al.*, 1996; Kytzia and Sandhoff, 1985), e.g., 4-MUGS (Bayleran *et al.*, 1984), and, most important, G_{M2} ganglioside. In the latter case only the α subunit in its heterodimeric Hex A form is functional *in vivo* (Meier *et al.*, 1991). In order for Hex A to hydrolyze G_{M2} ganglioside, it requires the small, heat-stable G_{M2} activator protein (activator). The activator interacts with both the carbohydrate and lipid portions of the ganglioside, solubilizing, or at least lifting a ganglioside molecule from the membrane, and "presenting" it to Hex A for hydrolysis (Meier *et al.*, 1991). *In vitro* assays demonstrate that detergents can be substituted for the activator; however, under these conditions Hex S, as well as Hex A, but not Hex B, can efficiently hydrolyze G_{M2} ganglioside. Interestingly, Hex B can hydrolyze G_{A2}, the neutral, asialo derivative of G_{M2}, in the presence of detergent, but not in the presence of activator alone. As well, the activator, even in the absence of G_{M2}, can inhibit the hydrolysis of MUGS by both Hex A and Hex S [reviewed by Fürst and Sandhoff (1992) and Sandhoff *et al.* (1989)]. These data indicate that the binding site for the complex is also located in the α subunit, but that elements of the β subunit [other than its active site (Hou *et al.*, 1996)] are necessary to somehow orientate correctly the complex and allow hydrolysis of the ganglioside. Further functions that have been identified for the β subunit are to greatly increase the stability of the resulting dimer and to facilitate the transport of the α subunit out of the endoplasmic reticulum (see later) [reviewed by Gravel *et al.* (1995) and Mahuran (1991)].

A. Catalytic site

The initial clue as to the location of the catalytic site came from the study of the B1 variant of Tay-Sachs disease (Kytzia *et al.*, 1983). Charlotte Brown, a Ph.D. student in Mahuran's lab (Brown and Mahuran, 1991; Brown *et al.*, 1989), analyzed the biochemical consequences of the B1 substitution in the α subunit, α-Arg^{178}His (Ohno and Suzuki, 1988a), by *in vitro* mutagenesis of the homologous codon in the β subunit, β-Arg^{211}His. In this study with the β analog we noted small changes in the stability and the rate of processing of the mutant protein which were totally eliminated when a more conservative substitution, β-Arg^{211}Lys (Figure 13.2), was introduced. Whereas this mutant Hex B retained an apparently normal K_m for MUG, its V_{max} was reduced by greater than 400-fold and its pH optimum shifted. From these and other data we concluded that α-Arg^{178} and β-Arg^{211} are active-site residues in Hex (Brown and Mahuran, 1991; Hou *et al.*, 1996). Recent molecular modeling of human Hex using the crystal structure of bacterial chitobiase suggests that these Arg are involved directly in substrate binding, interacting with OH3 and OH4 of the GlucNAc substrate (Tews *et al.*, 1996).

Yongmin Hou, a M.Sc. student with Mahuran, used Charlotte's β-Arg^{211}Lys encoding cDNA in co-transfection studies with another cDNA encoding the wild-type α. He produced a Hex A* with an inactive β subunit. This isozyme was used to characterize the substrate specificity of the α subunit of Hex A without the interferece of an active β subunit. Previously this type of work could only be done with the α-subunit homodimer, Hex S. Yongmin found that there were differences in the kinetics of the α subunit in it heterodimeric form as compared to homodimeric Hex S. The α-active site in Hex A* is sevenfold more catalytically active toward the neutral artifical substrate, MUG, than either of the α-active sites in Hex S. Heterodimeric α had a K_m for MUG similar to that of the β subunit. In Hex S the K_m for MUG is twice as high. Finally, he confirmed that an active β subunit did not enhance the ability of Hex A to hydrolyze G$_{M2}$ ganglioside in the presence of human activator protein (Hou *et al.*, 1996).

It is a widely held view that catalytic residues should be invariant in the aligned sequences from glycosyl hydrolases of the same family (family 20 for Hex B) (Henrissat, 1991). Students in Mahuran's lab, George Vavougios (M.Sc.) and Roderick Tse (Ph.D.), investigated several conserved acidic residues in the area of greatest homology between human Hex B and 15 Hex-related enzymes from different species, including bacteria. This area extends from β-Thr^{193} to β-Ser^{259} (Figure 13.2, dotted underlined). Two of these residues, β-Asp^{196} and β-Asp^{208}, as well as β-Arg^{211}/α-Arg^{178} (above), were invariant in all the aligned sequences. Of the acidic residues investigated, only the substitution of Asp^{196} by Asn produced normal levels of Hex B protein with a kcat of 0.2% of normal (Tse et al., 1996a). Mutagenesis of either β-Asp^{240} or β-Asp^{290} (conserved but not invariant in the aligned sequences) to Asn decreased kcat by 10- or 1.4-fold, but also raised the K_m of the enzyme 11- or 3-fold, respectively. The molecular modeling by Tews et al. (1996) supports the roles of Asp^{240} and Asp^{290} in substrate binding, but not the conclusion that Asp^{196} is a catalytic residue. They predict that α-Glu^{323}/β-Glu^{355} are the catalytic acidic residues, corresponding to Glu^{540} in chitobiase. Interestingly, β-Glu^{355} was labeled with a substrate analog containing a photoaffinity label in the aglycone position (Liessem et al., 1995). This position is not normally considered important for Hex binding and should be a considerable distance from the residue that interacts with the glycosidic linkage. Nevertheless, Fernandes et al. (1996) also implicated α-Glu^{323} in expression studies with mutated α subunits transfected into a neuroglial line from a Tay-Sachs fetus. More work needs to be done to reconcile our biochemical observations with the molecular modeling results.

B. Substrate-binding sites

Roderick, George, and a summer student, Yong Jian Wu, also constructed, expressed, and characterized two α/β fusion proteins (Tse et al., 1996b). First, a chimeric α/β chain was made by replacing the least well conserved amino-terminal section of the β chain with the corresponding α section (Figure 13.2). The biochemical characteristics of this protein were nearly identical to those of Hex B. Therefore, the most dissimilar regions in the subunits are not responsible for the subunits' dissimilar biochemical properties. A second fusion protein was made that also included the more homologous middle section of the α chain (Figure 13.2). This protein expressed the substrate specificity unique to isozymes containing an α subunit (A and S). We conclude that the region responsible for the ability of the α subunit to bind negatively charged substrates is located within residues α-132–283. Interestingly, the remaining carboxy-terminal section from the β chain, β-316–556, was sufficient to allow this chimera to hydrolyze G$_{M2}$ ganglioside with 10% the specific activity of heterodimeric hexosaminidase A. Thus the carboxy-terminal section of each subunit is likely involved in protein–protein interactions. Similar experiments have been conducted by Proia and Sandhoff and colleagues, which produced somewhat different conclusions (Pennybacker et al., 1996). The recently reported molecular modeling data (Tews et al., 1996) are not helpful in identifying candidate sites for sulfate or activator binding, because these functions are not present in chitobiase. However, they are bound to generate experiments to confirm or denying the role of other candidate α and/or β residues in substrate binding.

VII. MOLECULAR HETEROGENEITY IN TAY-SACHS AND SANDHOFF DISEASES

A very important consequence of the isolation of the cDNA clones was that it became possible to examine gene structure and mRNA expression in patient cell lines that heretofore had been limited to enzyme-based investigation. These studies were initiated with the finding that infantile Tay-Sachs disease in French Canadians and Ashkenazi Jews was due to different mutations, which in the former proved to be due to a partial gene deletion (Myerowitz and Hogikyan, 1986, 1987). Although this provided the first hint that numerous mutations would be involved, it was anticipated that a single mutation would likely be responsible for the infantile form of Ashkenazi Jewish Tay-Sachs disease. It was fitting that the mutation was announced at the symposium commemorating the 100th anniversary of Bernard Sachs' first description of the disease and saluted by this book. The story was presented by both Myerowitz and Gravel and with some trepidation because, in each study, a mutation was found but the patient's cells were heterozygous for it. There would have to be a second mutation. As it happened, both labs had investigated the same cell line and had found the same mutation. In the Gravel lab, these studies were led by Enrico Arpaia, a research associate. The results were published shortly after the meeting, along with a similar report by Suzuki (Arpaia et al., 1988; Myerowitz, 1988; Ohno and Suzuki, 1988b). The mutation was a G-to-C transversion of the first nucleotide in intron 12 (IVS12+1 G to C). This inactivated the obligatory GT dinucleotide of the donor splice junction, indicating that the mutation should result in aberrant splicing. Since the mutation resulted in the generation of a Dde I site, it was possible to distinguish between the normal and mutant alleles by Dde I digestion and polyacrylamide gel electrophoresis of PCR products (Figure 13.3). These experiments showed that the splice mutation accounted for a minority of mutant alleles in Ashkenazi Jews. Myerowitz and Costigan later found the major allele, a 4-bp insertion in exon 11 (Myerowitz and Costigan, 1988). These early mutations underscored the extent of genetic heterogeneity that was yet to be revealed in Tay-Sachs disease, as was proving to be true for many genetic diseases.

In subsequent work with Barbara Triggs-Raine, a postdoctoral fellow with Gravel, a multicenter study was made of the efficacy of enzyme versus DNA-based screening for Tay-Sachs carrier status among Ashkenazi Jews (Triggs-Raine et al., 1990). It demonstrated the superior precision of DNA-based testing but also confirmed the importance of retaining enzyme testing as part of screening

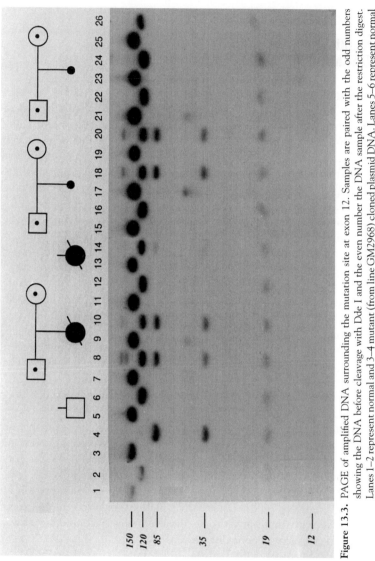

Figure 13.3. PAGE of amplified DNA surrounding the mutation site at exon 12. Samples are paired with the odd numbers showing the DNA before cleavage with Dde I and the even number the DNA sample after the restriction digest. Lanes 1–2 represent normal and 3–4 mutant (from line GM2968) cloned plasmid DNA. Lanes 5–6 represent normal DNA, and lanes 7–8 and 11–12 are from cell lines GM3051 and GM3052, parents of the patient represented by cell line GM2968 in lanes 9–10. In the even-numbered lanes, diagnostic bands for the mutation (new restriction site) are at 85 and 35 bp; an additional band at 120 bp indicates that the individual is a heterozygote.

programs, an issue that is now more relevant to the non-Jewish population because of the abundance of different mutations in the general population.

Studies on Sandhoff disease were also underway, particularly in the Mahuran lab. Brian O'Dowd initially used his β-subunit cDNA to survey fibroblasts from 11 infantile, 5 juvenile, and two adult chronic Sandhoff patients (O'Dowd *et al.*, 1986). While a great deal of molecular heterogeneity was found within this group of patients, two were found to be homozygous for a similar if not identical partial gene deletion. When this deletion was fully characterized by Kuldeep Neote, he found that it involved 16 kb of the *HEXB* gene, spanning from 2 kb upstream of exon 1 to within the first 1.5 kb of intron 5. More surprisingly, he found that it accounted for nearly 30% of all Sandhoff alleles he surveyed (Neote *et al.*, 1990b).

Beth McInnes, a postdoctoral fellow in Mahuran's lab, examined cells from one unusual, very mildly affected, French Canadian individual over 60 years of age with extremely low levels of serum Hex activity. The patient turned out to be heterozygous (along with the common 16-kb deletion mutation) for the same mutant allele as a homozygous Japanese patient with juvenile-onset Sandhoff disease, a $C^{1252}T$ in exon 11 encoding a $Pro^{417}Leu$ substitution. Since C^{1252} is B downstream from the intron 10/exon 11 junction (E^{+8}), it was surprising to discover that its detrimental effect on residual Hex levels was not caused by the missense mutation, but was caused by the large decrease in normally spliced β-mRNA (McInnes *et al.*, 1992; Wakamatsu *et al.*, 1992). Working in Gravel's lab as a postdoctoral fellow on studies initiated in Japan, Nobu Wakamatsu established PCR-based procedures for quantitating mRNA and showed that the patient's total β-mRNA was reduced to 8% of normal. Three species of RNA were characterized. Based on size, P1 was the normal transcript. P2 and P3 were alternatively spliced mRNA which used either a cryptic site in the middle of exon 11 or the normal 3' splice junction of exon 12. The ratio of these RNA species was P1:P2:P3, 11:2:1 (Wakamatsu *et al.*, 1992). Beth found these same RNA species in fibroblasts from the French Canadian patient (McInnes *et al.*, 1992). It remains to be explained why this patient has such a mild phenotype as compared to that found in the Japanese patient. However, another recent report has found this mutation in two heterozygous Italian patients. One of these patients also had the deletion allele and presented with a milder phenotype than did the second patient, with a missense mutation, $Cys^{292}Tyr$, as his second allele (Gomez-Lira *et al.*, 1995). These observations suggest a good deal of genetic heterogeneity in the RNA-splicing "machinery," even between individuals from the same ethnic group.

Since his move to Montreal, Gravel's lab has gone on to investigate non-Jewish Tay-Sachs mutations, including the common IVS-7 splice mutation (Akerman *et al.*, 1992) and a common pseudo-deficiency mutation, which has implications for carrier testing and prenatal diagnosis (Triggs-Raine *et al.*, 1992).

More recently, he has developed mouse models of Tay-Sachs and Sandhoff diseases (Phaneuf *et al.*, 1996), as have Proia and colleagues (Taniike *et al.*, 1995; Yamanaka *et al.*, 1994) and a collaborative group in Europe (Cohen-Tannoudji *et al.*, 1996). These latter experiments hold promise to address the pathophysiology of neuron death in these disorders and open the way to "approaches to therapy," part of the title to the original Medical Research Council program grant that initiated our adventure with Sandy Lowden back in 1976.

References

Akerman, B. R., Zielenski, J., Triggs-Raine, B. L., Prence, E. M., Natowicz, M. R., Lim-Steele, J. S., Kaback, M. M., *et al.* (1992). A mutation common in non-Jewish Tay-Sachs disease: Frequency and RNA studies. *Hum. Mutat.* **1**, 303–309.

Arpaia, E., Dumbrille-Ross, A., Maler, T., Neote, K., Tropak, M., Troxel, C., Stirling, J. L., *et al.* (1988). Identification of an altered splice site in Ashkenazic Tay-Sachs disease. *Nature* **333**, 85–86.

Bayleran, J., Hechtman, P., and Saray, W. (1984). Synthesis of 4-methylumbelliferyl-beta-D-N-acetylglucosamine-6-sulfate and its use in classification of GM2 gangliosidosis genotypes. *Clin. Chim. Acta* **143**, 73–89.

Bordier, C., and Crettol-Jarvinen, A. (1979). Peptide mapping of heterogeneous protein samples. *J. Biol. Chem.* **254**, 2565–2567.

Brown, C. A., and Mahuran, D. J. (1991). Active arginine residues in beta-hexosaminidase: Identification through studies of the B1 variant of Tay-Sachs disease. *J. Biol. Chem.* **266**, 15855–15862.

Brown, C. A., Neote, K., Leung, A., Gravel, R. A., and Mahuran, D. J. (1989). Introduction of the alpha subunit mutation associated with the B1 variant of Tay-Sachs disease into the beta subunit produces a beta-hexosaminidase B without catalytic activity. *J. Biol. Chem.* **264**, 21705–21710.

Cohen-Tannoudji, M., Marchand, P., Akli, S., and Puech, J. P. (1995). Disruption of murine Hexa gene leads to enzymeatic deficiency and to neuronal lysosomal storage, similar to that observed in Tay-Sachs disease. *Mammal Genome*, **6**, 844–849.

Fernandes, M. J. G., Leclerc, D., Henrissat, B., Vorgias, C. E., Gravel, R., Hechtman, P., and Kaplan, F. (1997). Identification of candidate active site residues in lysosomal beta-hexosaminidase A. *J. Biol. Chem.* **272**, 814–820.

Fürst, W., and Sandhoff, K. (1992). Activator proteins and topology of lysosomal sphingolipid catabolism. *Biochim. Biophys. Acta* **1126**, 1–16.

Geiger, B., Ben-Yoseph, Y., and Arnon, R. (1974). Purification of human hexosaminidase A and B by affinity chromatography. *FEBS Lett.* **45**, 276–281.

Gomez-Lira, M., Sangalli, A., Mottes, M., Perusi, C., Pignatti, P. F., Rizzuto, N., and Salviati, A. (1995). A common beta hexosaminidase gene mutation in adult Sandhoff disease patients. *Hum. Genet.* **96**, 417–422.

Gravel, R. A., Clarke, J. T. R., Kaback, M. M., Mahuran, D., Sandhoff, K., and Suzuki, K. (1995). The GM2 gangliosidoses. *In* "The Metabolic and Molecular Bases of Inherited Disease, Vol. 2," (C. R. Scriver, A. L. Beaudet, W. S. Sly, and D. Valle, eds.), pp. 2839–2879. McGraw-Hill, New York.

Hasilik, A., and Neufeld, E. F. (1980). Biosynthesis of lysosomal enzymes in fibroblasts: Synthesis as precursors of higher molecular weight. *J. Biol. Chem.* **255**, 4937–4945.

Henrissat, B. (1991). A classification of glycosyl hydrolases based on amino acid sequence similarities. *Biochem. J.* **280**, 309–316.

Henrissat, B., and Bairoch, A. (1993). New families in the classification of glycosyl hydrolases based on amino acid sequence similarities. *Biochem. J.* **293**, 781–788.

Hou, Y., Tse, R., and Mahuran, D. J. (1996). The direct determination of the substrate specificity of the alpha-active site in heterodimeric beta-hexosaminidase A. *Biochemistry* **35**, 3963–3969.

Hubbes, M., Callahan, J., Gravel, R., and Mahuran, D. (1989). The amino-terminal sequences in the pro-alpha and -beta polypeptides of human lysosomal beta-hexosaminidase A and B are retained in the mature isozymes. *FEBS Lett.* **249**, 316–320.

Korneluk, R. G., Mahuran, D. J., Neote, K., Klavins, M. H., O'Dowd, B. F., Tropak, M., Willard, H. F., et al. (1986). Isolation of cDNA clones coding for the alpha subunit of human beta-hexosaminidase: Extensive homology between the alpha and beta subunits and studies on Tay-Sachs disease. *J. Biol. Chem.* **261**, 8407–8413.

Kresse, H., Fuchs, W., Glossl, J., Holtfrerich, D., and Gilberg, W. (1981). Liberation of N-acetylglucosamine-6-sulfate by human beta-N-acetylhexosaminidase A. *J. Biol. Chem.* **256**, 12926–12932.

Kytzia, H.-J., and Sandhoff, K. (1985). Evidence for two different active sites on human beta-hexosaminidase A. *J. Biol. Chem.* **260**, 7568–7572.

Kytzia, H. J., Hinrichs, U., Maire, I., Suzuki, K., and Sandhoff, K. (1983). Variant of GM2-gangliosidosis with hexosaminidase A having a severely changed substrate specificity. *EMBO J.* **2**, 1201–1205.

Liessem, B., Glombitza, G. J., Knoll, F., Lehmann, J., Kellermann, J., Lottspeich, F., and Sandhoff, K. (1995). Photoaffinity labeling of human lysosomal beta-hexosaminidase B—Identification of Glu-355 at the substrate binding site. *J. Biol. Chem.* **270**, 23693–23699.

Little, L. E., Lau, M. M. L., Quon, D. V. K., Fowler, A. V., and Neufeld, E. F. (1988). Proteolytic processing of the alpha chain of the lysosomal enzyme beta-hexosaminidase, in normal human fibroblasts. *J. Biol. Chem.* **263**, 4288–4292.

Lowden, J. A., Zuker, S., Wilensky, A. J., and Skomorowski, M. A. (1974). Screening for carriers of Tay-Sachs disease: A community project. *Can. Med. Assoc. J.* **111**, 229–233.

Mahuran, D. J. (1990). Characterization of human placental beta-hexosaminidase I2: Proteolytic processing intermediates of hexosaminidase A. *J. Biol. Chem.* **265**, 6794–6799.

Mahuran, D. J. (1991). The biochemistry of HEXA and HEXB gene mutations causing GM2 gangliosidosis. *Biochim. Biophys. Acta* **1096**, 87–94.

Muhuran, D., and Gravel, R. (1988). The molecular biology of beta-hexosaminidase: Localization of the proteolytic processing and carbohydrate containing sites. In "NATO ASI Series A: Life Science, Vol. 150" (R. Salvayre, L. Douste-Blazy, and S. Gatt, eds.), pp. 225–236. Plenum, New York.

Mahuran, D. J., and Lowden, J. A. (1980). The subunit and polypeptide structure of hexosaminidase from human placenta. *Can. J. Biochem.* **58**, 287–294.

Mahuran, D. J., Tsui, F., Gravel, R. A., and Lowden, J. A. (1982). Evidence for two dissimilar polypeptide chains in the beta2 subunit of hexosaminidase. *Proc. Natl. Acad. Sci. (USA)* **79**, 1602–1605.

Mahuran, D., Novak, A., and Lowden, J. A. (1985). The lysosomal hexosaminidase isozymes. *Isozymes. Curr. Top. Biol. Med. Res.* **12**, 229–288.

Mahuran, D. J., Neote, K., Klavins, M. H., Leung, A., and Gravel, R. A. (1988). Proteolytic processing of human pro-beta hexosaminidase: Identification of the internal site of hydrolysis that produces the nonidentical beta a and beta b polypeptides in the mature beta-subunit. *J. Biol. Chem.* **263**, 4612–4618.

McInnes, B., Potier, M., Wakamatsu, N., Melancon, S. B., Klavins, M. H., Tsuji, S., and Mahuran, D. J. (1992). An unusual splicing mutation in the HEXB gene is associated with dramatically different phenotypes in patients from different racial backgrounds. *J. Clin. Invest.* **90**, 306–314.

Meier, E. M., Schwarzmann, G., Fürst, W., and Sandhoff, K. (1991). The human GM2 activator protein. A substrate specific cofactor of beta-hexosaminidase A. *J. Biol. Chem.* **266**, 1879–1887.

Myerowitz, R. (1988). Splice junction mutation in some Ashkenazi Jews with Tay-Sachs disease: Evidence against a single defect within this ethnic group. *Proc. Natl. Acad. Sci. (USA)* **85**, 3955–3959.

Myerowitz, R., and Costigan, F. C. (1988). The major defect in Ashkenazi Jews with Tay-Sachs disease is an insertion in the gene for the alpha-chain of beta-hexosaminidase. *J. Biol. Chem.* **263**, 18587–18589.

Myerowitz, R., and Hogikyan, N. D. (1986). Different mutations in Ashkenazi Jewish and non-Jewish French Canadians with Tay-Sachs disease. *Science* **232**, 1646–1648.

Myerowitz, R., and Hogikyan, N. D. (1987). A deletion involving Alu sequences in the beta-hexosaminidase alpha-chain gene of French Canadians with Tay-Sachs disease. *J. Biol. Chem.* **262**, 15396–15399.

Myerowitz, R., and Proia, R. L. (1984). cDNA clone for the alpha-chain of human beta-hexosaminidase: Deficiency of alpha-chain mRNA in Ashkenazi Tay-Sachs fibroblasts. *Proc. Natl. Acad. Sci. (USA)* **81**, 5394–5398.

Myerowitz, R., Piekarz, R., Neufeld, E. F., Shows, T. B., and Suzuki, K. (1985). Human beta-hexosaminidase alpha chain: Coding sequence and homology with the beta chain. *Proc. Natl. Acad. Sci. (USA)* **82**, 7830–7834.

Neote, K., Bapat, B., Dumbrille-Ross, A., Troxel, C., Schuster, S. M., Mahuran, D. J., and Gravel, R. A. (1988). Characterization of the human HEXB gene encoding lysosomal beta-hexosaminidase. *Genomics* **3**, 279–286.

Neote, K., Brown, C. A., Mahuran, D. J., and Gravel, R. A. (1990a). Translation initiation in the HEXB gene encoding the beta-subunit of human beta-hexosaminidase. *J. Biol. Chem.* **265**, 20799–20806.

Neote, K., McInnes, B., Mahuran, D. J., and Gravel, R. A. (1990b). Structure and distribution of an Alu-type deletion mutation in Sandhoff disease. *J. Clin. Invest.* **86**, 1524–1531.

O'Dowd, B., Quan, F., Willard, H., Lamhonwah, A. M., Korneluk, R., Lowden, J. A., Gravel, R. A., *et al.* (1985). Isolation of c-DNA clones coding for the beta-subunit of human beta-hexosaminidase. *Proc. Natl. Acad. Sci. (USA)* **82**, 1184–1188.

O'Dowd, B. F., Klavins, M. H., Willard, H. F., Gravel, R., Lowden, J. A., and Mahuran, D. J. (1986). Molecular heterogeneity in the infantile and juvenile forms of Sandhoff disease (O-variant GM2 gangliosidosis). *J. Biol. Chem.* **261**, 12680–12685.

O'Dowd, B. F., Cumming, D., Gravel, R. A., and Mahuran, D. J. (1988). Isolation and characterization of the major glycopeptides from human beta-hexosaminidase: Their localization within the deduced primary structure of the mature alpha and beta polypeptide chains. *Biochemistry* **27**, 5216–5226.

Ohno, K., and Suzuki, K. (1988a). Mutation in GM2-gangliosidosis B1 variant. *J. Neurochem.* **50**, 316–318.

Ohno, K., and Suzuki, K. (1988b). A splicing defect due to an exon-intron junctional mutation resulting in abnormal beta-hexosaminidase alpha chain mRNAs in Ashkenazi Jewish patients with Tay-Sachs disease. *Biochem. Biophys. Res. Commun.* **153**, 463–469.

Pennybacker, M., Liessem, B., Moczall, H., Tifft, C. J., Sandhoff, K., and Proia, R. L. (1996). Identification of domains in human beta-hexosaminidase that determine substrate specificity. *J. Biol. Chem.* **271**, 17377–17382.

Phaneuf, D., Wakamatsu, N., Huang, J. Q., Borowski, A., Peterson, A. C., Fortunato, S. R., Ritter, G., *et al.* (1996). Dramatically different phenotypes in mouse models of human Tay-Sachs and Sandhoff diseases. *Hum. Mol. Genet.* **5**, 1–14.

Proia, R. L. (1988). Gene encoding the human beta-hexosaminidase beta-chain: Extensive homology of intron placement in the alpha- and beta-genes. *Proc. Natl. Acad. Sci. (USA)* **85**, 1883–1887.

Proia, R. L., and Soravia, E. (1987). Organization of the gene encoding the human beta-hexosaminidase alpha chain. *J. Biol. Chem.* **262**, 5677–5681.

Quon, D. V. K., Proia, R. L., Fowler, A. V., Bleibaum, J., and Neufeld, E. F. (1989). Proteolytic processing of the beta-subunit of the lysosomal enzyme, beta-hexosaminidase, in normal human fibroblasts. *J. Biol. Chem.* **264,** 3380–3384.

Sagherian, C., Poroszlay, S., Vavougios, G., and Mahuran, D. J. (1993). Proteolytic processing of the probeta chain of beta-hexosaminidase occurs at basic residues contained within an exposed disulfide loop structure. *Biochem. Cell Biol.* **71,** 340–347.

Sandhoff, K., Conzelmann, E., Neufeld, E. F., Kaback, M. M., and Suzuki, K. (1989). The GM2 gangliosidoses. *In* "The Metabolic Basis of Inherited Disease" (C. V. Scriver, A. L. Beaudet, W. S. Sly, and D. Valle, eds.), Vol. 2, pp. 1807–1839. McGraw-Hill, New York.

Sonderfeld-Fresko, S., and Proia, R. L. (1989). Analysis of the glycosylation and phosphorylation of the lysosomal enzyme, beta-hexosaminidase B, by site-directed mutagenesis. *J. Biol. Chem.* **264,** 7692–7697.

Stirling, J., Leung, A., Gravel, R. A., and Mahuran, D. J. (1988). Localization of the pro-sequence within the total deduced primary structure of human beta-hexosaminidase B. *FEBS. Lett.* **231,** 47–50.

Taniike, M., Yamanaka, S., Proia, R. L., Langaman, C., Bone-Turrentine, T., and Suzuki, K. (1995). Neuropathology of mice with targeted disruption of Hexa gene, a model of Tay-Sachs disease. *Acta Neuropathol. (Berl.)* **89,** 296–304.

Tews, I., Perrakis, A., Oppenheim, A., Dauter, Z., Wilson, K. S., and Vorgias, C. E. (1996). Baterial chitobiase structure provides insight into catalytic mechanism and the basis of Tay-Sachs disease. *Nat. Struct. Biol.* **3,** 638–648.

Triggs-Raine, B. L., Feigenbaum, A. S. J., Natowitz, M., Skomorowski, M.-A., Schuster, S. M., Clarke, J. T. R., Mahuran, D. J., *et al.* (1990). Screening for carriers of Tay-Sachs disease among Ashkenazi Jews: Comparison of DNA-based and enzyme-based tests. *N. Engl. J. Med.* **323,** 6–12.

Triggs-Raine, B. L., Mules, E. H., Kaback, M. M., Lim-Steele, J. S. T., Dowling, C. E., Akerman, B. R., Natowicz, M. R., *et al.* (1992). A pseudodeficiency allele common in Non-Jewish Tay-Sachs carriers: Implications for carrier screening. *Am. J. Hum. Genet.* **51,** 793–801.

Tse, R., Vavougios, G., Hou, Y., and Mahuran, D. J. (1996a). Identification of an active acidic residue in the catalytic site of beta-hexosaminidase. *Biochemistry* **35,** 7599–7607.

Tse, R., Wu, Y. J., Vavougios, G., Hou, Y., Hinek, A., and Mahuran, D. J. (1996b). Identification of functional domains within the alpha and beta subunits of beta-hexosaminidase A through the expression of alpha-beta fusion proteins. *Biochemistry* **35,** 10894–10903.

Tsui, F., Mahuran, D. J., Lowden, J. A., Mosmann, T., and Gravel, R. A. (1983). Characterization of polypeptides serilogically and structurally related to hexosaminidase in cultured fibroblasts. *J. Clin. Invest.* **71,** 965–973.

von Heijne, G. (1986). A new method for predicting signal sequence cleavage sites. *Nucleic Acids Res.* **14,** 4683–4690.

Wakamatsu, N., Kobayashi, H., Miyatake, T., and Tsuji, S. (1992). A novel exon mutation in human beta-hexosaminidase beta subunit gene affecting the 3′ splice site selection. *J. Biol. Chem.* **267,** 2406–2413.

Weitz, G., and Proia, R. L. (1992). Analysis of the glycosylation and phosphorylation of the alpha-subunit of the lysosomal enzyme, beta-hexosaminidase A, by site-directed mutagenesis. *J. Biol. Chem.* **267,** 10039–10044.

Yamanaka, S., Johnson, O. N., Norflus, F., Boles, D. J., and Proia, R. L. (1994). Structure and expression of the mouse beta-hexosaminidase genes, HEXA and HEXB. *Genomics* **21,** 588–596.

14

Biosynthesis of Normal and Mutant β-Hexosaminidases

Elizabeth F. Neufeld*
Department of Biological Chemistry
UCLA School of Medicine
Los Angeles, California 90095

Alessandra d'Azzo
Department of Genetics
St. Jude's Children's Research Hospital
Memphis, Tennessee 38105

I. The Normal Biosynthetic Pathway
II. Biosynthesis of Mutant β-Hexosaminidases
References

I. THE NORMAL BIOSYNTHETIC PATHWAY

When Andrej Hasilik joined our laboratory at the National Institutes of Health (NIH) in 1978, it was with the express purpose of studying the biosynthesis of lysosomal enzymes. He planned to label cultured human fibroblasts with radioactive precursors and then isolate the enzymes of interest by immunoprecipitation and polyacrylamide gel electrophoresis. This procedure, which soon became standard practice, required antibodies highly specific to the enzymes of interest. Andrej purified three lysosomal enzymes from human placenta and used them to immunize goats. In the legacy that he left us upon returning to Germany were large amounts

*To whom correspondence should be addressed. Email: eneufeld@mednet.ucla.edu. Fax: (310) 206-1929. Telephone: (310) 825-7149.

Advances in Genetics, Vol. 44

of excellent antisera against β-hexosaminidase and its isolated subunits, which have served us and our colleagues for nearly two decades.

Hasilik discovered that proteins destined for lysosomes, including both subunits of β-hexosaminidase, underwent modifications on the way to or after reaching their final destination. They were made as precursors which were slowly converted to mature forms of smaller size (Hasilik and Neufeld, 1980a). The precursors acquired the mannose 6-phosphate recognition marker necessary for targeting to lysosomes (Hasilik and Neufeld, 1980b). These early findings were confirmed and extended by a number of investigators to many other lysosomal enzymes; the reader is referred to extensive reviews on the subject (von Figura and Hasilik, 1986; Neufeld, 1991; Kornfeld, 1992; Hille-Rehfeld, 1995).

The path of β-hexosaminidase to lysosomes is shown schematically in Figure 14.1, with the reactions occurring in the various sites identified in the legend. The α and β subunits are synthesized independently; they lose the signal peptide and acquire N-linked oligosaccharides and intrachain disulfide bonds upon translocation into the endoplasmic reticulum (Sonderfeld-Fresko and Proia, 1988). They then acquire the mannose 6-phosphate signal for targeting to

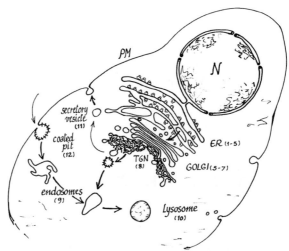

Figure 14.1. The natural history of β-hexosaminidase. The numbered reactions occurring in the designated organelles are as follows: (1) translation of α- and β-subunit mRNAs on membrane-bound polysomes; (2) cleavage of signal peptide; (3) N-glycosylation and formation of disulfide bonds; (4) first step of synthesis of mannose 6-phosphate targeting signal; (5) subunit association; (6) completion of synthesis of mannose 6-phosphate targeting signal; (7) formation of hybrid and complex carbohydrate structures; (8) binding to mannose 6-phosphate receptors; (9) dissociation of enzyme from receptors; (10) enzymatic function of β-hexosaminidase, also limited proteolysis and carbohydrate trimming; (11) secretion of β-hexosaminidase; (12) endocytosis mediated by mannose 6-phosphate receptors.

lysosomes. The acquisition of this structure involves two enzymatic steps, the transfer of phospho-N-acetylglucosamine from UDPGlcNAc to high-mannose oligosaccharides and the subsequent uncovering of the diesterified phosphate groups. Both reactions were initially thought to take place in the Golgi apparatus, but later the first reaction was shown to be initiated in the endoplasmic reticulum (ER) and to be completed in the Golgi (Lazzarino and Gabel, 1988).

A reaction of particular importance to the biology and pathology of the β-hexosaminidase system is association of the α and β subunits to form the heterodimeric Hex A (αβ) isozyme, which is active against all naturally occurring substrates including G_{M2} ganglioside. It was Rick Proia who found that αβ subunit association could be monitored by judicious use of the antisera raised by Hasilik: an antiserum that recognized only monomeric α subunits (anti-α), one that recognized the α subunit only when it was associated with the β subunit (anti-B), and one that recognized the α subunit in all forms, whether monomeric or associated into an αα or αβ dimer (anti-A), as shown in Figure 14.2 (Proia et al., 1984). An additional antiserum (against monomeric β subunits) was available to study association of the β subunit into ββ dimers. Our studies showed that the association of α and β subunits occurred only after acquisition of the mannose 6-phosphate marker (as measured by ^{32}P incorporation), and we therefore thought that it occurred in the Golgi. But because the transfer of PGlcNAc

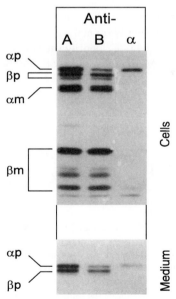

Figure 14.2. Use of specific antisera to differentiate between free and associated forms of the α subunit. See text for explanation; αp, βp = precursor α and β subunit, αm and βm = mature α and β subunits.

begins in the endoplasmic reticulum (see above), subunit association may also start in that organelle. This uncertainty leads us to place the association step in both the ER and the Golgi in Figure 14.1. The β subunits have been found to associate with each other in the endoplasmic reticulum in a cell-free translation system (Sonderfeld-Fresko and Proia, 1988).

As the precursor enzyme proceeds through the Golgi, some of the oligo-saccharides which did not acquire mannose 6-phosphate may be further processed to complex or hybrid structures. The pathway from the distal compartment of trans-Golgi network (TGN) is mediated by either of the two mannose 6-phosphate receptors (see above reviews). Transport to lysosomes proceeds through an inter-mediate acidified compartment, where the enzyme dissociates from the receptors, allowing the receptors to cycle back to the TGN and be available for further rounds of transport. A small amount of enzyme is secreted even though it has the mannose 6-phosphate signal, probably because binding in the TGN is not com-plete. However, the secreted enzyme can bind to mannose 6-phosphate receptors on the cell surface, reenter the cell through coated pits, and proceed to lysosomes through intermediate organelles. Late endosomes are probably the organelles in which the endocytic and endogenous pathways converge (Figure 14.1).

Several proteolytic cleavages occur during the synthesis of the enzyme, which were identified very precisely after the amino acid sequence of the subunits became known from cDNAs. We can distinguish by size the pre-precursor forms of the subunits that can only be seen in cell-free translation, the precursor forms produced after cleavage of the signal peptide, and the mature forms formed after the enzymes have reached their lysosomal destination (Figure 14.2). The amino termini of the various polypeptides identify the sites of proteolysis (Little *et al.*, 1988; Quon *et al.*, 1989; Mahuran *et al.*, 1988; Hubbes *et al.*, 1989). Cleaved segments remain associated by disulfide bonds. A schematic representation of the processing of the two β-hexosaminidase subunits is shown in Figure 14.3. These

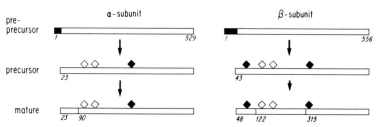

Figure 14.3. Maturation of α and β subunits of β-hexosaminidase. The polypeptides are represented by bars, with the N-terminal amino acid identified by number; the solid segments of the bars are the signal peptides. The squares above the bars indicate N-glycosylation sites; the solid squares indicate the carbohydrates which also contain the mannose 6-phosphate targeting signal. Modified from Neufeld (1989) to include data from Hubbes *et al.* (1989) and Weitz and Proia (1992).

proteolytic cleavages of β-hexosaminidase are not necessary for acquisition of enzymatic activity, as evidenced by full activity of the secreted enzyme, which is in the precursor form (Figure 14.2). Rather, they represent reactions which, along with dephosphorylation and trimming of carbohydrate chains, occur in the degradative environment of lysosomes without destroying the catalytic activity of β-hexosaminidase.

II. BIOSYNTHESIS OF MUTANT β-HEXOSAMINIDASES

Biosynthetic studies featured prominently in sorting out mutations before the *HEXA* gene was cloned. Our first clue that absence of β-hexosaminidase A activity in Tay-Sachs disease could have multiple causes came in a study of mRNA translation followed by immunoprecipitation of the translated product with the anti-A serum (Proia and Neufeld, 1982). The mRNA from cells of a Tay-Sachs patient of Ashkenazi Jewish origin did not support translation of the α subunit, whereas mRNA prepared from cells of a Tay-Sachs patient of Italian origin (GM1110) could do so. We were surprised because the latter had been reported not to make any α-subunit protein (Hasilik and Neufeld, 1980a). It turned out that the GM1110 cells synthesized a normal amount of precursor α subunit which was insoluble under the milder extraction conditions of the earlier study. This abnormal α polypeptide was N-glycosylated but failed to undergo further processing, such as phosphorylation or proteolytic cleavage. Thus it appeared not to be transported to lysosomes. Furthermore, it was not secreted, indicating that it was trapped in the endoplasmic reticulum. The mutation responsible for the synthesis of this defective protein is an amino acid substitution, Glu482Ser (Nakano *et al.*, 1988). A different mutant α-polypeptide that was shown to be trapped in the endoplasmic reticulum was derived from cells of another Italian patient, WG1051. This mutant polypeptide was truncated because the deletion of a nucleotide (ΔC1510) in the penultimate exon caused frameshift and premature termination (Lau *et al.*, 1989). It is likely that of the approximately 40 missense mutations known to occur in the *HEXA* gene (Myerowitz, 1987; see also Chapter 17), a substantial fraction must involve retention of the mutant protein in the endoplasmic reticulum, a compartment that exercises quality control by keeping misfolded proteins from progressing to other organelles (Lodish, 1988).

A different but very interesting biochemical defect of the α subunit is faulty association with the β subunit (d'Azzo *et al.*, 1984). The defective α subunit acquires mannose 6-phosphate but remains monomeric. Such a defect was found in fibroblasts from patients with milder forms of Hex A deficiency (chronic or adult G_{M2} gangliosidoses). Complementation analysis showed the defect to be in the α subunit. The failure of the two subunits to associate in these fibroblasts was not absolute; a very low level of association was detected, in keeping with the milder clinical phenotype of the patients. The underlying mutation was shown to be Gly269Ser (Paw et al., 1989; Navon and Proia, 1989). The α subunit also remained monomeric in cells from a juvenile G_{M2} gangliosidosis patient, whose mutation (Arg504His) must be less permissive of residual activity (Paw et al., 1990). These studies underscored the importance of association of the subunits for acquisition of catalytic activity as well as transport to lysosomes.

There are also amino acid substitutions that do not interfere with post-translational events but interfere with catalytic activity. These are of two types. The B1 mutations cause Hex A to have the catalytic properties of Hex B because the α subunit is catalytically inert (Kytzia et al., 1983; Ohno and Suzuki, 1988). Most of these are caused by three mutations at the same codon, Arg178His/Cys/Leu. The pseudo-deficiency mutations, Arg247Trp and Arg249Trp, allow residual activity toward G_{M2} ganglioside but not toward the synthetic substrates commonly used for diagnostic assays (Triggs-Raine et al., 1992; see also Chapter 17). Neither the B1 mutations nor the pseudo-deficiency mutations would be expected to produce abnormal processing intermediates.

In the current molecular era, studies of mutations at the level of protein synthesis and processing have largely given way to characterization at the level of DNA. The next stage, which awaits the crystallization of the β-hexosaminidase isoenzymes, will be to explain the effects of the mutations on folding and function of the proteins. A start has been made in that direction, using the distantly related bacterial chitobiase for modeling the catalytic domain of the α subunit of human β-hexosaminidase (Tews et al., 1996).

References

D'Azzo, A., Proia, R. L., Kolodny, E. H., Kaback, M. M., and Neufeld, E. F. (1984). Faulty association of α- and β-subunits in some forms of β-hexosaminidase A deficiency. J. Biol. Chem. **259**, 11070–11074.

Hasilik, A., and Neufeld, E. F. (1980a). Biosynthesis of lysosomal enzymes in fibroblasts. Synthesis as precursors of higher molecular weight. J. Biol. Chem. **255**, 4937–4945.

Hasilik, A., and Neufeld, E. F. (1980b). Biosynthesis of lysosomal enzymes in fibroblasts. Phosphorylation of mannose residues. J. Biol. Chem. **235**, 4946–4950.

Hille-Rehfeld, A. (1995). Mannose 6-phosphate receptors in sorting and transport of lysosomal enzymes. Biochim. Biophys. Acta. **1241**, 177–194.

Hubbes, M., Callahan, J., Gravel, R., and Mahuran, D. (1989). The amino-terminal sequences in the pro-alpha and -beta polypeptides of human lysosomal beta-hexosaminidase A and B are retained in the mature isozymes. FEBS Lett. **249**, 316–320.

Kornfeld, S. (1992). Structure and function of the mannose 6-phosphate/insulin-like growth factor II receptors. Annu. Rev. Biochem. **61**, 307–330.

Kytzia, H.-J., Hinrichs, U., Maire, I., Suzuki, K., and Sandhoff, K. (1983). Variant of G_{M2} gangliosidosis with hexosaminidase A having a severely changed substrate specificity. EMBO J. **2**, 1201–1205.

Lau, M. M. H., and Neufeld, E. F. (1989). A frameshift mutation in a patient with Tay-Sachs disease causes premature termination and defective intracellular transport of the α-subunit of β-hexosaminidase. J. Biol. Chem. **264**, 21376–21380.

Lazzarino, D. A., and Gabel, C. A. (1988). Biosynthesis of the mannose 6-phosphate marker in transport-impaired mouse lymphoma cells. *J. Biol. Chem.* **263,** 10118–10126.

Little, L. E., Lau, M. M. H., Quon, D. V. K., Fowler, A. V., and Neufeld, E. F. (1988). Proteolytic processing of the lysosomal enzyme, β-hexosaminidase, in normal human fibroblasts. *J. Biol. Chem.* **263,** 4288–4292.

Lodish, H. (1988). Transport of secretory and membrane glycoproteins from the endoplasmic reticulum to the Golgi. *J. Biol. Chem.* **263,** 2107–2110.

Mahuran, D. J., Neote, K., Klavins, M. H., Leung, A., and Gravel, R. A. (1988). Proteolytic processing of the pro-α and pro-β precursors from human β-hexosaminidase. *J. Biol. Chem.* **263,** 4612–4618.

Myerowitz, R. (1987). Tay-Sachs disease-causing mutations and neutral polymorphisms in the Hex A gene. *Hum. Mutat.* **9,** 195–208.

Nakano, T., Muscillo, M., Ohno, K., Hoffman, A. J., and Suzuki, K. (1988). A point mutation in the coding sequence of the β-hexosaminidase α gene results in defective processing of the enzyme protein in an unusual GM2-gangliosidosis variant. *J. Neurochem.* **51,** 984–987.

Navon, R., and Proia, R. L. (1989). The mutation in Ashkenazi Jews with G_{M2} gangliosidosis, the adult form of Tay-Sachs disease. *Science* **243,** 1471–1474.

Neufeld, E. F. (1989). Natural history and inherited disorders of a lysosomal enzyme, β-hexosaminidase. *J. Biol. Chem.* **264,** 10927–10930.

Neufeld, E. F. (1991). Lysosomal storage diseases. *Annu. Rev. Biochem.* **60,** 257–280.

Paw, B. H., Kaback, M. M., and Neufeld, E. F. (1989). Molecular basis of adult-onset and chronic G_{M2} gangliosidoses in patients of Ashkenazi Jewish origin: Substitution of serine for glycine at position 269 of the α-subunit of β-hexosaminidase. *Proc. Natl. Acad. Sci. (USA)* **86,** 2413–2417.

Paw, B. H., Moskowitz, S. M., Uhrhammer, N., Wright, N., Kaback, M. M., and Neufeld, E. F. (1990). Juvenile G_{M2} gangliosidosis caused by substitution of histidine for arginine at position 499 or 504 of the α-subunit of β-hexosaminidase. *J. Biol. Chem.* **265,** 9452–9457.

Proia, R. L., and Neufeld, E. F. (1982). Synthesis of β-hexosaminidase in cell-free translation and in intact fibroblasts: An insoluble precursor α-chain in a rare form of Tay-Sachs disease. *Proc. Natl. Acad. Sci. (USA)* **79,** 6360–6364.

Proia, R. L., D'Azzo, A., and Neufeld, E. F. (1984). Association of the α- and β-subunit during the biosynthesis of β-hexosaminidase in cultured human fibroblasts. *J. Biol. Chem.* **259,** 3350–3354.

Quon, D. V. K., Proia, R. L., Fowler, A. V., Bleibaum, J., and Neufeld, E. F. (1989). Proteolytic processing of the β-subunit of β-hexosaminidase, in normal human fibroblasts. *J. Biol. Chem.* **264,** 3380–3384.

Sonderfeld-Fresko, S., and Proia, R. L. (1988). Synthesis and assembly of a catalytically active lysosomal enzyme, β-hexosaminidase B, in a cell-free system. *J. Biol. Chem.* **263,** 13463–13469.

Tews, I., Perrakis, A., Oppenheim, A., Dauter, Z., Wilson, K. S., and Vorgias, C. (1996). Bacterial chitobiase structure provides insight into catalytic mechanisms and the basis of Tay-Sachs disease. *Nature Struct Biol.* **3,** 638–648.

Triggs-Raine, B. L., and Gravel, R. A. Chapter 17 (this volume).

Triggs-Raine, B. L., Mules, A. H., Kaback, M. M., Lim-Steele, J. S. T., Dowling, C. E., Akerman, B. R., Natowicz, M. R., Grebner, E. E., Navon, R., Welch, J. P., Greenberg, C. R., Thomas, G. H., and Gravel, R. A. (1992). A pseudodeficiency allele common in non-Jewish Tay-Sachs carriers: Implications for carrier screening. *Am. J. Hum. Genet.* **51,** 793–801.

von Figura, K., and Hasilik, A. (1986). Lysosomal enzymes and their receptors. *Annu. Rev. Biochem.* **55,** 167–193.

Weitz, G., and Proia, R. L. (1992). Analysis of the glycosylation and phosphorylation of the α-subunit of the lysosomal enzyme, β-hexosaminidase A, by site-directed mutagenesis. *J. Biol. Chem.* **267,** 10039–10044.

15 Recognition and Delineation of β-Hexosaminidase α-Chain Variants: A Historical and Personal Perspective

Kunihiko Suzuki

Neuroscience Center
Departments of Neurology and Psychiatry
University of North Carolina School of Medicine
Chapel Hill, North Carolina 27599

I. At the Beginning
II. Increasing Complexity
III. Era of Molecular Genetics
IV. Evolution of B1 Variant
   A. History
   B. Enzymology of B1 Variant
   C. Molecular Genetics of B1 Variant
V. Genotype–Phenotype Correlation
   Acknowledgments
   References

I. AT THE BEGINNING

Nearly 40 years have passed since I started working on Tay-Sachs and related diseases during my neurology residency years. When I arrived in the Bronx as a first-year resident from Japan in 1960, immediately after my internship, the Albert Einstein College of Medicine was a new start-up with the typical enthusiasm and informality of youth. Some of the first Einstein graduates were in my residency class. The late Saul Korey, then Chairman of Neurology, and Bob Terry, Head of Neuropathology, had just initiated the first multidisciplinary efforts on Tay-Sachs and related diseases (Gomez *et al.*, 1963). A few years earlier, Terry and Korey (1960) had discovered the membranous cytoplasmic bodies (MCB) in neurons of Tay-Sachs patients. The classical form of the disease prevalent in Ashkenazi Jews

Advances in Genetics, Vol. 44

was then the only known ganglioside storage disease. The term amaurotic idiocy was still widely used. The term also included all of the phenotypic variants of what is now known as neuronal ceroid-lipofuscinosis. This innocent and simplistic state changed drastically within a few years of my entering a scientific career, when existence of an entirely different category of ganglioside storage disease was uncovered, in which G_{M1}-ganglioside, instead of G_{M2}-ganglioside, was stored abnormally (O'Brien *et al.*, 1965; Gonatas and Gonatas, 1965; Jatzkewitz and Sandhoff, 1963). This led to our proposal in 1967 for a systematic nomenclature of genetic ganglioside storage diseases (G_{M1}-gangliosidosis, G_{M2}-gangliosidosis, etc.), that has since been widely accepted (Suzuki and Chen, 1967). During the intervening 35 years, the field has moved from analytical biochemistry to enzymology, and most recently, to molecular biology. Each progressive step has brought additional degrees of complexity that could hardly have been imagined when I started my career.

II. INCREASING COMPLEXITY

By the end of the 1960s, three major classes were recognized among G_{M2}-gangliosidoses; N-acetyl-β-hexosaminidase (Hex) A deficiency, B deficiency, and the form in which both isozymes appeared normal (Sandhoff, 1969; Okada and O'Brien, 1969). This last form was later characterized as being due to a defect in a genetically unrelated G_{M2} activator protein (Conzelmann and Sandhoff, 1978). Tay-Sachs, Sandhoff, and the AB variant are commonly used to denote these categories. As we found out more about the enzymology of β-hexosaminidases and about the metabolism of G_{M2}-ganglioside and related compounds, it became clear that the above three categories of G_{M2}-gangliosidosis are not only enzymologically distinct but also are caused by abnormalities in three different genes—Hex α, Hex β, and the G_{M2} activator protein. Regardless of the nomenclature used, these three categories are conceptually unambiguous.

Genetic heterogeneity within each of the three major categories of G_{M2}-gangliosidosis was predictable. Our laboratory contributed descriptions of some of the early cases of atypical phenotypes. Collaboration with Isabelle Rapin was critically important for many of these unusual cases. She is an astute clinician with an uncanny ability to sort out potentially significant patients for in-depth studies. An unusual Ashkenazi Jewish family was found in which all three sibs were affected by a chronic adult form of the disease (Rapin *et al.*, 1976). The clinical onset was in the first few years of life; the patients survived into adulthood with slowly progressive dystonia and spinocerebellar degeneration, but without obvious intellectual deterioration. They were partially deficient in Hex A, with disproportionately severe deficiencies in the serum. The parents gave the heterozygote levels of Hex A activities. The youngest sib died of phenothiazine intoxication at age 17, but the other two lived into their fourth decade. As to Hex B-deficient variants, we described one of the earliest patients with the juvenile form in the mid-1970s (Goldie *et al.*, 1977).

Numerous phenotypic variants of Hex α and β deficiencies have been described since the early 1970s, some with genetic and/or enzymologic characterizations (for a comprehensive review and references, see Sandhoff, 1969). Before the molecular biology era, availability of specific antibodies directed toward the respective subunits made it possible to further examine the nature of the enzyme proteins at steady state as well as during biosynthesis and processing. The classical Tay-Sachs disease in the Ashkenazi Jewish population was found to be cross-reacting material (CRM)-negative (Proia and Neufeld, 1982). On the other hand, unusual patients were described in whom Hex α subunit protein is synthesized but abnormal in such ways that it is smaller than normal (Zokaeem *et al.*, 1987), abnormally insoluble and thus not processed properly (Proia and Neufeld, 1982), or unable to associate normally with the β subunit (d'Azzo *et al.*, 1984).

III. ERA OF MOLECULAR GENETICS

Cloning of cDNA clones encoding the Hex α and β subunits (Korneluk *et al.*, 1986; O'Dowd *et al.*, 1986; Myerowitz *et al.*, 1985 Myerowitz and Proia, 1984) and elucidation of the genomic organization of the respective genes (Proia and Soravia, 1987; O'Dowd *et al.*, 1986; Proia, 1988) in the mid-1980s and corresponding progress in the G_{M2} activator protein several years later (Klima *et al.*, 1991; Xie *et al.*, 1991; Schröder *et al.*, 1989) finally made it possible to examine directly the mutations in all forms of genetic G_{M2}-gangliosidosis. Whereas presence or absence of CRM generally parallels the presence or absence of the corresponding mRNA, Northern analyses have shown further complexities in the quantity and size of mRNA in both α (Budde-Steffen *et al.*, 1996; Ohno and Suzuki, 1988a,b; Myerowitz and Proia, 1984) and β mutations (O'Dowd *et al.*, 1986). Even within the mRNA-negative types, a major deletion at the 5' end of the Hex α gene has been demonstrated in the French-Canadian form, whereas the Hex α gene appears grossly normal in the Ashkenazi form (Myerowitz and Hogikyan, 1986). The gene in the French-Canadian type is unlikely to be transcribed, whereas apparently normal transcription was demonstrated for the Ashkenazi Jewish form, despite the essentially undetectable mRNA (Paw and Neufeld, 1988). Subsequent to the identification of the two major mutations in the classical Jewish infantile Tay-Sachs disease (Myerowitz and Costigan, 1988; Ohno and Suzuki, 1988a; Arpaia *et al.*, 1988; Myerowitz, 1988), activities in mutation analysis have become widespread and continue to this day in many laboratories. Approximately 80 disease-causing mutations have been identified so far in the β-hexosaminidase α-subunit gene alone (Gravel *et al.*, 1995). Among the more common variants, the mutation underlying the adult form of β-hexosaminidase α deficiency occurring in Israel and the surrounding regions has been characterized by Navon and Proia (1989).

IV. EVOLUTION OF B1 VARIANT

A. History

One of those patients first uncovered by Isabelle Rapin was a Puerto Rican girl with a juvenile form of the disease (Suzuki *et al.*, 1970). Her early development appeared perfectly normal. Soon after she started elementary school, she developed clumsiness and fell easily. She started losing her mental capacity slowly from about 6 years, developed seizures at 8 years, and cerebellar signs at 9 years. Pyramidal signs and dystonia became prominent between $9\frac{1}{2}$ and 11 years, and dysphagia was difficult to manage by 14 years. She became areflexic at 14 years and died at $14\frac{1}{2}$ years. Some of the symptoms and signs commonly seen in the infantile disease never developed. Analysis of the postmortem brain showed the total amount of cortical ganglioside to be twice normal, with G_{M2}-ganglioside comprising 50% of the total, compared with the four-to-five times total normal amount and 85–90% G_{M2} in classical infantile Tay-Sachs patients. These findings were initially insufficient to convince us that the patient had a genetically defined G_{M2}-gangliosidosis. The conclusive evidence came when we isolated a pure fraction of the membranous cytoplasmic bodies and found that their ganglioside content was exclusively G_{M2}. Very shortly afterwards, the discovery of Hex A deficiency in the classical Tay-Sachs disease was made and we were able to show that this patient was partially deficient in Hex A activity (Suzuki and Suzuki, 1970). More recent advances have established that this so-called juvenile G_{M2}-gangliosidosis is one of the important variants of genetic β-hexosaminidase α deficiencies. From the beginning, we pointed out that the partial Hex A deficiency observed in this patient with chromogenic or fluorogenic artificial substrates was indistinguishable from that in carriers of classical infantile Tay-Sachs disease and that there must be an enzymologic difference between them. Whereas the reduction in Hex A activity was approximately 50% with the conventional fluorogenic substrates, it was disproportionately severe when assayed with the natural lipid substrate, G_{M2}-ganglioside (Zerfowski and Sandhoff, 1974). From the clinical, enzymologic, and biochemical findings, it is highly likely that this patient represents the first detailed description of patients homozygous for the typical B1 variant mutation,

533G→A, R178H ("DN-allele," see below). Loss of all tissue materials from this patient unfortunately precludes molecular genetic confirmation.

B. Enzymology of B1 variant

The first confirmed patient with this variant, now commonly referred to as B1 variant, was described in a Puerto Rican boy. We initially thought that he had an atypical case of AB variant (Goldman et al., 1980), because β-hexosaminidase A and B activities were normal when tested with the conventional fluorogenic substrate and yet there was a major accumulation of G_{M2}-ganglioside in the brain. He was later found to have an abnormality in the Hex α subunit (Kytzia et al., 1983; Li et al., 1981). This was an understandable sequence of events, given the knowledge of this group of disorders at that time. More detailed enzymologic studies soon clarified that this patient did not have G_{M2} activator protein deficiency but had an enzymologically unique form of β-hexosaminidase α deficiency (Kytzia et al., 1983). This led to the enzymologic definition of B1 variant. Total β-hexosaminidase, as well as the proportion of the A and B isozyme activities in B1 variant patients, appear either completely normal or only moderately decreased enzymatically, if the conventional chromogenic or fluorogenic artificial substrates are used for the assay. Even when reduced, the degree of the reduction is nowhere near the level that would allow diagnosis of Hex α deficiency. However, the catalytic activity toward the natural substrate, G_{M2}-ganglioside, and the recently developed artificial 6-sulfated substrates, such as 4-MU glcNAc 6-sulfate, is profoundly deficient. These are the enzymologic criteria that define the B1 variant. It should be emphasized that the diagnosis of B1 variant should not be made solely on the basis of relatively high residual activities of β-hexosaminidase A determined by the unsulfated substrate and A/B differential inactivation. The diagnosis of B1 variant can be made only when the activity against either G_{M2}-ganglioside or the sulfated substrate is *disproportionately* low relative to the isozyme A activity determined by the conventional procedure. Unfortunate confusions have arisen in some cases in the literature because this important criterion was not well understood. In retrospect, the other Puerto Rican patient whom we had described a dozen years earlier as having juvenile G_{M2}-gangliosidosis satisfies the criteria for diagnosis of B1 variant (Suzuki et al., 1970).

Our initial interpretation of the phenomenology of B1 variant was that the underlying mutation occurred in the α chain in such a way that it inactivated only the catalytic activity toward the natural substrate, G_{M2}-ganglioside, without affecting the activity toward the nonsulfated artificial substrates. This was quickly revised with the findings reported by Kytzia and Sandhoff (1985), who showed clearly that each of the two subunits has its own catalytic site. The site on the α subunit is active toward the G_{M2}-ganglioside with the help of the

G_{M2} activator protein but inactive toward the conventional artificial substrates, such as 4-methylumbelliferyl N-acetylglucosaminide, which can be cleaved only by the catalytic site on the β subunit. Therefore, if a mutation inactivates the catalytic activity of the α chain without affecting its stability and its ability to associate with the β subunit to form a stable β-hexosaminidase A isozyme, the enzymologic phenotype of B1 variant will result.

C. Molecular genetics of B1 variant

I spent my sabbatical year in the laboratory of Elizabeth Neufeld in 1984–1985 and actively participated in the project of cloning and sequencing of human β-hexosaminidase α subunit cDNA as the primary warm-body sequencer (Myerowitz et al., 1985). Upon my return to the Albert Einstein College of Medicine and subsequent move to North Carolina, mutation analysis of our B1 variant patient seemed to be an ideal vehicle for me to apply the newly learned trade and to expand my experience. The likelihood that this variant would generate stable enzyme protein was particularly attractive to me, as I had gone through the phases of analytical biochemistry and enzymology of this group of disorders. Should a mutation cause complete absence of the enzyme protein, there is not much an old-fashioned enzymologist can do. The mutation identified in 1988 in our original patient with the B1 variant, 533G→A, R178H, turned out to be the first point mutation in the coding sequence of the mature Hex α subunit responsible for any form of G_{M2}-gangliosidosis (Ohno and Suzuki, 1988b). Little did we imagine then that this particular mutation had still more interesting future developments.

A series of studies by Akemi Tanaka, then working in our laboratory, demonstrated that the R178H mutation is widespread among patients from the general Mediterranean region (Italian, French, Spanish, English/Italian/ Hungarian, English/French/Azores, and, the original Puerto Rican, who can be considered as originally from the region (Tanaka et al., 1988; 1990). This was further expanded when we learned that B1 variant is the prevalent form of all β-hexosaminidase α deficiencies in Northern Portugal (Ribeiro et al., 1991; Maia et al., 1990) and when Maria Rosário Dos Santos, from the laboratory of Clara Sá Miranda, brought specimens from eleven B1 variant patients, all from the area surrounding Porto, Portugal. Ten patients were homoallelic and one patient was heteroallelic for R178H (533G→A)(Dos Santos et al., 1991). We also learned that a major wave of emigration occurred from Portugal to Puerto Rico a few hundred years ago. Subsequent studies with additional patients have further confirmed the very high prevalence of this mutation in the Northern Portuguese population. Of 18 B1 variant patients, 14 were homoallelic and four were heteroallelic. These series of B1 variant patients carrying this mutation at least in one allele suggest that the genotype is reflected in the clinical as well as biochemical phenotype.

Patients homoallelic for this mutation are invariably of the juvenile phenotype, whereas a few late-infantile B1 variant patients have been found to have this mutation on one allele and an mRNA-negative infantile mutation on the other. The patient for whom the "DN-allele" was named was of a late-infantile phenotype and had a genotype of the DN/Jewish infantile 4-base deletion. An adult patient with enzymatically definable B1 variant (Specola *et al.*, 1990) has been found to be heteroallelic with one DN-allele (M. T. Vanier, personal communication). Although the mutation on the other allele remains to be characterized, it is not unreasonable to anticipate that it may be a mutation which, if homoallelic by itself, would have resulted in an adult form of the disease. Not only is the genotype reflected on the clinical phenotype, there also appears to be a correlation with the degree of G_{M2}-ganglioside accumulation in the brain. The genotypes of patients were, in the decreasing order of G_{M2}-ganglioside accumulation, infantile/infantile, DN/infantile, DN/DN homoallelic, and DN/probable adult allele.

The functional importance of Arg 178 was demonstrated by Brown and Mahuran (1991), who took advantage of there being a homologous arginine in the β subunit. We took the computer-assisted molecular modeling approach to assess the functional importance of Arg178 (Suzuki and Vanier, 1991). When peptides of approximately 40 amino acids surrounding Arg178 in the α subunit and its homologous counterpart in the β subunit were energy minimized, both α and β subunits formed similar three-dimensional structures of a deep pocket with the arginine residues at the bottom, despite the fact that the two peptides share only two-thirds of the amino acids. In contrast, the three-dimensional structure of the α subunit peptide completely "fell apart" when the single Arg178 was replaced by histidine.

During her work on the series of B1 variant patients, Akemi Tanaka identified the second point mutation underlying B1 variant as 532C→T (R178C) in a patient from Czechoslovakia (Tanaka *et al.*, 1990). The mutation is in the same Arg178 as in the DN mutation, with the frequent mutation at a CpG site occurring on the other strand. This finding delighted us because it was consistent with what I had anticipated. From the unique enzymologic phenotype of B1 variant, we had thought that mutations responsible for the variant were likely to be located at or near the catalytic or binding site of the enzyme protein critical for its function. Another but different point mutation occurring in the same codon also underlying the B1 variant phenotype provided further support for the hypothesis. Another mutation, 533G→T, R178L, is known in the same codon in a patient who showed clinical and enzymologic characteristics intermediate between the classical infantile disease and the B1 variant (Triggs-Raine *et al.*, 1991). Hex A activity was lower than the usual B1 variant, although there was a requisite disproportionately low activity against the sulfated substrate. Two additional mutations have since been reported to be responsible for the enzymologic phenotype of B1 variant, V192L (Ainsworth and Coulter-Mackie, 1992) and D258H (Fernandes *et al.*, 1992). However, validity of V192L as an underlying molecular defect of the B1

Table 15.1. Hex α Subunit Mutations Responsible for B1 Variant Phenotype

Mutation[a]	References	Note
533G→A, R178H	(Dos Santos et al., 1991; Tanaka et al., 1988, 1990; Ohno and Suzuki, 1988b)	By far the most common; originated in Northern Portugal?
532T→C, R178C	(Tanaka et al., 1990)	Czech patient
533G→T, R178L	(Triggs-Raine et al., 1991)	Relatively early, rapidly progressive course; Hex A activity with nonsulfated substrate lower than usual B1 variant
G574→C, Val192L	(Ainsworth and Coulter-Mackie, 1992)	But see (Hou et al., 1996)
772G→C, D258H	(Fernandes et al., 1992)	Portuguese patient
755G→A, R252H	Ribeiro et al., unpublished	Portuguese patient

[a]The first three mutations occur in the same codon.

variant phenotype has been questioned recently (Hou et al., 1996). Additionally, a new B1 mutation has been identified, R252H, that satisfied the enzymologic criteria for B1 variant in an expression study (Ribeiro et al., 1996). Thus, altogether, six point mutations have been ascribed to B1 variant form of G_{M2}-gangliosidosis, one, R178H is by far the most common, and one of them, V192L, may not hold up (Table 1).

V. GENOTYPE–PHENOTYPE CORRELATION

As in any other genetic disorders, the traditional classification of genetic β-hexosaminidase α deficiencies that was based primarily on the clinical phenotype now faces fundamental as well as pragmatic complications. The fundamental logical question is how to define a disease or its variant. In autosomal recessive disorders, the random combinations of two abnormal alleles of n mutations is $n(n + 1)/2$. Over 60 disease-causing mutations are already known for genetic G_{M2}-gangliosidosis due to β-hexosaminidase α subunit defects. Thus, at least 1830 different genotypes are theoretically possible among patients even within the already known mutations, and this number is constantly increasing. Except for consanguineous cases and in isolated populations, compound heterozygosity is the norm rather than the exception. Thus it becomes impossible to define a disease logically on the basis of the genotype for any given patient. Nevertheless, for pragmatic purposes, one hopes that the traditional clinical classification can be useful. Prediction, however, is not very reassuring. In autosomal recessive disorders, two abnormal alleles contribute to the phenotypic expression. If one makes a reasonable assumption that the clinical phenotype depends on the residual activity of the mutant enzyme protein *in vivo*, a mutant enzyme with higher residual activity tends to override that with less residual activity. Thus it is likely that all patients who have at least one allele with an adult mutation are of the adult phenotype, regardless of the nature of the other allele. In contrast, one juvenile mutant allele will define the juvenile phenotype only if the other allele carries either a juvenile or an infantile mutation. If the other allele carries an adult mutation, the patient will be of an adult phenotype. Then patients with an infantile phenotype must carry two infantile mutations. As already discussed, patients who are homoallelic for the common B1 variant mutation, R178H, invariably exhibit the juvenile phenotype, and the phenotype is modified by the nature of the second mutation when heteroallelic. This type of analysis holds only if there are *adult*, *juvenile*, and *infantile* mutations with discontinuous ranges of residual activities. Probability of random mutations dictates that it is far more likely that the degree of residual activities of all individual mutations will be continuous between normal activity and zero activity. Any degree of discontinuity among individual mutations will further be smoothed out by the additive nature of the contributions by the two mutant alleles. For the same reason, it is doubtful that more detailed enzymological profiles, based on, for example, pI, K_m, etc., can be very useful. The diverse natural substrates and possible differential effect of any mutation on substrate specificities of the enzyme further complicate the eventual phenotypic expressions. The range of the final outcome of the residual activity contributed by the two mutant genes in any given patient defies any hope for discrete clinical phenotypes, and consequently, any orderly clinical classification.

I believe we are at the stage where we will have to recognzie more clearly the logical distinction between "definitions" and "names" of individual diseases. One can describe and define a disease in sufficient detail. Once it is defined without ambiguity, then the name of the disease has to be only unique to the defined entity but does not even have to be descriptive. Eponyms have no descriptive meaning and yet they are often useful as long as the disease under the name is unambiguously defined.

Trying to classify and give names to all variants of G_{M2}-gangliosidosis is likely to be futile. There is no *a priori* basis to designate any particular form as the prototype and others as variants. Even the Ashkenazi Jewish form of Tay-Sachs disease with classical infantile phenotype has been shown to be genetically heterogeneous. Perhaps we should recognize only three G_{M2}-gangliosidoses as disease entities—Hex α, Hex β, and the activator gene deficiencies. Each patient should then be defined by the mutation(s). Descriptive terms may be used for convenience, but not as names of any particular "variants." This is psychologically unsettling, but after all, we are at a great stage of flux and we will probably have to endure it.

Acknowledgments

Work from our own laboratory on G_{M2}-gangliosidosis has been supported in recent years by National Institutes of Health research grant RO1 NS28997. Drs. Kousaku Ohno, Takeshi Nakano, Akemi Tanaka, Rose-Mary Boustany, Maria Rosáro Dos Santos, Maria Gil Ribeiro, Eiji Nanba, and Junji Nishimura all contributed significantly to our own work on genetic β-hexosaminidase deficiencies.

References

Ainsworth, P. J., and Coulter-Mackie, M. B. (1992). A double mutation in exon 6 of the β-hexosaminidase α subunit in a patient with the B1 variant of Tay-Sachs disease. *Am. J. Hum. Genet.* **51**, 802–809.

Arpaia, E., Dumbrille-Ross, A., Maler, T., Neote, K., Tropak, M., Troxel, C., Stirling, J. L., *et al.* (1988). Identification of an altered splice site in Ashkenazi Tay-Sachs disease. *Nature.* **333**, 85–86.

Brown, C. A., and Mahuran, D. J. (1991). Active arginine residues in β-hexosaminidase. Identification through studies of the B1 variant of Tay-Sachs disease. *J. Biol. Chem.* **266**, 15855–15862.

Budde-Steffen, C., Steffen, M., Siegel, D. A., and Suzuki, K. (1996). Presence of β-hexosaminidase α-chain mRNA in two different variants of GM2-gangliosidosis. *Neuropediatrics* **19**, 59–61.

Conzelmann, E., and Sandhoff, K. (1978). AB variant of infantile GM2 gangliosidosis: Deficiency of a factor necessary for stimulation of hexosaminidase A-catalyzed degradation of ganglioside GM2 and glycolipid GA2. *Proc. Natl. Acad. Sci. (USA)* **75**, 3979–3983.

d'Azzo, A., Proia, R. L., Kolodny, E. H., Kaback, M. M., and Neufeld, E. F. (1984). Faulty association of alpha- and beta-subunits in some forms of beta-hexosaminidase A deficiency. *J. Biol. Chem.* **259**, 11070–11074.

Dos Santos, M. R., Tanaka, A., Sá Miranda, M. C., Ribeiro, M. G., Maia, M., and Suzuki, K. (1991). GM2-gangliosidosis B1 variant: Analysis of β-hexosaminidase α gene mutations in 11 patients from a defined region in Portugal. *Am. J. Hum. Genet.* **49**, 886–890.

Fernandes, M., Kaplan, F., Natowicz, M., Prence, E., Kolodny, E., Kaback, M., and Hechtman, P. (1992). A new Tay-Sachs disease B1 allele in exon 7 in two compound heterozygotes each with a second novel mutation. *Hum. Mol. Genet.* **1**, 759–761.

Goldie, W. D., Holtzman, D., and Suzuki, K. (1977). Chronic hexosaminidase A and B deficiency: A case report. *Ann. Neurol.* **2**, 156–158.

Goldman, J. E., Yamanaka, T., Rapin, I., Adachi, M., Suzuki, K., and Suzuki, K. (1980). The AB variant of GM2-gangliosidosis: Clinical, biochemical and pathological studies of two patients. *Acta Neuropathol.* **52**, 189–202.

Gomez, C. J., Gonatas, J., Korey, S. R., Samuels, S., Stein, A., Suzuki, K., Terry, R. D., *et al.* (1963). Study in Tay-Sachs disease. *J. Neuropathol. Exp. Neurol.* **22**, 2–104.

Gonatas, N. K., and Gonatas, J. (1965). Ultrastructural and biochemical observations on a case of systemic late infantile lipidosis and its relationship to Tay-Sachs disease and gargoylism. *J. Neuropathol. Exp. Neurol.* **24**, 318–340.

Gravel, R. A., Clarke, J. T. R., Kaback, M. M., Mahuran, D., Sandhoff, K., and Suzuki, K. (1995). The gangliosidosis. *In* "The Metabolic and Molecular Basis of Inherited Disease" (Scriver, C. R., Beaudet, A. L., Sly, W. S., and Valle, D., eds.), pp. 2839–2879. McGraw-Hill, New York.

Hou, Y., Vavougios, G., Hinek, A., Wu, K. K., Hechtman, P., Kaplan, F., and Mahuran, D. J. (1996). The Val[192] Leu mutation in the α-subunit of β-hexosaminidase A is not associated with the B1-variant from of Tay-Sachs disease. *Am. J. Hum. Genet.* **59**, 52–58.

Jatzkewitz, H., and Sandhoff, K. (1963). On a biochemically special form of infantile amaurotic idiocy. *Biochim. Biophys. Acta.* **70**, 354–356.

Klima, H., Tanaka, A., Schnabel, D., Nakano, T., Schröder, M., Suzuki, K., and Sandhoff, K. (1991). Characterization of full-length cDNAs and the gene coding for the human GM2 activator protein. *FEBS Lett.* **289,** 260–264.

Korneluk, R. G., Mahuran, D. J., Neote, K., Klavins, M. H., O'Dowd, B. F., Tropak, M., Willard, H. F., *et al.* (1986). Isolation of cDNA clones coding for the alpha-subunit of human beta-hexosaminidase. Extensive homology between the alpha- and beta-subunits and studies on Tay-Sachs disease. *J. Biol. Chem.* **261,** 8407–8413.

Kytzia, H.-J., and Sandhoff, K. (1985). Evidence for two different active sites on human beta-hexosaminidase A. Interaction of GM2 activator protein with beta-hexosaminidase A. *J. Biol. Chem.* **260,** 7568–7572.

Kytzia, H.-J., Hinrichs, U., Maire, I., Suzuki, K., and Sandhoff, K. (1983). Variant of GM2-gangliosidosis with hexosaminidase A having a severely changed substrate specificity. *EMBO J.* **2,** 1201–1205.

Li, S.-C., Hirabayashi, Y., and Li, Y.-T. (1981). A new variant of type-AB GM2-gangliosidosis. *Biochem. Biophys. Res. Commun.* **101,** 479–485.

Maia, M., Alves, D., Ribeiro, G., Pinto, R., and Sa Miranda, M. C. (1990). Juvenile GM2 gangliosidoses variant B1: Clinical and biochemical study in seven patients. *Neuropediatrics* **21,** 18–23.

Myerowitz, R. (1988). Splice junction mutation in some Ashkenazi Jews with Tay-Sachs disease: Evidence against a single defect within this ethnic group. *Proc. Natl. Acad. Sci. (USA)* **85,** 3955–3958.

Myerowitz, R., and Costigan, F. C. (1988). The major defect in Ashkenazi Jews with Tay-Sachs disease is an insertion in the gene for the α-chain of β-hexosaminidase. *J. Biol. Chem.* **263,** 18587–18589.

Myerowitz, R., and Hogikyan, N. D. (1986). Different mutations in Ashkenazi Jewish and non-Jewish French Canadians with Tay-Sachs disease. *Science* **232,** 1646–1648.

Myerowitz, R., and Proia, R. L. (1984). cDNA clone for the alpha-chain of human beta-hexosaminidase: Deficiency of alpha-chain mRNA in Ashkenazi Tay-Sachs fibroblasts. *Proc. Natl. Acad. Sci. (USA)* **81,** 5394–5398.

Myerowitz, R., Piekarz, R., Neufeld, E. F., Shows, T. B., and Suzuki, K. (1985). Human β-hexosaminidase α chain: Coding sequence and homology with the β chain. *Proc. Natl. Acad. Sci. (USA)* **82,** 7830–7834.

Navon, R., and Proia, R. L. (1989). The mutation in Ashkenazi Jews with adult GM2 gangliosidosis, the adult form of Tay-Sachs disease. *Science* **243,** 1471–1474.

O'Brien, J. S., Stern, M. B., Landing, B. H., O'Brien, J. K., and Donnell, G. N. (1965). Generalized gangliosidosis. Another inborn error of ganglioside metabolism? *Am. J. Dis. Child.* **109,** 338–346.

O'Dowd, B. F., Klavins, M. H., Willard, H. F., Gravel, R., Lowden, J. A., and Mahuran, D. J. (1986). Molecular heterogeneity in the infantile and juvenile forms of Sandhoff disease (O variant GM2-gangliosidosis). *J. Biol. Chem.* **261,** 12680–12685.

Ohno, K., and Suzuki, K. (1988a). A splicing defect due to an exon-intron junctional mutation results in abnormal β-hexosaminidase α chain mRNAs in Ashkenazi Jewish patients with Tay-Sachs disease. *Biochem. Biophys. Res. Commun.* **153,** 463–469.

Ohno, K., and Suzuki, K. (1988b). Mutation in GM2-gangliosidosis B1 variant. *J. Neurochem.* **50,** 316–318.

Okada, S., and O'Brien, J. S. (1969). Tay-Sachs disease: Generalized absence of a β-D-N-acetylhexosaminidase component. *Science* **165,** 698–700.

Paw, B. H., and Neufeld, E. F. (1988). Normal transcription of the beta-hexosaminidase alpha chain gene in the Ashkenazi Tay-Sachs mutation. *J. Biol. Chem.* **263,** 3012–3015.

Proia, R. L. (1988). Gene encoding the human beta-hexosaminidase beta chain: Extensive homology of intron placement in the alpha- and beta-genes. *Proc. Natl. Acad. Sci. (USA)* **85,** 1883–1887.

Proia, R. L., and Neufeld, E. F. (1982). Synthesis of beta-hexosaminidase in cell-free translation and in intact fibroblasts: An insoluble precursor alpha chain in a rare form of Tay-Sachs disease. *Proc. Natl. Acad. Sci. (USA)* **79**, 6360–6364.

Proia, R. L., and Soravia, E. (1987). Organization of the gene encoding the human β-hexosaminidase α-chain. *J. Biol. Chem.* **262**, 5677–5681.

Rapin, I., Suzuki, K., Suzuki, K., and Valsamis, M. P. (1976). Adult (chronic) GM2-gangliosidosis. Atypical spinocerebellar degeneration in a Jewish sibship. *Arch. Neurol.* **33**, 120–130.

Ribeiro, M. G., Pinto, R. A., Dos Santos, M. R., Maia, M., and Sá Miranda, M. C. (1991). Biochemical characterization of β-hexosaminidase in different biological specimens from eleven patients with GM2-gangliosidosis B1 variant. *J. Inherit. Metab. Dis.* **14**, 715–720.

Ribeiro, M. G., Sonin, T., Pinto, R. A., Fontes, A., Ribeiro, H., Pinto, E., Palmeira, M. M., and Sá Miranda, M. C. (1996). Clinical, enzymatic, and molecular characterization of a Portuguese family with a chronic form of G_{M2}-gangliosidosis B1 variant. *J. Med. Genet.* **33**, 341–343.

Sandhoff, K. (1969). Variation of β-hexosaminidase pattern in Tay-Sach disease. *FEBS Lett.* **4**, 351–354.

Schröder, M., Klima, H., Nakano, T., Kwon, H., Quintern, L. E., Gaertner, S., Suzuki, K., *et al.* (1989). Isolation of a cDNA encoding the human GM2 activator protein. *FEBS Lett.* **251**, 197–200.

Specola, N., Vanier, M. T., Goutieres, F., Mikol, J., and Aicardi, J. (1990). The juvenile and chronic forms of GM2 gangliosidosis—clinical and enzymatic heterogeneity. *Neurology* **40**, 145–150.

Suzuki, K., and Chen, G. C. (1967). Brain ceramide hexosides in Tay-Sachs disease and generalized gangliosidosis (GM1-gangliosidosis). *J. Lipid Res.* **8**, 105–113.

Suzuki, Y., and Suzuki, K. (1970). Partial deficiency of hexosaminidase component A in juvenile GM2-gangliosidosis. *Neurology* **20**, 848–851.

Suzuki, K., and Vanier, M. T. (1991). Biochemical and molecular aspects of late-onset GM2-gangliosidosis: B1 variant as a prototype. *Dev. Neurosci.* **13**, 288–294.

Suzuki, K., Rapin, I., Suzuki, Y., and Ishii, N. (1970). Juvenile GM2-gangliosidosis: Clinical variant of Tay-Sachs disease or a new disease. *Neurology* **20**, 190–204.

Tanaka, A., Ohno, K., and Suzuki, K. (1988). GM2-gangliosidosis B1 variant: A wide geographic and ethnic distribution of the specific β-hexosaminidase α chain mutation originally identified in a Puerto Rican patient. *Biochem. Biophys. Res. Commun.* **156**, 1015–1019.

Tanaka, A., Ohno, K., Sandhoff, K., Maire, I., Kolodny, E. H., Brown, A., and Suzuki, K. (1990). GM2-gangliosidosis-B1 variant—Analysis of β-hexosaminidase α gene abnormalities in 7 patients. *Am. J. Hum. Genet.* **46**, 329–339.

Terry, R. D., and Korey, S. R. (1960). Membranous cytoplasmic granules in infantile amaurotic idiocy. *Nature.* **188**, 1000–1002.

Triggs-Raine, B. L., Akerman, B. R., Clarke, J. T. R., and Gravel, R. A. (1991). Sequence of DNA flanking the exons of the HEXA gene, and identification of mutations in Tay-Sachs disease. *Am. J. Hum. Genet.* **49**, 1041–1054.

Xie, B., McInnes, B., Neote, K., Lamhonwah, A.-M., and Mahuran, D. (1991). Isolation and expression of a full-length cDNA encoding the human GM2 activator protein. *Biochem. Biophys. Res. Commun.* **177**, 1217–1223.

Zerfowski, J., and Sandhoff, K. (1974). Juvenile GM2-Gangliosidose mit veränderter Substrat-Spezifität der Hexosaminidase A. *Acta Neuropathol.* **27**, 225–232.

Zokaeem, G., Bayleran, J., Kaplan, P., Hechtman, P., and Neufeld, E. (1987). A shortened beta-hexosaminidase alpha-chain in an Italian patient with infantile Tay-Sachs disease. *Am. J. Hum. Genet.* **40**, 537–547.

16

Late-Onset G_{M2} Gangliosidosis and Other Hexosaminidase Mutations among Jews

Ruth Navon
Department of Human Genetics and Molecular Medicine
Tel-Aviv University, Sackler School of Medicine
Ramat Aviv, Tel-Aviv 69978, Israel
and
Laboratory of Molecular Genetics
Sapir Medical Center, Kfar Sava
44281, Israel

I. Adult G_{M2} Gangliosidosis
   A. Biochemical Investigations
   B. Pathology Findings
   C. Clinical Evaluation of Hex A-Deficient Adults
   D. The Molecular Basis
   E. Effects of Antidepressant Drugs
   F. Summary of AGG
II. Tay-Sachs Disease among Moroccan Jews
III. Heat-Labile β-Hexosaminidase B and the Genotyping
   of Tay-Sachs Disease
   Acknowledgments
   References

I. ADULT G_{M2} GANGLIOSIDOSIS

Early in 1970 we received our second request for prenatal diagnosis regarding Tay-Sachs disease (TSD) at the genetic clinic of the Sheba Medical Center, Israel. It was just a few months after the announcement by Okada and O'Brien that TSD is caused by absence of β-hexosaminidase (Hex) A, following which we had developed our own heat inactivation method (HIM) for quantifying Hex A using the synthetic substrate 4-MUG (Padeh and Navon, 1971). This couple, who were Ashkenazi Jews, had lost their 2-year-old TSD son, and the woman was 17 weeks pregnant. Even though there was no doubt that these parents were both carriers of TSD, I asked for their blood samples and left their separated leukocytes in the refrigerator. All my efforts were directed toward gaining as much experience as possible with the HIM on cultured amniotic cells, and to establishing the normal range of Hex A levels. Only 6 weeks later, after having found 30% Hex A activity in this fetus's cells, did I remember its parents' stored leukocytes and tested them for Hex A. The mother had the expected intermediate enzymatic levels, but the father (SR) had no Hex A activity at all. I was, of course puzzled, but attributed it to an artifact, probably related to the lability of Hex A and the relatively long "incubation" of the cells at 4°C. I did not pursue this finding until about a year later, when the father's sister (PG) asked to be tested, together with her fiance. By then, I had become quite experienced and confident of the HIM results. PG's leukocytes proved to be devoid of Hex A, and I immediately recalled her brother's puzzling result of the previous year. He agreed to be retested, and his fresh leukocytes were again totally Hex A deficient. This time the identical finding in the two sibs could not be explained away by trivial reasons, and it was obvious that a thorough investigation was called for.

Our geneticist, Avinoam Adam, joined me on a visit to the parents of SR and PG, during which we questioned them about all six of their children. It turned out that PG had been mentally ill since the age of 17. The other four sisters had developed normally. We then discovered that besides our propositus (RS), who had a TSD child, one of his sisters (HS) had two children who had died of TSD. She divorced because of that, remarried, and had three healthy children. Avinoam drew the family pedigree, and within a few minutes he offered an explanation for the findings: "two different alleles are segregating in this family." He suggested that the two sibs who had TSD children were compound heterozygotes for an allele resulting in the classical infantile TSD, and for another allele resulting in an apparent Hex A deficiency, and that the grandparents were heterozygotes, each for a different allele. Avinoam also said that HS's three children must be carriers for one of these mutated alleles, and we should find intermediate levels of Hex A in all three of them. I checked the three generations of that family for Hex A activity. In addition to the propositus RS and PG, HS and another sister had no enzyme, and as predicted, the three healthy children of HS had intermediate levels of Hex A (Figure 16.1). The four Hex A-deficient sibs did not show overt clinical symptoms, except for the mentally ill sister, whose condition seemed at that time irrelevant to her Hex A deficiency.

It took me about a year to convince the skeptical head of our Human Genetics Unit, Baruch Padeh, to publish these observations as "Apparent Deficiency of Hexosaminidase A in Healthy Members of a Family with Tay-Sachs Disease" (Navon et al., 1973). Only several years later did we realize that all four Hex

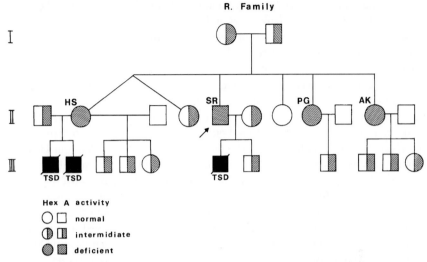

Figure 16.1. The R family. The propositus SR is marked with an arrow.

A-deficient sibs had slowly progressive adult-onset G$_{M2}$ gangliosidosis (AGG) (see below).

A. Biochemical investigations

In the early 1970s I thought of three hypotheses that might resolve the paradoxical finding of TSD-like enzymic deficiency in apparently healthy adults: (1) in the absence of Hex A, these people degrade the G$_{M2}$ ganglioside through some alternative enzymic pathway; (2) they have a variant Hex A that hydrolyzes the natural substrate but not the synthetic MUG used in the HIM; (3) they have minute amounts of Hex A that cannot be detected by the HIM, but are sufficient to degrade the G$_{M2}$ ganglioside, at least to some extent.

Following our 1973 paper, R. Brady of the (U.S.) National Institutes of Health (NIH) offered to collaborate, and soon J. Tallman of his laboratory arrived in Israel with a labeled ^{3}H-G$_{M2}$ ganglioside. We found that our Hex A-deficient adults could degrade the natural substrate-like TSD heterozygous controls (Tallman *et al.*, 1974). These results were compatible with the first two hypotheses, but later proved to be wrong because we had used a detergent in the reaction mixture.

Concentrating on the development of Hex A assays that would be more sensitive than the HIM, I turned to Ruth Arnon, then Head of Immunology at the Weizmann Institute of Science. Together with B. Geiger and Y. Ben Yoseph, we developed antibodies to the α subunit of Hex A and came up with a radioimmunoassay that revealed minute amounts of Hex A in the four Hex A-deficient adults of the R family. Mario Rattazzi, then in Buffalo, New York, found the same when he offered to test the leukocytes and skin fibroblasts of those adults by cellulose acetate electrophoresis. Using these more sensitive methods, all with the synthetic substrate, a very low residual activity of Hex A, 5–10% of normal, was detected in our Hex A-deficient adults (Navon et al., 1976).

Years later it became clear that, in vivo, the interaction between Hex A and the G_{M2} ganglioside substrate is accomplished by a nonenzymatic protein, the G_{M2} activator. Then, in Konrad Sandhoff's laboratory in Bonn, with E. Conzelmann and H.-J. Kytzia, we used radioactive G_{M2} ganglioside together with the G_{M2} activator, and found that the residual activity of Hex A in our Hex A-deficient adults was between 2% and 3.6%, compared with 0.1% in TSD patients (Conzelmann et al., 1983). These results confirmed the findings with the sensitive methods, and differed from our previous result with Tallman. The discrepancy was resolved when it became clear that the use of detergents in the assay leads to incorrect results and must be avoided. At any rate, we proved that the Hex A-deficient adults did have residual Hex A activity that was significantly higher than in TSD patients, but significantly lower than in TSD heterozygotes.

A different approach to studying the degree of hydrolysis of G_{M2} ganglioside was first used by E. Kolodny and colleagues: they fed cultured skin fibroblasts with radiolabeled ganglioside G_{M2} and quantified the gangliosides in the extracted lipids separated on thin-layer chromatography (Raghavan et al., 1985). In 1994, Shimon Gatt in Jerusalem and his co-workers reported introducing a nonradioactive substrate—lissamine rhodamine dodecanoic acid (LR12) sphingomyelin—into skin fibroblasts and defining different diseases by measuring intracellular degradation. I suggested trying the same method on the different G_{M2} gangliosidosis variants, including the AGG patients. I had hoped that by feeding cells with the G_{M1} ganglioside (rather than G_{M2}), we could better estimate the degree of G_{M2} degradation and come up with a way to predict the severity of AGG disease. We used sulforhodamine G_{M1} (SR12-GM1) to feed skin fibroblasts as well as various blood cells including lymphocytes, monocytes, and macrophage cells of different patients and carriers, and found distinctly different rates of degradation by AGG patients and controls (normals and carriers) or TSD patients (Agmon et al., 1996).

In 1984, Elizabeth Neufeld and co-workers showed that fibroblasts from adult Hex A-deficient patients synthesized a defective α subunit that failed to associate with the β subunit (d'Azzo et al., 1984). The same year we showed that the α-subunit precursor was not converted to the mature form in AGG patients (Frisch et al., 1984), a fact that served later as the basis for our strategy to identify the AGG mutation.

B. Pathology findings

When HS of the R family had to undergo hysterectomy, I asked the surgeon to secure a piece of her tissue for electron microscopy. Since he routinely also

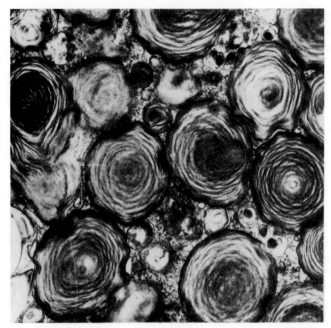

Figure 16.2. Membranous cytoplasmatic bodies (MCB) found in neurons of the myenteric plexus of HS's appendix.

performed appendectomy in such cases, we sent over this tissue to pathologist Uri Sandbank at Beilinson Hospital. The finding of extensive neuronal storage bodies (Figure 16.2) was the final proof that we were dealing with a disease. I still remember how impressed R. Desnick was when, on a visit to Israel in 1978, I showed him the electron micrographs with the storage bodies, which were practically identical to those of a TSD brain.

Sandbank's electron microscopy, along with a variety of other tests, later enabled us to diagnose AGG for the first (and probably the only) time in a fetus (Navon *et al.*, 1986b).

C. Clinical evaluation of Hex A-deficient adults

Meticulous neurological examinations of the four Hex A-deficient sibs of the R family by Z. Argov of Jerusalem revealed that, contrary to the title of our 1973 paper, all of them suffered from a neurologic disorder (Navon *et al.*, 1981a). Over the years we found additional unrelated AGG families in Israel and abroad, and other investigators, mostly from the United States, reported more Ashkenazim as well as non-Jewish cases (referenced in Gravel *et al.*, 1995). The first report was by Rapin *et al.* (1976) on two adult Ashkenazi siblings with progressive deterioration of gait as well as dysarthria. It is now well established that AGG is manifested by an extremely variable clinical spectrum that includes psychosis (Argov and Navon, 1984; Navon *et al.*, 1986a). The great clinical variability in AGG patients (see Section I.F) is not yet understood, and presents serious problems for genetic counseling. However, it is generally agreed that the severity of the different G_{M2} gangliosidosis variants correlates with the extent of accumulation of the undegraded G_{M2} ganglioside, which depends on the activity of Hex A. It seems that both the formation and effectiveness of a marginal amount of Hex A can be influenced by additional factors.

D. The molecular basis

With the help of R. Proia, who introduced me to the field of molecular genetics at the NIH, we set out to identify the molecular defect in AGG. We obtained skin biopsies and blood samples from the affected and nonaffected members of our R family, and concluded that the AGG members were compound heterozygotes for an infantile TSD allele and for another allele. We assumed that the latter allele must be responsible for the AGG phenotype, and that it produced the precursor of the α-chain polypeptide of Hex A. When I sequenced the entire α chain cDNA of the propositus RS, I found only one change: a single nucleotide substitution of G to A, which changed the amino acid glycine (GGT) to serine (AGT) (Navon and Proia, 1989). The same was found almost at the same time in E. Neufeld's laboratory (Paw *et al.*, 1989). The substitution is in the last base of exon 7 of the *HEXA* gene, and must be responsible for the presence, albeit at a low level, of Hex A found in AGG patients. Testing the mutation by allele-specific oligonucleotide (ASO) hybridization in genomic DNA of the R family revealed that the mutation was present heterozygously, as expected.

This finding raised a number of questions: Is the Gly269Ser mutation indeed responsible for the AGG phenotype? Do all Ashkenazi patients carry this mutation? Are all Ashkenazi patients compound heterozygotes with an Ashkenazi infantile mutation? Is the Gly269Ser mutation also responsible for AGG in non-Jews? Do individuals who are homozygous for the Gly269Ser mutation manifest clinical symptoms?

Examination of more Ashkenazi AGG patients by us and others showed that all of them had the Gly269Ser mutation and all were compound heterozygotes, each carrying one of the two Ashkenazi TSD alleles known at the time (see Chapter 12). As far as I know, there is only one Ashkenazi AGG patient who does not carry one of the two Ashkenazi TSD alleles, an 80-year-old woman with only a Gly269Ser mutation and only a severe tremor in her hands. Unfortunately, our efforts to identify her second mutation have not been successful.

Samples from six non-Jewish patients supplied by H. Mitsumuto from Cleveland, and by E. Kolodny, now in New York, enabled us to determine that all non-Jewish AGG patients carry the same Gly269Ser mutations. We also found that two of the non-Jewish affected siblings were homozygotes for this mutation, and the rest were compound heterozygotes with one of the infantile Ashkenazi mutations. When I told E. Kolodny about the finding of two homozygotes, he was not at all surprised: "I know these siblings for some time—one is a 53-year-old policeman who drives a motorcycle and his brother is 61, and their clinical symptoms are so mild they cannot be compared to other AGG patients I know." Thus, we could conclude that the Gly269Ser mutation is the basis of AGG among both Ashkenazim and non-Jews (Navon *et al.*, 1990). However, Kolodny and co-workers later described an AGG patient of Ashkenazi origin who was negative for the Gly269Ser mutation (De Gasperi *et al.*, 1995).

E. Effects of antidepressant drugs

Psychotic AGG patients—like other psychotics—are often treated with antide-pressant drugs, most of them tricyclic amine compounds and lysosomotropic in structure. Seeing such pills at the bedsides of two demented AGG patients led me to suspect that these antidepressant drugs, with a common amphiphilic structure, exacerbate their disease. We hypothesized that these antidepressants deplete the cells of psychotic AGG patients of their already minute amount of Hex A, which probably accelerates the accumulation of the G$_{M2}$ gangliosisde in their neurons and causes further deterioration of their condition. We confirmed this through *in vitro* experiments: when imipramine (10^{-4} M) was added to normal cultured fibroblasts, the cells lost 50% of their initial Hex A activity within 24 h. The effect was reversible: removing the imipramine brought the cells back to their normal Hex A levels (Navon and Baram, 1987; Palmeri *et al.*, 1992). The effect of drugs on lysosomal enzymes, documented before, is probably analogous to that of NH$_4$Cl and chloroquine (Lullman-Rauch *et al.*, 1978). Logically, the conse-quence of antidepressant treatment of AGG patients would be reduction in their limited ability to degrade the accumulating G$_{M2}$ ganglioside. In recent years it has become increasingly accepted that psychotic AGG patients have more severe neurologic manifestations than nonpsychotic ones. Reports that their condition improved when they were taken off the drugs corroborate our *in vitro* finding (Rosbush *et al.*, 1995, and references therein).

F. Summary of AGG

AGG is a slowly progressing neurodegenerative disorder. The patients have very low Hex A activity and usually carry the Gly269Ser mutation in compound het-erozygosity with an infantile TSD mutation. Homozygotes for the Gly269Ser mutation are only mildly affected.

    The age of onset is heterogenous and very difficult to determine, because the disorder progresses very slowly and patients are usually diagnosed at a relatively advanced stage of the disease, usually around age 20.

Neurologic manifestations: The symptoms and severity are remarkably variable, even among siblings, but in all cases there is involvement of one or another part of the central nervous system. The neurologic manifestations may include proximal muscle weakness, muscle cramping and wasting, lack of coordination, unsteady gait, ataxia, hand tremors, dysarthria, and dystonia and/or dyskinesia. Sensory impairment, chorea, and seizures are less common. Some patients were first misdiagnosed with Friedrich ataxia, Kugelberg-Welander syndrome disease, or amyotrophic lateral sclerosis.

Psychiatric abnormalities seem to be an integral part of AGG and often precede the neurologic manifestations. They include acute hebephrenic schizophrenia, agitation, delusions, hallucinations, paranoia, and recurrent depression. Disorganization of thoughts is a common sign, but dementia is generally not a prominent feature.

Incidence: About 4% of Ashkenazi and 3% of the non-Jewish people diagnosed enzymatically as carriers of TSD are in fact carriers of the AGG G269S mutation (Paw et al., 1990). This means that about 1 in every 750 Ashkenazim is a heterozygote for this mutation. So far no cases of AGG have been found in non-Ashkenazi Jews.

Treatment: To date there is no treatment for most AGG patients. Only the psychiatric patients can be treated to control or alleviate their suffering. For the prospects of enzyme or gene therapy, see Chapters 18 and 25 in this volume.

II. TAY-SACHS DISEASE AMONG MOROCCAN JEWS

S. Moses, head of pediatrics at Soroka Medical Center in Be'er Sheva, Israel, told me in 1974 about what appeared to be a classical case of TSD that he had diagnosed in a Moroccan Jewish infant. This was quite a surprise, since TSD was seen in those years exclusively among Ashkenazim (we were both unaware then of reports from the 1950s of a few non-Ashkenazi cases, including Moroccan Jews). Moses learned that two sibs in that family had died of a similar disease. When the HIM test showed no Hex A in this infant's blood, and typical intermediate levels in his parents, we were satisfied that this was indeed a genuine classical TSD. However, at the suggestion of Maimon Cohen (now in the United States), G. Bach and M. Zeigler of the genetics department at Hadassah Hospital in Jerusalem conducted biochemical tests comparing Ashkenazi TSD cases and Moses' TSD child, and found some unusual kinetic and electrophoretic features of Hex A in his family. These findings led us to suggest that TSD in Moroccan Jews might be caused by a different mutation than that causing the disease in Ashkenazim (Bach et al., 1976), which was then assumed to be a single mutation. Bach's group later noted that TSD was not as rare among Moroccan Jews as was previously thought, and in fact, during my 15 years' experience with the Israeli TSD screening prevention program I encountered seven Moroccan TSD families.

Rachel Myerowitz's surprising finding in 1987 of more than one TSD mutation among Ashkenazim, reports of molecular heterogeneity of mutations in a variety of other Mendelian disorders, and the observation that different mutations appear often in different populations, naturally raised the questions: Does such a heterogeneity existed among Moroccan Jews? Does this community share any of the Ashkenazi mutations? Answers to these questions were especially pertinent to controversies over issues such as the genetic affinities between the diverse Jewish communities, and the relative roles of drift versus selection for TSD heterozygotes as factors responsible for the relatively high frequencies of the disease in some populations.

I prepared DNA samples from eight Moroccan obligate heterozygotes and one TSD child in Israel, and a skin biopsy from that child, and brought these materials to the NIH for investigation. We soon learned that all nine Moroccan samples were negative for the two Ashkenazi mutations! The TSD infant proved to be a compound heterozygote for two different mutations, a 3-bp deletion resulting in a deletion of phenylalanine (ΔF), and an unknown mutation (Navon and Proia, 1991). By 1991, having continued the search in Israel, we could account in molecular terms for all our obligate Moroccan carriers, who had between them ΔF and two other mutations identified by one of my students (Drucker et al., 1992).

When we analyzed samples of Moroccan carriers, who were defined enzymatically as such by the national screening program in Israel (i.e., with no known cases of TSD in their families), we were surprised to find that 10 of 44 such carriers were negative for all known Moroccan and Ashkenazi mutations. We suspected that they might carry one or two of pseudo-deficiency mutations that were found mainly in non-Jews. However, another student in my laboratory proved that this was not the case, and instead she identified two mutations among seven of them. One is a substitution E482K identified previously in one Italian case and one Japanese TSD case, whereas, the other is a novel IVS5-2(a→g) mutation, resulting in a skipping of the whole of exon 6 of the *HEXA* gene—in all probability a disease allele (Kaufman et al., 1997).

To date, it is estimated that 1 in 140 Moroccans is a TSD carrier, and we have identified seven different TSD mutations in this community. Table 16.1 summarizes their distribution among 44 unrelated enzyme-defined carriers and 10 unrelated representatives of eight TSD families.

Thus, although the Moroccan Jewish population of approx. 750,000 is much smaller than the Ashkenazi population, it harbors more identified TSD mutations. A possible explanation could be their much longer history, which probably includes traces of proselytes from before the Christian era as well as waves of immigrations, the latest of which occurred with the expulsion of Jews from Iberia at the end of the fifteenth century. The finding of Ashkenazi mutations in two

Table 16.1. HEXA Mutations among Moroccan Jews

Mutations	Carriers		Remarks
	Enzyme-defined	Obligate	
$\Delta F^{304/305}$	17	5	Found also in France among non-Jews
R170Q	17	2	Reported once in a Japanese TSD case
IVS5-2(a→g)	6	—	New
Y180X	—	1	The only case reported
E482K	1	—	Reported also in Italian and Japanese TSD cases
1278+TATC	—	1	Common in Ashkenazim
IVS12+1(g→c)	—	1	Common in Ashkenazim
Unidentified	3	—	
Totals	44	10	

of the Moroccan carriers cannot be taken yet as evidence for a common origin. Further investigations, such as the characterization of intragenic and flanking haplotypes, may clarify this point.

III. HEAT-LABILE β-HEXOSAMINIDASE B AND THE GENOTYPING OF TAY-SACHS DISEASE

The enzyme β-hexosaminidase (Hex) B does not degrade the G_{M2} ganglioside in humans, and therefore has no direct involvement in TSD. However, it is a major participant in the routine enzymatic assay for identification of TSD genotypes because Hex B, like Hex A, hydrolyzes the synthetic substrate 4-MUG, and, unlike Hex A, it is heat stabile. These two properties form the basis of the widely used HIM.

The presence of a rare, mutant, heat-labile Hex B (HLB) would undermine the HIM and lead to erroneous conclusions regarding the proportions of Hex A. The recognition (Navon et al., 1981b, 1985) and subsequent identification of a harmless HLB mutation in our laboratory (Ginat et al., 1995), as well as the finding that HLB is not extremely rare among some Jewish communities (Navon and Adam, 1990), were the late outcomes of an investigation that began in 1979 of Israeli–Arab first cousins whose child was hospitalized at Beilinson Hospital for several weeks with an undiagnosed neurodegenerative disorder. The "exclusively Jewish" TSD had not been seriously considered in an Arab patient until pediatrician-geneticist Sara Kaffe (now in New York) was asked to consult on this case. The child's early normal development and the red spot in her retina led S. Kaffe to suggest TSD as the most probable diagnosis.

Hex A assays performed on this family in our laboratory at the Sheba Medical Center turned up some amazing results. The proposita had no Hex A and the mother had the typical carrier level, thus establishing S. Kaffe's TSD diagnosis. However, the Hex A level of the mother's father (who was also the uncle of the child's father) was above normal! When we finally persuaded the child's father to be examined, we found that instead of the intermediate Hex A level expected in a parent of a TSD patient, he too had a "supernormal" Hex A level. Whereas some colleagues questioned his paternity, I was convinced that we were faced with a genetic variant of Hex. Indeed, when we separated Hex B from Hex A, we found that the Hex B of these two men was heat labile. We developed a method to identify carriers of both TSD and HLB (Navon et al., 1981b), and eventually discovered that the HLB in this Arab family must be caused by a harmless Hex B mutation, since we found two healthy HLB homozygotes in the same family (Navon et al., 1985).

After the findings of TSD among Moroccan Jews, the national TSD screening program was offered to all Israelis. When one of the first HIM-tested non-Ashkenazim was an Iraqi Jew with supernormal Hex A, who proved to be an HLB carrier, we became alarmed that perhaps some TSD carriers might have been missed and were considered normal because they were also HLB carriers.

We checked results of suspiciously high Hex A levels during the previous 5 years and found two HLB carriers among some 38,000 Ashkenazim, and no HLB among a few hundred Moroccans. A check of some 2600 non-Ashkenazi blood donors including an additional 1500 Moroccan and other North African Jews turned up no HLB. On the other hand, there were 18 HLB carriers (about 1.7%) out of more than 1000 Jews originating from Iraq, Iran, Turkey, Syria, and Lebanon, where TSD is extremely rare or nonexistent (Navon and Adam, 1990).

A student in my laboratory recently characterized the molecular basis of HLB in our populations, identifying a 1627 G→A transition in exon 14 of the Hex B gene that causes a 543 Ala→Thr substitution in the β subunit of Hex. She demonstrated complete correspondence between the presence of this mutation and the enzymatic genotyping of HLB in the original Arab family. The same mutation was found in 9 of 10 examined Ashkenazi and non-Ashkenazi HLB carriers. Proia and Pennybacker then proved at the NIH that this mutation indeed causes HLB (Ginat et al., 1997).

At the present time, everyone in the screening laboratories in Israel and abroad is cognizant of HLB. In the Israeli TSD program, they now find more than 100 HLB Jewish carriers every year (Peleg and Goldman, personal communication). Thus, the small risk of misleading HIM diagnoses of TSD genotypes because of HLB has been practically eliminated.

Acknowledgments

I would like to acknowledge all the physicians and the investigators in the different laboratories, in Israel and around the world, the patients and their families, and especially Avinoam Adam, who went with me all this long way since 1970. Without the dedication and hard work of every one of them I would not be able to make this contribution.

References

Agmon, V., Khosravi, R., Marchesini, S., Dinur, T., Dagan, A., Gatt, S., and Navon, R. (1996). Intracellular degradation of sulforhodamine-G_{M1}: Use for a fluorescence-based characterization of G_{M2} gangliosidosis variants in fibroblasts and white blood cells. *Clin. Chim. Acta* **247,** 105–120.

Argov, Z., and Navon, R. (1984). Clinical and genetic variations in the syndrome of GM2 gangliosidosis due to hexosaminidase A deficiency. *Ann. Neurol.* **16,** 14–20.

Bach, G., Navon, R., Zeigler, M., Legth, Y., Porter, E., and Cohen, M. M. (1976). Tay-Sachs disease in a Moroccan Jewish family. A possible new mutation. *Israel J. Med. Sci.* **12,** 1432–1439.

Conzelmann, E., Kytzia, H. J., Navon, R., and Sandhoff, K. (1983). Ganglioside G_{M2} N-acetyl-β-D-galactosaminidase activity in cultured fibroblasts of late infantile and adult G_{M2} gangliosidosis and of healthy probands with low hexosaminidase level. *Am. J. Hum. Genet.* **35,** 900–913.

d'Azzo, A., Proia, R. L., Kolodny, E. H., Kaback, M. M., and Neufeld, E. F. (1984). Faulty association of the α- and β-subunits in some forms of β-hexosaminidase deficiency. *J. Biol. Chem.* **259,** 11070–11074.

De Gasperi, R., Gama Sosa, M. A., Battistini, S., Yeretsian, J., and Kolodny, E. H. (1995). Late onset G_{M2} gangliosidosis in two siblings of Ashkenazi Jewish ancestry results from mutation in the HEXA gene causing abnormal thermolability of hexosaminidase A. *Am. J. Hum. Genet.* **57,** A238.

Drucker, L., Proia, R. L., and Navon, R. (1992). Identification and rapid detection of three Tay-Sachs mutations in the Moroccan Jewish population. *Am. J. Hum. Genet.* **51,** 371–377.

Frisch, A., Baram, D., and Navon, R. (1984). Hexosaminidase A deficient adult: Presence of α chain precursor in cultured skin fibroblasts. *Biochem. Biophys. Res. Commum.* **119,** 101–107.

Ginat, N., Adam, A., Jaber, L., Pennybacker, M., Proia, R. L., and Navon, R. (1997). Molecular basis of heat labile hexosaminidase B among Jews and Arabs. *Hum. Mutat.* **10,** 424–429.

Gravel, R. A., Clark, J. T. R., Kaback, M. M., Mahuran, D. J., Sandhoff, K., and Suzuki, K. (1995). The GM2 gangliosidoses. *In* "The Metabolic and Molecular Bases of Inherited Disease." (C. R. Scriver, A. L. Beaudet, W. S., Sly, and D. Valle, eds.), pp. 2839–2879. McGraw-Hill, New York.

Kaufman, M., Grinshpun-Cohen, J., Kapati, M., Peleg, L., Goldman, B., Akstein, E., Adam, A., and Navon, R. (1997). Mutations involved in Tay-Sachs disease among Moroccan Jews. *Hum. Mutat.* **10,** 295–300.

Lullman-Rauch, H., Lullman-Rauch, R., and Wassermann, D. (1978). Lipidosis induced by amphiphilic cationic drugs. *Biochem. Pharmacol.* **27,** 1103–1108.

Navon, R., and Adam, A. (1990). Thermolabile hexosaminidase (Hex) B: Diverse frequencies among Jewish communities and implications for screening of sera for Hex A deficiency. *Hum. Hered.* **40,** 99–104.

Navon, R., and Baram, D. (1987). Depletion of cellular β-hexosaminidase by imipramine is prevented by dexamethasone: Implications for treating psychotic hexosaminidase-A deficient patients. *Biochem. Biophys. Res. Commun.* **148,** 1098–1103.

Navon, R., and Proia, R. L. (1989). The mutations in Ashkenazi Jews with adult GM2 gangliosidosis, the adult form of Tay-Sachs disease. *Science* **243,** 1471–1474.

Navon, R., and Proia, R. L. (1991). Tay-Sachs disease in Moroccan Jews: Deletion of a phenylalanine in the α subunit of β-hexosaminidase. *Am. J. Hum. Genet.* **48,** 412–419.

Navon, R., Padeh, B., and Adam, A. (1973). Apparent deficiency of hexosaminidase A in healthy members of a family with Tay-Sachs disease. *Am. J. Hum. Genet.* **25,** 287–292.

Navon, R., Geiger, B., Ben-Yoseph, Y., and Rattazzi, M. C. (1976). Low levels of β-hexosaminidase A in healthy individuals with apparent deficiency of this enzyme. *Am. J. Hum. Genet.* **28,** 339–349.

Navon, R., Argov, Z., Brandt, N., and Sandbank, U. (1981a). Adult G_{M2} gangliosidosis in association with Tay-Sachs disease: A new phenotype. *Neurology* **31,** 1397–1401.

Navon, R., Nutman, J., Kopel, R., Gaber, L., Gadoth, N., Goldman, B., and Nitzan, M. (1981b). Hereditary heat labile hexosaminidase B: Its implication of recognizing Tay-Sachs genotypes. *Am. J. Hum. Genet.* **33,** 907–915.

Navon, R., Kopel, R., Nutman, J., Frisch, A., Conzelmann, E., Sandhoff, K., and Adam, A. (1985). Hereditary heat-labile hexosaminidase B: A variant whose homozygotes synthesize a functional Hex A. *Am. J. Hum. Genet.* **37,** 138–146.

Navon, R., Frisch, A., and Argov, Z. (1986a). Hexosaminidase A deficiency in adults. *J. Med. Genet.* **24,** 179–196.

Navon, R., Sandbank, U., Frisch, A., Baram, D., and Adam, A. (1986b). Adult onset GM2 gangliosidosis diagnosed in a fetus. *Prenatal Diag.* **6,** 169–176.

Navon, R., Kolodny, E. H., Mitsumoto, H., Thomas, G. H., and Proia, R. L. (1990). Ashkenazi Jewish and non-Jewish adult GM2 gangliosidosis patients share a common genetic defect. *Am. J. Hum. Genet.* **46,** 817–821.

Padeh, B., and Navon, R. (1971). Diagnosis of Tay-Sachs disease by hexosaminidase activity in leukocytes and amniotic fluid cells. *Israel J. Med. Sci.* **7,** 259–263.

Palmeri, S., Mangano, L., Battisti, C., Malandrini, A., and Federico, A. (1992). Imipramine induced lipidosis and dexamethasone effect: Morphological and biochemical study in normal and chronic GM2 gangliosidosis fibroblasts. *J. Neurol. Sci.* **110,** 215–221.

Paw, B. H., Kaback, M. M., and Neufeld, E. F. (1989). Molecular basis of adult-onset and chronic GM2 gangliosidosis in patients of Ashkenazi Jewish origin: Substitution of serine for glycine at position 269 of the α subunit of β-hexosaminidase. *Proc. Natl. Acad. Sci. (USA)* **86,** 2413–2417.

Paw, B. H., Tieu, P. T., Kaback, M. M., Lim, J., and Neufeld, E. F. (1990). Frequency of three Hex A mutant alleles among Jewish and non-Jewish carriers identified in a Tay-Sachs screening program. *Am. J. Hum. Genet.* **47,** 698–705.

Proia, R. L., Kolodny, E. H., and Navon, R. (1990). Hexosaminidase pseudodeficiency. *Am. J. Hum. Genet.* **47,** 880–881.

Raghavan, S., Krusell, A., Krusell, J., Lyerla, T. A., and Kolodny, E. H. (1985). G$_{M2}$-ganglioside metabolism in hexosaminidase A deficiency states: Determination in situ using labeled G$_{M2}$ added to fibroblast cultures. *Am. J. Hum. Genet.* **37,** 1071–1082.

Rapin, I., Suzuki, K., Suzuki, K., and Valsamis, M. P. (1976). Adult (chronic) G$_{M2}$ gangliosidosis. A typical spinocerebellar degeneration in a Jewish sibship. *Arch. Neurol.* **33,** 120–130.

Rosbush, P. I., MacQueen, G. M., Clarke, J. T., Callahan, J. W., Strasberg, P. M., and Mazurek, M. F. (1995). Late-onset Tay-Sachs disease presenting as catatonic schizophrenia: Diagnostic and treatment issues. *J. Clin. Psychiatry* **56,** 347–353.

Tallman, J. F., Brady, R. O., Navon, R., and Padeh, B. (1974). Ganglioside catabolism in hexosaminidase A deficient adults. *Nature* **252,** 254–255.

17

Naturally Occurring Mutations in G_{M2} Gangliosidosis: A Compendium

Barbara Triggs-Raine*
Department of Biochemistry and Medical Genetics
University of Manitoba
Winnipeg, Canada R3E 0W3

Don J. Mahuran and Roy A. Gravel
The Research Institute and the Department of Laboratory Medicine
and Pathobiology
University of Toronto, The Hospital for Sick Children
Toronto, Canada M5G 1X8

I. Introduction
II. β-Hexosaminidase A Mutations
   A. Genotype/Phenotype Relationships
   B. Epidemiology of Tay-Sachs Disease
III. β-Hexosaminidase B Mutations
   A. Genotype/Phenotype Relationships
   B. Epidemiology of Sandhoff Disease
IV. GM2A Mutations
V. Structure/Function Relationships of β-Hexosaminidase
   References

I. INTRODUCTION

The molecular basis of the G_{M2} gangliosidoses has been extensively investigated over the past decade. The characterization of cDNA clones (Myerowitz and Proia,

* To whom correspondence should be addressed. E-mail: traine@ms.umanitoba.ca. Fax: (204) 789-3900. Telephone: (204) 789-3218.

1984; Myerowitz *et al.*, 1985; O'Dowd *et al.*, 1985; Korneluk *et al.*, 1986) and subsequently the genes (Proia, 1988; Proia and Soravia, 1987; Neote *et al.*, 1988) encoding the subunits of β-hexosaminidase (Hex) set the stage for delineating the mutation basis of these disorders. Early efforts focused on the French Canadian and Ashkenazi Jewish populations because of the elevated incidence of Tay-Sachs disease in these populations. Guided by elegant biochemical and genetic studies, these analyses rapidly progressed to other populations and individuals with variant forms of both Tay-Sachs and Sandhoff diseases. The more recent cloning of the cDNA and gene encoding the G_{M2} activator (Schroder *et al.*, 1989; Xie *et al.*, 1991; Klima *et al.*, 1991; Nagarajan *et al.*, 1992) has also opened this disease to mutation analysis.

   All three disorders result from a functional deficiency of Hex A activity, required for the lysosomal degradation of G_{M2} ganglioside principally in neuronal tissues. They are caused by mutations in both alleles of any one of three genes. Two of the genes, *HEXA* and *HEXB*, encode the α and β subunits, respectively, of Hex A (structure, αβ) and Hex B (ββ). The third gene, GM2A, encodes the G_{M2} activator. Only Hex A, acting on the G_{M2} ganglioside–G_{M2} activator complex, can hydrolyze the terminal N-acetylgalactosamine of G_{M2} ganglioside. Mutations in the *HEXA* gene, leading to deficiency of Hex A activity, cause Tay-Sachs disease or less severely affected variants. Mutations in the *HEXB* gene, leading to deficiency of both Hex A and Hex B activities, cause Sandhoff disease or its variants. Finally, mutations in the GM2A gene, leading to defects of the G_{M2} activator while retaining functional Hex A and Hex B, result in the AB variant of G_{M2} gangliosidosis. The nomenclature used in this chapter is from Gravel *et al.* (1995), which retains the names Tay-Sachs and Sandhoff diseases and G_{M2} activator deficiency to refer to infantile acute forms of G_{M2} gangliosidosis due to *HEXA*, *HEXB*, and GM2A mutations, respectively. Milder clinical variants are designated according to the severity of the disease. These include subacute (juvenile-onset) and chronic (adult-onset) forms.

   Before the cloning of the *HEXA* and *HEXB* genes, genetic and biochemical evidence indicated that Hex A and Hex B enzyme deficiencies had diverse molecular origins. A wide spectrum of G_{M2} gangliosidosis phenotypes was evident, suggesting that numerous *HEXA* and *HEXB* alleles would be identified. Allelic heterogeneity was also implied by the observation of unique biochemical phenotypes among some patients. The most dramatic is illustrated by the severely affected, B1 variant of Tay-Sachs disease. Affected patients were found to have substantial Hex A activity when the enzyme was assayed with a synthetic substrate (4-methylumbelliferyl-β-D-N-acetyl-glucosaminide, 4-MUG), while little or no activity was detected using a sulfated derivative (4-MUG-sulfate, 4-MUGS) or G_{M2} ganglioside (Kytzia and Sandhoff, 1985). In a contrasting example, asymptomatic individuals were found who had deficient Hex A activity when assayed with synthetic substrates but who had sufficient activity toward G_{M2} ganglioside (Hex A pseudo-deficiency or Hex A-minus normal) (Vidgoff *et al.*, 1973;

Kelly *et al.*, 1976; Thomas *et al.*, 1982; Grebner *et al.*, 1986; O'Brien *et al.*, 1978; Navon *et al.*, 1986; Raghavan *et al.*, 1985b). In most cases, the activity of all three substrates would be expected to behave similarly. Significantly, only about 10% of residual Hex A activity appears to be required to remain healthy; all clinical phenotypes result from levels of Hex A activity that fall between zero and 5% of normal (Conzelmann *et al.*, 1983; Raghavan *et al.*, 1985a; Leinekugel *et al.*, 1992). Within this narrow range, disease severity has been found to correlate inversely with the capacity of residual Hex A activity in patient fibroblasts to cleave G$_{M2}$ ganglioside. Many other studies have described genetic and biochemical heterogeneity that today has been largely accounted for through the discovery of allelic diversity.

II. β-HEXOSAMINIDASE A MUTATIONS

A *HEXA* locus mutation database has been established by F. Kaplan, P. M. Nowacki, and P. Hechtman, McGill University. It can be found at http://www.debelle.mcgill.ca/hexa/hexadb.htm.

To date, 74 mutations in *HEXA* have been identified responsible for Tay-Sachs disease or one of its variants (Table 17.1; see also Gravel *et al.* [1995] for details of biochemical phenotypes). These mutations are diverse in nature. They include one large and several small deletions, one single and one multiple nucleotide insertion, many missense and nonsense mutations, and several mutations in or nearby intron/exon junctions that lead to abnormal splicing. These mutations are scattered throughout the gene (Figure 17.1), with a large cluster of mutations in exon 5.

In addition to the deleterious *HEXA* mutations, polymorphisms with no apparent effect on Hex A activity have also been identified (Table 17.2). In some cases, amino acid substitutions have been identified in enzyme-defined carriers of Tay-Sachs disease, but their affect on Hex A activity remains unknown (Table 17.2). Future analyses of these mutations in Tay-Sachs disease patients, their expression in *ex vivo* systems, and/or interpretations from model enzyme structures will be required to assess the effect of such changes on Hex A activity.

A. Genotype/phenotype relationships

Distinct correlations between genotype and phenotype can be made for the majority of the *HEXA* mutations that have been identified. Such correlations are straightforward if a mutation is common and therefore found in the homozygous state or in compound heterozygosity with other alleles clearly associated with an acute phenotype. In contrast, the determination of phenotypes associated with uncommon mutations is more difficult if the patient has a delayed-onset form of disease.

Table 17.1. HEXA Mutations in Tay-Sachs Disease and Its Variants

Mutation	Location	Expected result	Phenotype	Origin	References
del7.6kb	5' of gene to IVS-1	No mRNA	Acute	French Canadian	Myerowitz and Hogikyan, 1986; Myerowitz and Hogikyan, 1987; De Braekeleer et al., 1992
1A→G	Exon 1	Met1Val	Acute	Black American	Mules et al., 1992b
2T→C	Exon 1	Met1Thr	Severe subacute, second allele Pro25Ser	English	Harmon et al., 1993
73C→T	Exon 1	Pro25Ser	Severe subacute, second allele Met1Thr	English	Harmon et al., 1993
78G→A	Exon 1	Trp26Stop	Acute	English; Israeli Arab	Triggs-Raine et al., 1991; Drucker and Navon, 1993
116T→G	Exon 1	Leu39Arg	Acute	Polish	Akli et al., 1993a
346+1G→C	IVS-2	Abnormal splicing	Acute	German/Irish/English	Triggs-Raine et al., 1991
346+1G→A	IVS-2	Abnormal splicing	Acute	French	Akli et al., 1991
380T→G	Exon 3	Leu127Arg	Acute	Italian	Akli et al., 1993a
409C→T	Exon 3	Arg137Stop	Acute	French; French Canadian; Irish	Akli et al., 1991; Mules et al., 1992b
412+1G→T	IVS-3	Abnormal splicing	Acute	Japanese	Tanaka et al., 1994
423delTT	Exon 4	Frameshift	Acute	French	Akli et al., 1993a
436delG	Exon 4	Frameshift	Acute	Black American	Mules et al., 1992b
459+5G→A	IVS-4	Abnormal splicing	Acute	French	Akli et al., 1991
460-1G→T	IVS-4	Abnormal splicing	Acute	Black American	Mules et al., 1991
477delTG	Exon 5	Frameshift	Acute	English	Triggs-Raine et al., 1991
496C→G	Exon 5	Arg166Gly	Severe subacute, second allele 498delC	Syrian	Peleg et al., 1995
498delC	Exon 5	Frameshift	Severe subacute due to second allele Arg166Gly	Syrian	Peleg et al., 1995

508C→T	Exon 5	Arg170Trp	Acute	French Canadian; Italian	Fernandes et al., 1992; Akli et al., 1993a; Triggs-Raine et al., 1991; Triggs-Raine et al., 1995
509G→A	Exon 5	Arg170Gln	Acute	Japanese; Scottish; Moroccan Jew	Nakano et al., 1990; Drucker et al., 1992; Akli et al., 1993a
532C→T	Exon 5	Arg178Cys	Acute (B1 variant)	Czechoslovakian	Tanaka et al., 1990a
533G→A	Exon 5	Arg178His	Severe subacute or subacute (B1 variant)	Portuguese; diverse	Kytzia et al., 1983; Ohno and Suzuki, 1988a; Tanaka et al., 1988, 1990a; Dos Santos et al., 1991
533G→T	Exon 5	Arg178Leu	Acute (B1 variant)	English	Triggs-Raine et al., 1991
538T→C	Exon 5	Tyr180His	Chronic	Ashkenazi Jewish	De Gasperi et al., 1996
540C→G	Exon 5	Tyr180Stop	Acute	Moroccan Jewish	Drucker et al., 1992
547insA	Exon 5	Frameshift	Acute	Chinese	Akalin et al., 1992
570G→A	Exon 5	Abnormal splicing	Severe subacute	Tunisian	Akli et al., 1990
570+1G→ A	IVS-5	Abnormal splicing	Acute	Turkish	Ozkara et al., 1995
571-1G→T	IVS-5	Abnormal splicing	Acute	Japanese	Tanaka et al., 1993
574G→C	Exon 6	Val192Leu	Acute	German/Romanian	Gordon et al., 1988; Ainsworth and Coulter-Mackie, 1992; Hou et al., 1996
598G→A		Val200Met	Acute		
590A→C	Exon 6	Lys197Thr	Chronic, second allele Arg499His	Dutch	Akli et al., 1993a
611A→G	Exon 6	His204Arg	Acute	German	Akli et al., 1993a
629C→T	Exon 6	Ser210Phe	Acute	Algerian; North African	Akli et al., 1991; Poenaru and Akli, 1994
632T→C	Exon 6	Phe211Ser	Acute	Italian	Akli et al., 1993a
672+1G→A	IVS-6	Abnormal splicing	Subacute/chronic due to second unknown allele?	American	Akli et al., 1993a

(continues)

Table 17.1. (continued)

Mutation	Location	Expected result	Phenotype	Origin	References
739C→T	Exon 7	Arg247Trp	Pseudo-deficiency	Diverse	Triggs-Raine et al., 1992, 1995; Mules et al., 1992a; Kaback et al., 1993; Tomczak et al., 1993;
745C→T	Exon 7	Arg249Trp	Pseudo-deficiency	Diverse	Cao et al., 1993; Kaback et al., 1993; Triggs-Raine et al., 1991, 1995
749G→A	Exon 7	Gly250Asp	Subacute	Lebanese	Hechtman et al., 1989; Trop et al., 1992
755G→A	Exon 7	Arg252His	Chronic, second allele Arg178His	Portuguese	Ribeiro et al., 1996
772G→C	Exon 7	Asp258His	Acute (B1 variant)	Scottish-Irish	Bayleran et al., 1987; Fernandes et al., 1992a
805G→A	Exon 7	Gly269Ser	Chronic	Diverse	d'Azzo et al., 1984; Navon and Proia, 1989; Paw et al., 1989; Navon et al., 1990
805+1G→A	IVS-7	Abnormal splicing	Acute	French Canadian	Hechtman et al., 1992; Triggs-Raine et al., 1991, 1995
805+1G→C	IVS-7	Abnormal splicing	Acute	Portuguese	Ribeiro et al., 1995
902T→G	Exon 8	Met301Arg	Acute	Yugoslav	Akli et al., 1993a
910del3	Exon 8	ΔPhe304	Acute	Moroccan Jewish; French; Italian/Portuguese	Akli et al., 1991; Drucker et al., 1992; Navon and Proia, 1991; Poenaru and Akli, 1994
927delCT	Exon 8	Frameshift	Acute	German Scandinavian	Fernandes et al., 1992a
958del3	Exon 8	ΔGly320	Severe subacute	Irish	Mules et al., 1992b

Mutation	Location	Effect	Phenotype	Population	Reference
986+3A→G	IVS-8	Abnormal splicing	Severe subacute, second allele Arg178His	German/Hungarian/Irish	Richard et al., 1995
987G→A	Exon 9	Trp329Stop	Severe subacute	German/English	Mules et al., 1992b
1003A→T	Exon 9	Ile335Phe	Affected prenatal	non-Jewish	Tomczak and Grebner, 1994
1039del18	Exon 9	Insertion of 6 amino acids	Affected prenatal	Ashkenazi Jewish	Tomczak and Grebner, 1994
1073+1G→A	IVS-9	Abnormal splicing	Acute	Diverse; Celtic origin	Landels et al., 1992, 1993; Akerman et al., 1992; Mules et al., 1992a; Akli et al., 1993b
1074-8del5	IVS-9	Abnormal splicing	Acute	Polish	Triggs-Raine et al., 1991
1074-1G→T	IVS-9	Abnormal splicing	Acute	Irish/French	Brown et al., 1995
1171G→A	Exon 11	Val391Met	Chronic, second allele Arg178His	Greek	Navon et al., 1995
1176G→A	Exon 11	Trp392Stop	Acute	Ashkenazi Jewish	Shore et al., 1992
1177C→T	Exon 11	Arg393Stop	Acute	French; Turkish	Akli et al., 1991; Ozkara et al., 1995
1182delG	Exon 11	Frameshift	Acute	non-Jewish	Tomczak and Grebner, 1994
1260G→C	Exon 11	Trp420Cys	Acute	German	Tanaka et al., 1990b
1278ins4	Exon 11	Frameshift	Acute	Diverse	Myerowitz and Costigan, 1988; Paw and Neufeld, 1988; Corthorn et al., 1991
1360G→A	Exon 12	Gly454Ser	Acute	Italian	Akli et al., 1993a
1373G→A	Exon 12	Cys458Tyr	Acute	Japanese	Tanaka et al., 1994
1421+1G→C	IVS-12	Abnormal splicing	Acute	Ashkenazi Jewish	Arpaia et al., 1988; Myerowitz, 1988; Ohno and Suzuki, 1988b

(continues)

Table 17.1. (continued)

Mutation	Location	Expected result	Phenotype	Origin	References
1444G→A	Exon 13	Glu482Lys	Acute	Italian/Chinese	Nakano et al., 1988; Proia and Neufeld, 1982; Poenaru and Akli, 1994
1451T→C	Exon 13	Leu484Pro	Acute	Japanese	Tanaka et al., 1994
1453T→C	Exon 13	Trp485Arg	Acute	Chinese	Akalin et al., 1992
1495C→T	Exon 13	Arg499Cys	Acute	European	Mules et al., 1992b; Akli et al., 1993a
1496G→A	Exon 13	Arg499His	Subacute	European	Paw et al., 1990a; Triggs-Raine et al., 1991; Akli et al., 1993a
1510delC	Exon 13	Frameshift	Acute	Italian	Zokaeem et al., 1987; Lau and Neufeld, 1989
1510C→T	Exon 13	Arg504Cys	Acute	Diverse	Paw et al., 1991; Akli et al., 1991, 1993c; Tanaka et al., 1994
1511G→A	Exon 13	Arg504His	Subacute	Diverse	Paw et al., 1990a; Boustany et al., 1991; Akli et al., 1993a; Tanaka et al., 1994

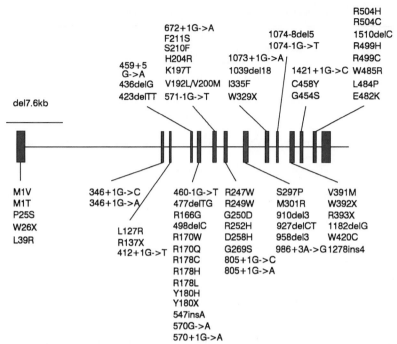

Figure 17.1. Schematic of the mutations currently identified in the *HEXA* gene. The solid boxes represent the 14 exons (not to scale). The lines indicate the exon in which the mutation is located or is found nearby.

Most of the identified *HEXA* mutations are associated with the acute form of Tay-Sachs disease. Many, such as deletions, frameshifts, and splice mutations, are "null" because they prevent the expression of α subunits. However, some point mutations that result in amino acid substitutions are also functionally null because the resulting Hex A is inactive or fails to reach the lysosome. A small number of mutations have been associated with severe subacute forms of Tay-Sachs disease, and once again, these include various types of mutations. In a few cases, where both mutations in an individual are novel, it has been difficult to predict which allele is responsible for the residual enzyme activity associated with the delayed onset of the disease. There are several examples in Table 17.1 of individuals with a severe subacute phenotype who are compound heterozygotes for different amino acid substitutions. In some cases, it is evident which is likely the milder mutation (e.g., Met1Thr/Pro25Ser, since the first ablates the initiation codon [Harmon *et al.*, 1993]). In a different type of ambiguity of phenotype, the B1-type Arg178His mutation, identified in numerous patients, is clearly associated with a severe subacute form of disease when in compound heterozygosity with a null allele. When the Arg178His allele is present in the homozygous form,

Table 17.2. *HEXA* Mutations with No Known Disease Association

Mutation	Effect	Location	Probable phenotype	Origin	Reference
9C→T	Ser3Ser	Exon 1	None	Diverse	Mules *et al.*, 1992b; Grinshpun *et al.*, 1995
587A→G	Asn196Ser	Exon 6	Unknown	French Canadian	Triggs-Raine *et al.*, 1995
672+30T→G	Unknown	IVS-6	None	French Canadian	Triggs-Raine *et al.*, 1995
748G→A	Gly250Ser	Exon 7	Unknown	French Canadian	Triggs-Raine *et al.*, 1995
749G→A	Val253Val	Exon 7	None	Pennsylvania Dutch	Mules *et al.*, 1992a
1146+18A→G	Unknown	IVS-10	None	French Canadian	Triggs-Raine *et al.*, 1995
1164C→G	Ile388Met	Exon 11	Unknown	French Canadian	Triggs-Raine *et al.*, 1995
1195A→G	Asn399Asp	Exon 11	None	Black American	Mules *et al.*, 1992b
1306G→A	Val436Ile	Exon 11	None	Black American	Mules *et al.*, 1992b
1338T→C	Pro446Pro	Exon 12	None	French Canadian	Triggs-Raine *et al.*, 1995
1518G→A	Glu506Glu	Exon 13	None	Not reported	Paw *et al.*, 1991
1527-6T→C	Unknown	IVS-13	None	Ashkenazi Jewish	Kaplan *et al.*, 1993
1684T→G	Unknown	Exon 14	None	French Canadian	Triggs-Raine *et al.*, 1995

the presentation is delayed and in some cases classified as subacute (reviewed by Dos Santos *et al.*, 1991). This illustrates the additivity of the Hex A residual activity associated with the B1 mutation and the impact on ameliorating the phenotype that it can have when in double dose.

Subacute and chronic forms of G_{M2} gangliosidosis are more variable in age of onset and severity. Three mutations (Gly250Asp, Arg499His, and Arg504His) have been associated with subacute G_{M2} gangliosidosis (Trop *et al.*, 1992; Paw *et al.*, 1990a). One additional mutation, 672+1G→A, was identified in a subacute/chronic (i.e., juvenile/adult-onset) patient (Akli *et al.*, 1993a). However, since most mutations in the +1 positions of introns are associated with acute phenotypes, it is likely the second allele, which has not been identified, that is responsible for the milder phenotype in that patient. The chronic form of G_{M2} gangliosidosis has been associated with six mutations: Tyr180His, Lys197Thr, Arg252His, Gly269Ser, 958del3, and Val391Met (Table 17.1). Once again, the determination of the phenotype associated with the less common mutations is difficult and the phenotype may vary among patients. The Gly269Ser mutation is the most common mutation in chronic patients, and it is clearly associated with a chronic phenotype in either the homozygous or compound heterozygous forms (Navon et al., 1990).

At the very end of the spectrum of phenotypes resulting from Hex A deficiency is the asymptomatic phenotype, Hex A pseudo-deficiency, which results from benign mutations that cause Hex A deficiency but do not cause disease. Two mutations, Arg247Trp and Arg249Trp, when in combination with a second deleterious allele, result in Hex A pseudo-deficiency (Triggs-Raine et al., 1992; Cao et al., 1993).

B. Epidemiology of Tay-Sachs disease

The Tay-Sachs disease carrier frequency among Ashkenazi Jews is about 1 in 30, approximately 10 times higher than that estimated in the non-Jewish population by Hardy-Weinberg analysis (reviewed by Kaback et al., 1993). Three mutations, 1278ins4, 1421+1G→C, and Gly269Ser, account for approximately 98% of the disease-causing alleles in this population (Triggs-Raine et al., 1990; Paw et al., 1990b; Grebner and Tomczak, 1991). In enzyme-defined Ashkenazi Jewish heterozygotes for Tay-Sachs disease, these mutations account for a slightly reduced proportion of alleles (Fernandes et al., 1992b; Grebner and Tomczak, 1991; Paw et al., 1990b; Triggs-Raine et al., 1990; Yoo et al., 1993) because of the presence of a small number of benign mutations (~2%) and the necessary inclusion of some false positive samples in establishing a sensitive cutoff for the heterozygote range for Hex A activity.

Tay-Sachs disease has an elevated incidence in Eastern Quebec (Andermann et al., 1977), and unique mutations account for most of the alleles among French Canadians (Fernandes et al., 1992a; Hechtman et al., 1992; De Braekeleer et al., 1992). The primary mutation in this population is a 7.6-kb deletion that starts upstream of exon 1 and extends into intron 1 (Myerowitz and Hogikyan, 1987). One additional mutation, 805+1G→A, appears unique to French Canadians (Hechtman et al., 1992), and a third, 508C→T (Arg170Trp), has been found predominantly among French Canadians (Fernandes et al., 1992a). The 7.6-kb deletion accounts for about 80% of the mutant alleles in this population (Hechtman et al., 1990).

An elevated incidence of Tay-Sachs disease has also been reported among the Pennsylvania Dutch (Kelly et al., 1975), the Cajun population (McDowell et al., 1992), and Moroccan Jews (Vecht et al., 1983). The molecular basis of the disease in these populations in now relatively well defined. The 1073+1G→A mutation accounts for the majority of disease-causing alleles among the Pennsylvania Dutch (Mules et al., 1992a), the 1278ins4 mutation for most mutant alleles among the Cajuns (McDowell et al., 1992), and 910del3 for most mutant alleles among Moroccan Jews (Navon and Proia, 1991).

In non-Jewish populations, a diversity of mutations has been identified among Tay-Sachs disease patients and variants. Of these, only one has been found to be relatively common, 1073+1G→A, which accounts for about 10% of mutant alleles (Akerman *et al.*, 1992; Akli *et al.*, 1993c; Landels *et al.*, 1992, 1993). The two benign mutations, Arg247Trp and Arg249Trp, account for about 36% of alleles in non-Jewish enzyme-defined carriers, making their detection extremely important in enzyme screening programs designed for the prevention of Tay-Sachs disease. These benign or pseudo-deficiency mutations could result in incorrect or misleading counselling and diagnosis.

III. β-HEXOSAMINIDASE B MUTATIONS

The molecular analysis of Sandhoff disease and its variants has been less intense. As shown in Table 17.3, only 20 *HEXB* mutations have been reported. Once again, these mutations are diverse in nature and are distributed throughout the gene (Figure 17.2). Among null-type mutations, there are two large deletions, several small deletions or insertions, and a splice junction mutation. Half the mutations are missense mutations, resulting in amino acid substitutions. Although their number is small, some clustering has become apparent in exons 7 and 11. Only one point mutation has been reported for exon 5, the major site of point mutations in the homologous *HEXA* gene.

The two large deletions have been the subject of some ambiguity in the literature. One of these, a 16-kb deletion extending from the promoter region to intron 5 appears to have been derived from recombination between two Alu sequences, with the breakpoint at the midpoint in each sequence (Neote *et al.*,

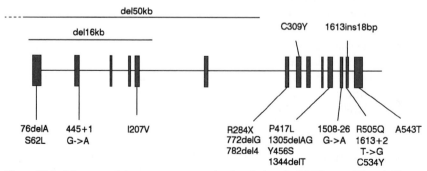

Figure 17.2. Schematic of the mutations currently identified in the *HEXB* gene. The solid boxes represent the 14 exons (not to scale). The lines indicate the exon in which the mutation is located or is found nearby. The boundaries of the deletions are not precisely placed. A broken line is used to represent the undefined 5′ boundary of the 50-kb deletion.

1990b). This mutation was independently reported as 50 kb in length, although diagnostic assays confirmed that they were the same deletion (Bikker *et al.*, 1989; Bolhuis and Bikker, 1992). The ambiguity mentioned above lies in the identification of a second deletion, also estimated at 50 kb in length, which extends from beyond the 5′ end of the gene to intron 6 (Zhang *et al.*, 1995). Its 5′ boundary has not been determined. Comparison of the two mutations in side-by-side samples by pulsed field electrophoresis has confirmed the different lengths of the two deletions (Zhang *et al.*, 1995).

There has also been confusion concerning the convention for numbering the nucleotide and amino acid positions of mutations in the *HEXB* gene (Table 17.3). Most reports have numbered mutations from the "A" of the currently accepted initiation codon (Neote *et al.*, 1990a). However, the pre-pro-β polypeptide has three Met residues upstream of the signal peptide cleavage site, and one report of the structure of the *HEXB* gene begins from the second ATG (Proia, 1988). Indeed, while all three Met residues can support translation initiation *in vitro* (Sonderfeld-Fresko and Proia, 1988; Neote *et al.*, 1990a), only the first of the three initiates translation *in vivo* (Neote *et al.*, 1990a). Correct numbering of the human Hex β subunit sequence can be found in (Neote *et al.*, 1988).

A. Genotype/Phenotype relationships

As with Tay-Sachs disease and variants, the severity of disease due to *HEXB* mutations also correlates with the residual level of Hex A activity. This is because it is the level of Hex A, independent of the level of Hex B, that is relevant to G$_{M2}$ ganglioside catabolism in the CNS. Consequently, all mutations leading to complete absence of Hex A activity, including the large and small deletions, the latter of which all produce frameshifts, and the splice junction mutation, are associated with the infantile acute phenotype and early death.

Among the point mutations, a broad spectrum of phenotypes is beginning to emerge. Many have occurred as compound heterozygotes with obvious null second mutations. In these instances, as with *HEXA* mutations, the phenotype appears to be determined by the milder of the two mutations. However, where both mutations produce amino acid substitutions, it is difficult to be specific about the contribution of each allele to the phenotype if the patient has a late onset or mild disease. Only four mutations associated with the subacute or chronic phenotype have been described. Two mutations, 1514G→A (Arg505Gln) (Bolhuis *et al.*, 1993) and 1627G→A (Ala543Thr) (De Gasperi *et al.*, 1995), are associated with a heat-labile Hex B. Both occur near the C terminus of the β subunit and were identified along with more severe second mutations. Another, 1252C→T (Pro417Leu), results in a disease of highly variable severity. Wakamatsu *et al.* (1992) described the mutation in a homozygous Japanese patient who had a childhood-onset disease with mental retardation and local panatrophy. Although

Table 17.3. *HEXB* Mutations in Sandhoff Disease and Its Variants

Mutation	Location	Expected result	Phenotype	Origin	Reference
del16kb	Promoter to IVS-5	No mRNA	Acute	French; others	Neote et al., 1990b; the following are also the same allele: Bikker et al., 1989, 1990; Bolhuis and Bikker, 1992;
del50kb	5' of promoter to IVS-6	No mRNA	Acute		Zhang et al., 1995
76delA	Exon 1	Frameshift	Acute	Cyprus Maronite	Hara et al., 1994
185C→T	Exon 1	Ser62Leu	Acute		Zhang et al., 1995
445+1G→A	IVS-2	Abnormal splicing	Acute	Argentinean	de Kremer et al., 1985, 1987; Brown et al., 1992; Kleiman et al., 1994
619A→G	Exon 5	Ile207Val	Benign polymorphism (formerly described as chronic)		Banerjee et al., 1991, 1994; see Redonnet-Vernhet et al., 1996, for explanation of polymorphism
772delG	Exon 7	Frameshift	Acute	Caucasian; Indian	O'Dowd et al., 1986; McInnes et al., 1992a; Zhang et al., 1994
782del4	Exon 7	Frameshift	Acute	Argentinean	Brown et al., 1992; Kleiman et al., 1994
850C→T	Exon 7	Arg284Stop	Acute		Zhang et al., 1994
926G→A[a]	Exon 8	Cys309Tyr	Chronic (second allele Pro417Leu)	Italian	Gomez-Lira et al., 1995b

1252C→T[a]	Exon 11	Pro417Leu (affects splicing)	Subacute/chronic	Japanese; Italian; French Canadian	Wakamatsu et al., 1992; McInnes et al., 1992b; Gomez-Lira et al., 1995b
1305delAG	Exon 11	Frameshift	Acute	Caucasian; Indian	McInnes et al., 1992a; Zhang et al., 1994
1344delT	Exon 11	Frameshift	Acute		Zhang et al., 1994
1367A→C	Exon 11	Tyr456Ser	Probably acute		Banerjee et al., 1991, 1994
1508-26G→A	IVS-12	Insertion of 8 amino acids due to cryptic splicing	Subacute		Nakano and Suzuki, 1989; Dlott et al., 1990; Mitsuo et al., 1990
1514G→A	Exon 13	Arg505Gln	Chronic		Oonk et al., 1979; Bolhuis et al., 1987, 1993
1613ins18bp	IVS-13/exon 14	Insertion of 6 amino acids	Asymptomatic, "hexosaminidase Paris"	French	Dreyfus et al., 1977; Dlott et al., 1990
1601G→A[a]	Exon 13	Cys534Tyr	Acute	Japanese	Kuroki et al., 1995
1613+2T→G[a]	IVS-13	Abnormal splicing	Acute		Gomez-Lira et al., 1995a
1627G→A	Exon 14	Ala543Thr	Chronic	Oriental Jews	De Gasperi et al., 1995

[a]Position of mutation is incorrect in at least one cited references (based on counting from second ATG); the number in the table corresponds to the correct position.

the mutation is located in an exon 8 bp from a splice junction, these authors showed that it acts by disrupting splicing rather than affecting enzyme function per se. This same mutation has been identified, along with more severe mutations, in individuals with late-onset, chronic diseases (McInnes et al., 1992b; Gomez-Lira et al., 1995b). In one case, the disease was very mild, with symptoms suggestive primarily of impairment of the autonomic nervous system (McInnes et al., 1992b). The same genotype was also observed in siblings who were free of disease manifestations. In another report, a patient showed an isolated lower motor disease in one case and a more profound late-onset disease with motor neuron and cerebellar involvement in another (Gomez-Lira et al., 1995b).

Two complex mutations have been described which are associated with variable phenotypes including pseudo-deficiency. One is a point mutation in intron 12, IVS12-26G→A (Nakano and Suzuki, 1989; Mitsuo et al., 1990; Dlott et al., 1990). It generates a cryptic acceptor site which results in an in-frame insertion of eight amino acids between exons 12 and 13. The second mutation is a duplication of 18 bp from −16 to +2 at the IVS-13/exon 14 junction (Dlott et al., 1990). It results in the insertion of six amino acids at the beginning of exon 14. In both cases, the modified β subunit fails to fold properly and is not processed. However, a small amount of correct splicing appears to be responsible for the formation of a small but highly variable amount of normal Hex A. This has resulted in a variable clinical phenotype, ranging from a subacute (juvenile) disease in the case of the IVS-12 splice mutation to an asymptomatic presentation in a case of the IVS-13 insertion (Nakano and Suzuki, 1989; Mitsuo et al., 1990; Dlott et al., 1990). [In the former, an asymptomatic case has also been described, but the phenotype was likely accounted for by the second unidentified mutation (Nakano and Suzuki, 1989)]. In all of these individuals, Hex B was virtually absent. This biochemical phenotype, Hex A-plus/Hex B-minus, was originally termed "Hexosaminidase Paris" (Dreyfus et al., 1977). These results suggest that the C-terminal sequence of the β subunit, like the α subunit, appears to be important in its proper folding or initial processing.

B. Epidemiology of Sandhoff disease

The most common HEXB mutation identified thus far is the 16-kb deletion. It has been identified mainly in individuals of French or French-Canadian origin and may account for as many as one-third of HEXB alleles (Bikker et al., 1990; Neote et al., 1990b). The highest recorded frequency of Sandhoff disease carriers, 1/26, is in the Cordoba region of Argentina in a population of mixed Creole, Spanish, and native peoples (de Kremer et al., 1987). Most of the carriers share the same mutation, IVS2+1G→A (30/31 alleles examined) (Brown et al., 1992; Kleiman et al., 1994). A second mutation, seen in only one individual, is 784del4 (Brown et al., 1992). A potentially common mutation that has yet to be evaluated for prevalence is Pro417Leu. It has been described in a Japanese (Wakamatsu et al.,

Table 17.4. GM2A Mutations in GM2 Activator Deficiency

Mutation	Location	Expected result	Phenotype	Origin	Reference
262del3	Exon 3	ΔLys88	Acute	Saudi Arabian	Schepers *et al.*, 1996
410delA	Exon 3	Frameshift	Acute	Spanish	Schepers *et al.*, 1996
412T→C	Exon 3	Cys138Arg[a]	Acute	Black American	Schroder *et al.*, 1991; Xie *et al.*, 1992
506G→C	Exon 4	Arg169Pro	Acute	Indian	Schroder *et al.*, 1993

[a]Given as Cys107Arg by Schroder *et al.* (1991).

1992), French-Canadian (McInnes *et al.*, 1992b), and several Italian patients (Gomez-Lira *et al.*, 1995b). Other populations that have reported an elevated frequency of Sandhoff disease include an inbred community of Metis Indians in northern Saskatchewan (Lowden *et al.*, 1978) and in Lebanon (Der Kaloustian *et al.*, 1981). There may also be an elevated carrier frequency among Hispanics of Mexican or Central African origin (Cantor *et al.*, 1987).

IV. *GM2A* MUTATIONS

G$_{M2}$ activator deficiency (AB variant) is a very rare disease with only the acute phenotype documented in the literature (Gravel *et al.*, 1995). Thus far, four mutations have been described (Table 17.4). Two are small deletions. One is of a single nucleotide, 410delA, resulting in a frameshift (Schepers *et al.*, 1996), and the other is a 3-bp deletion, 262del3, resulting in the deletion of a single amino acid, Lys88. Two other mutations result in amino acid substitutions: 412T→C (Cys138Arg), identified in the same patient in two independent studies (Schroder *et al.*, 1991; Xie *et al.*, 1992), and 506G→C (Arg169Pro) (Schroder *et al.*, 1993). All of these mutations are associated with the expression of very little protein, due to the generation of incompletely processed or unstable proteins that are rapidly degraded.

V. STRUCTURE/FUNCTION RELATIONSHIPS OF β-HEXOSAMINIDASE

Until 1996, studies of structure/function relationships involving *HEXA* or *HEXB* mutations were primarily limited to correlating levels of expressed enzyme activity to specific amino acids changes. Recently, the resolution of the crystal structure of the *Serratia marcescens* chitobiase, a member of a family of β-hexosaminidases that includes human lysosomal Hex, has allowed a predicted model for the human Hex α subunit to be proposed (Tews *et al.*, 1996). Care must be taken in assessing the impact of mutations to recognize that those that affect folding or stability may never reach the lysosome. Nevertheless, many of the mutations associated with acute disease were localized to the interior of the modeled protein, while those associated with less severe forms of disease were more external. For example, the B1 mutation site, Arg178, was located in the vicinity of the proposed active site, while Gly269, associated with the chronic phenotype, is in the exterior of the molecule. Significantly, the B1 mutation sites, Arg178 and Asp258, were positioned in the proposed substrate binding pocket, while other residues, such as Gly250 and Ser279 (mutations leading to subacute variants), Gly269 (leading to chronic variant), or Arg247 and Arg249 (pseudo-deficiency mutations) were located far from the active site, where they may be more likely to affect folding, subunit association, or enzyme processing. One caveat in assessing the role of mutations in models based on distantly related enzymes is that a high degree of sequence homology is normally required. Although the *S. marcescens* is only 26% homologous to the Hex α subunit in the active-site region, the authors were still able to obtain a structure. Therefore, while necessarily speculative, the proposed Hex α structure is an important addition to the repertoire of tools by which human disease mutations can be understood.

References

Ainsworth, P. J., and Coulter-Mackie, M. B. (1992). A double mutation in exon 6 of the beta-hexosaminidase alpha subunit in a patient with the B1 variant of Tay-Sachs disease. *Am. J. Hum. Genet.* **51,** 802–809.

Akalin, N., Shi, H. P., Vavougios, G., Hechtman, P., Lo, W., Scriver, C. R., Mahuran, D., *et al.* (1992). Novel Tay-Sachs disease mutations from China. *Hum. Mutat.* **1,** 40–46.

Akerman, B. R., Zielenski, J., Triggs-Raine, B. L., Prence, E. M., Natowicz, M. R., Lim-Steele, J. S., Kaback, M. M., *et al.* (1992). A mutation common in non-Jewish Tay-Sachs disease: Frequency and RNA studies. *Hum. Mutat.* **1,** 303–309.

Akli, S., Chelly, J., Mezard, C., Gandy, S., Kahn, A., and Poenaru, L. (1990). A "G" to "A" mutation at position −1 of a 5′ splice site in a late infantile form of Tay-Sachs disease. *J. Biol. Chem.* **265,** 7324–7330.

Akli, S., Chelly, J., Lacorte, J. M., Poenaru, L., and Kahn, A. (1991). Seven novel Tay-Sachs mutations detected by chemical mismatch cleavage of PCR-amplified cDNA fragments. *Genomics* **11,** 124–134.

Akli, S., Boue, J., Sandhoff, K., Kleijer, W., Vamos, E., Young, E., Gatti, R., *et al.* (1993a). Collaborative study of the molecular epidemiology of Tay-Sachs disease in Europe. *Eur. J. Hum. Genet.* **1,** 229–238.

Akli, S., Chelly, J., Kahn, A., and Poenaru, L. (1993b). A null allele frequent in non-Jewish Tay-Sachs patients. *Hum. Genet.* **90,** 614–620.

Akli, S., Chomel, J. C., Lacorte, J. M., Bachner, L., Kahn, A., and Poenaru, L. (1993c). Ten novel mutations in the HEXA gene in non-Jewish Tay-Sachs patients. *Hum. Mol. Genet.* **2,** 61–67.

Andermann, E., Scriver, C. R., Wolfe, L. S., Dansky, L., and Andermann, F. (1977). Genetic variants of Tay-Sachs disease: Tay-Sachs disease and Sandhoff's disease in French Canadians, juvenile Tay-Sachs disease in Lebanese Canadians, and a Tay-Sachs screening program in the French-Canadian population. *Prog. Clin. Biol. Res.* **18,** 161–188.

Arpaia, E., Dumbrille Ross, A., Maler, T., Neote, K., Tropak, M., Troxel, C., Stirling, J. L., *et al.* (1988). Identification of an altered splice site in Ashkenazi Tay-Sachs disease. *Nature* **333,** 85–86.

Banerjee, P., Siciliano, L., Oliveri, D., McCabe, N. R., Boyers, M. J., Horwitz, A. L., Li, S. C., *et al.* (1991). Molecular basis of an adult form of beta-hexosaminidase B deficiency with motor neuron disease. *Biochem. Biophys. Res. Commun.* **181,** 108–115.

Banerjee, P., Boyers, M. J., Berry-Kravis, E., and Dawson, G. (1994). Preferential beta-hexosaminidase (Hex) A (alpha beta) formation in the absence of beta-Hex B (beta beta) due to heterozygous point mutations present in beta-Hex beta-chain alleles of a motor neuron disease patient. *J. Biol. Chem.* **269,** 4819–4826.

Bayleran, J., Hechtman, P., Kolodny, E., and Kaback, M. (1987). Tay-Sachs disease with hexosaminidase A: Characterization of the defective enzyme in two patients. *Am. J. Hum. Genet.* **41,** 532–548.

Bikker, H., van den Berg, F. M., Wolterman, R. A., de Vijlder, J. J., and Bolhuis, P. A. (1989). Demonstration of a Sandhoff disease-associated autosomal 50-kb deletion by field inversion gel electrophoresis. *Hum. Genet.* **81,** 287–288.

Bikker, H., van den Berg, F. M., Wolterman, R. A., Kleijer, W. J., de Vijlder, J. J., and Bolhuis, P. A. (1990). Distribution and characterization of a Sandhoff disease-associated 50-kb deletion in the gene encoding the human beta-hexosaminidase beta-chain. *Hum. Genet.* **85,** 327–329.

Bolhuis, P. A., and Bikker, H. (1992). Deletion of the 5′-region in one or two alleles of HEXB in 15 out of 30 patients with Sandhoff disease. *Hum. Genet.* **90,** 328–329.

Bolhuis, P. A., Oonk, J. G., Kamp, P. E., Ris, A. J., Michalski, J. C., Overdijk, B., and Reuser, A. J. (1987). Ganglioside storage, hexosaminidase lability, and urinary oligosaccharides in adult Sandhoff's disease. *Neurology* **37,** 75–81.

Bolhuis, P. A., Ponne, N. J., Bikker, H., Baas, F., and Vianney de Jong, J. M. (1993). Molecular basis of an adult form of Sandhoff disease: Substitution of glutamine for arginine at position 505 of the beta-chain of beta-hexosaminidase results in a labile enzyme. *Biochim. Biophys. Acta* **1182,** 142–146.

Boustany, R. M., Tanaka, A., Nishimoto, J., and Suzuki, K. (1991). Genetic cause of a juvenile form of Tay-Sachs disease in a Lebanese child. *Ann. Neurol.* **29,** 104–107.

Brown, C. A., McInnes, B., de Kremer, R. D., and Mahuran, D. J. (1992). Characterization of two HEXB gene mutations in Argentinean patients with Sandhoff disease. *Biochim. Biophys. Acta* **1180,** 91–98.

Brown, D. H., Triggs-Raine, B. L., McGinniss, M. J., and Kaback, M. M. (1995). A novel mutation at the invariant acceptor splice site of intron 9 in the HEXA gene [IVS9-1 G→T] detected by a PCR-based diagnostic test. *Hum. Mutat.* **5,** 173–174.

Cantor, R. M., Roy, C., Lim, J. S., and Kaback, M. M. (1987). Sandhoff disease heterozygote detection: A component of population screening for Tay-Sachs disease carriers. II. Sandhoff disease gene frequencies in American Jewish and non-Jewish populations. *Am. J. Hum. Genet.* **41,** 16–26.

Cao, Z., Natowicz, M. R., Kaback, M. M., Lim-Steele, J. S., Prence, E. M., Brown, D., Chabot, T., *et al.* (1993). A second mutation associated with apparent beta-hexosaminidase A pseudodeficiency: Identification and frequency estimation. *Am. J. Hum. Genet.* **53,** 1198–1205.

Conzelmann, E., Kytzia, H. J., Navon, R., and Sandhoff, K. (1983). Ganglioside GM2 N-acetyl-beta-D-galactosaminidase activity in cultured fibroblasts of late-infantile and adult GM2 gangliosidosis patients and of healthy probands with low hexosaminidase level. *Am. J. Hum. Genet.* **35,** 900–913.

Corthorn, J., Cantin, M., and Thibault, G. (1991). Rat atrial secretory granules and pro-ANF processing enzyme. *Mol. Cell Biochem.* **103,** 31–39.

d'Azzo, A., Proia, R. L., Kolodny, E. H., Kaback, M. M., and Neufeld, E. F. (1984). Faulty association of alpha- and beta-subunits in some forms of beta-hexosaminidase A deficiency. *J. Biol. Chem.* **259,** 11070–11074.

De Braekeleer, M., Hechtman, P., Andermann, E., and Kaplan, F. (1992). The French Canadian Tay-Sachs disease deletion mutation: Identification of probable founders. *Hum. Genet.* **89,** 83–87.

De Gasperi, R., Gama Sosa, M. A., Grebner, E. E., Mansfield, D., Battistini, S., Sartorato, E. L., Raghavan, S. S., *et al.* (1995). Substitution of alanine543 with a threonine residue at the carboxy terminal end of the beta-chain is associated with thermolabile hexosaminidase B in a Jewish family of Oriental ancestry. *Biochem. Mol. Med.* **56,** 31–36.

De Gasperi, R., Sosa, M. A. G., Battistini, S., Yeretsian, J., Raghavan, S., Zelnik, N., Leshinsky, E., *et al.* (1996). Late-onset GM2 Gangliosidosis—Ashkenazi Jewish family with an exon 5 mutation (Tyr180→His) in the Hex A alpha-chain gene. *Neurology* **47,** 547–552.

de Kremer, R. D., Boldini, C. D., Capra, A. P., Levstein, I. M., Bainttein, N., Hidalgo, P. K., and Hliba, H. (1985). Sandhoff disease: 36 cases from Cordoba, Argentina. *J. Inher. Metab. Dis.* **8,** 46.

de Kremer, R. D., Depetris de Boldini, C., Paschini de Capra, A., Pons de Veritier, P., Goldenhersch, H., Corbella, L., Sembaj, A., *et al.* (1987). Estimation of heterozygote frequency of Sandhoff disease in a high-risk Argentinian population. Predictive assignment of the genotype through statistical analysis. *Medicina* **47,** 455–463.

Der Kaloustian, V. M., Khoury, M. J., Hallal, R., Idriss, Z. H., Deeb, M. E., Wakid, N. W., and Haddad, F. S. (1981). Sandhoff disease: A prevalent form of infantile GM2 gangliosidosis in Lebanon. *Am. J. Hum. Genet.* **33,** 85–89.

Dlott, B., d'Azzo, A., Quon, D. V., and Neufeld, E. F. (1990). Two mutations produce intron insertion in mRNA and elongated beta-subunit of human beta-hexosaminidase. *J. Biol. Chem.* **265,** 17921–17927.

Dos Santos, M. R., Tanaka, A., Sa Miranda, M. C., Ribeiro, M. G., Maia, M., and Suzuki, K. (1991). GM2-gangliosidosis B1 variant: Analysis of beta-hexosaminidase alpha gene mutations in 11 patients from a defined region in Portugal. *Am. J. Hum. Genet.* **49,** 886–890.

Dreyfus, J. C., Poenaru, L., Vibert, M., Ravise, N., and Boue, J. (1977). Characterization of a variant of beta-hexosaminidase: "hexosaminidase Paris." *Am. J. Hum. Genet.* **29,** 287–293.

Drucker, L., and Navon, R. (1993). Tay-Sachs disease in an Israeli Arab family: Trp26→stop in the alpha-subunit of hexosaminidase A. *Hum. Mutat.* **2,** 415–417.

Drucker, L., Proia, R. L., and Navon, R. (1992). Identification and rapid detection of three Tay-Sachs mutations in the Moroccan Jewish population. *Am. J. Hum. Genet.* **51,** 371–377.

Fernandes, M., Kaplan, F., Natowicz, M., Prence, E., Kolodny, E., Kaback, M., and Hechtman, P. (1992a). A new Tay-Sachs disease B1 allele in exon 7 in two compound heterozygotes each with a second novel mutation. *Hum. Mol. Genet.* **1,** 759–761.

Fernandes, M. J., Kaplan, F., Clow, C. L., Hechtman, P., and Scriver, C. R. (1992b). Specificity and sensitivity of hexosaminidase assays and DNA analysis for the detection of Tay-Sachs disease gene carriers among Ashkenazic Jews. *Genet. Epidemiol.* **9,** 169–175.

Gomez-Lira, M., Perusi, C., Brutti, N., Farnetani, M. A., Margollicci, M. A., Rizzuto, N., Pignatti, P. F., *et al.* (1995a). A 48-bp insertion between exon 13 and 14 of the HEXB gene causes infantile-onset Sandhoff disease. *Hum. Mutat.* **6,** 260–262.

Gomez-Lira, M., Sangalli, A., Mottes, M., Perusi, C., Pignatti, P. F., Rizzuto, N., and Salviati, A. (1995b). A common beta hexosaminidase gene mutation in adult Sandhoff disease patients. *Hum. Genet.* **96,** 417–422.

Gordon, B. A., Gordon, K. E., Hinton, G. G., Cadera, W., Feleki, V., Bayleran, J., and Hechtman, P. (1988). Tay-Sachs disease: B1 variant. *Ped. Neurol.* **4,** 54–57.

Gravel, R. A., Clarke, J. T. R., Kaback, M. M., Mahuran, D., Sandhoff, K., and Suzuki, K. (1995). The GM2 gangliosidoses. *In* "The Metabolic and Molecular Bases of Inherited Disease" (C. R. Scriver, A. L. Beaudet, W. S. Sly, and D. Valle, eds.), pp. 2839–2879. McGraw-Hill, New York.

Grebner, E. E., and Tomczak, J. (1991). Distribution of three alpha-chain beta-hexosaminidase A mutations among Tay-Sachs carriers. *Am. J. Hum. Genet.* **48,** 604–607.

Grebner, E. E., Mansfield, D. A., Raghavan, S. S., Kolodny, E. H., d'Azzo, A., Neufeld, E. F., and Jackson, L. G. (1986). Two abnormalities of hexosaminidase A in clinically normal individuals. *Am. J. Hum. Genet.* **38,** 505–514.

Grinshpun, J., Khosravi, R., Peleg, L., Goldman, B., Kaplan, F., Triggs-Raine, B., and Navon, R. (1995). An Alu1-polymorphism in the HEXA gene is common in Ashkenazi and Sephardic Jews, Israeli Arabs, and French Canadians of Quebec and Northern New England. *Hum. Mutat.* **6,** 89–90.

Hara, Y., Ioannou, P., Drousiotou, A., Stylianidou, G., Anastasiadou, V., and Suzuki, K. (1994). Mutation analysis of a Sandhoff disease patient in the Maronite community in Cyprus. *Hum. Genet.* **94,** 136–140.

Harmon, D. L., Gardner-Medwin, D., and Stirling, J. L. (1993). Two new mutations in a late infantile Tay-Sachs patient are both in exon 1 of the beta-hexosaminidase alpha subunit gene. *J. Med. Genet.* **30,** 123–128.

Hechtman, P., Boulay, B., Bayleran, J., and Andermann, E. (1989). The mutation mechanism causing juvenile-onset Tay-Sachs disease among Lebanese. *Clin. Genet.* **35,** 364–375.

Hechtman, P., Kaplan, F., Bayleran, J., Boulay, B., Andermann, E., and De Braekeleer, M., *et al.* (1990). More than one mutant allele causes infantile Tay-Sachs disease in French-Canadians. *Am. J. Hum. Genet.* **47,** 815–822.

Hechtman, P., Boulay, B., De Braekeleer, M., Andermann, E., Melancon, S., Larochelle, J., Prevost, C., *et al.* (1992). The intron 7 donor splice site transition: A second Tay-Sachs disease mutation in French Canada. *Hum. Genet.* **90,** 402–406.

Hou, Y., Vavougios, G., Hinek, A., Wu, K. K., Hechtman, P., Kaplan, F., and Mahuran, D. J. (1996). The Val192Leu mutation in the α-subunit of β-hexosaminidase is not associated with the B1-variant form of Tay-Sachs disease. *Am. J. Hum. Genet.* **59,** 52–58.

Kaback, M., Lim-Steele, J., Dabholkar, D., Brown, D., Levy, N., and Zeiger, K. (1993). Tay-Sachs disease—Carrier screening, prenatal diagnosis, and the molecular era. An international perspective, 1970 to 1993. The International TSD Data Collection Network. *J. Am. Med. Assoc.* **270,** 2307–2315.

Kaplan, F., Kapoor, S., Lee, D., Fernandes, M., Vienozinskis, M., Mascisch, A., Scriver, C. R., *et al.* (1993). A Pst+ polymorphism in the HEXA gene with an unusual geographic distribution. *Eur. J. Hum. Genet.* **1,** 301–305.

Kelly, T. E., Chase, G. A., Kaback, M. M., Kumor, K., and McKusick, V. A. (1975). Tay-Sachs disease: High gene frequency in a non-Jewish population. *Am. J. Hum. Genet.* **27,** 287–291.

Kelly, T. E., Reynolds, L. W., and O'Brien, J. S. (1976). Segregation within a family of two mutant alleles for hexosaminidase A. *Clin. Genet.* **9,** 540–543.

Kleiman, F. E., de Kremer, R. D., de Ramirez, A. O., Gravel, R. A., and Argarana, C. E. (1994). Sandhoff disease in Argentina: High frequency of a splice site mutation in the HEXB gene and correlation between enzyme and DNA-based tests for heterozygote detection. *Hum. Genet.* **94,** 279–282.

Klima, H., Tanaka, A., Schnabel, D., Nakano, T., Schroder, M., Suzuki, K., and Sandhoff, K. (1991). Characterization of full-length cDNAs and the gene coding for the human GM2 activator protein. *FEBS Lett.* **289,** 260–264.

Korneluk, R. G., Mahuran, D. J., Neote, K., Klavins, M. H., O'Dowd, B. F., Tropak, M., Willard, H. F., *et al.* (1986). Isolation of cDNA clones coding for the alpha-subunit of human beta-hexosaminidase. Extensive homology between the alpha- and beta-subunits and studies on Tay-Sachs disease. *J. Biol. Chem.* **261,** 8407–8413.

Kuroki, Y., Itoh, K., Nadaoka, Y., Tanaka, T., and Sakuraba, H. (1995). A novel missense mutation (C522Y) is present in the beta-hexosaminidase beta-subunit gene of a Japanese patient with infantile Sandhoff disease. *Biochem. Biophys. Res. Commun.* **212,** 564–571.

Kytzia, H. J., and Sandhoff, K. (1985). Evidence for two different active sites on human beta-hexosaminidase A. Interaction of GM2 activator protein with beta-hexosaminidase. A. *J. Biol. Chem.* **260,** 7568–7572.

Kytzia, H. J., Hinrichs, U., Maire, I., Suzuki, K., and Sandhoff, K. (1983). Variant of GM2-gangliosidosis with hexosaminidase A having a severely changed substrate specificity. *EMBO J.* **2,** 1201–1205.

Landels, E. C., Green, P. M., Ellis, I. H., Fensom, A. H., and Bobrow, M. (1992). Beta-hexosaminidase splice site mutation has a high frequency among non-Jewish Tay-Sachs disease carriers from the British Isles. *J. Med. Genet.* **29,** 563–567.

Landels, E. C., Green, P. M., Ellis, I. H., Fensom, A. H., Kaback, M. M., Lim-Steele, J., Zeiger, K., *et al.* (1993). Further investigation of the HEXA gene intron 9 donor splice site mutation frequently found in non-Jewish Tay-Sachs disease patients from the British Isles. *J. Med. Genet.* **30,** 479–481.

Lau, M. M., and Neufeld, E. F. (1989). A frameshift mutation in a patient with Tay-Sachs disease causes premature termination and defective intracellular transport of the alpha-subunit of beta-hexosaminidase. *J. Biol. Chem.* **264,** 21376–21380.

Leinekugel, P., Michel, S., Conzelmann, E., and Sandhoff, K. (1992). Quantitative correlation between the residual activity of beta-hexosaminidase A and arylsulfatase A and the severity of the resulting lysosomal storage disease. *Hum. Genet.* **88,** 513–523.

Lowden, J. A., Ives, E. J., Keene, D. L., Burton, A. L., Skomorowski, M. A., and Howard, F. (1978). Carrier detection in Sandhoff disease. *Am. J. Hum. Genet.* **30,** 38–45.

McDowell, G. A., Mules, E. H., Fabacher, P., Shapira, E., and Blitzer, M. G. (1992). The presence of two different infantile Tay-Sachs disease mutations in a Cajun population. *Am. J. Hum. Genet.* **51,** 1071–1077.

McInnes, B., Brown, C. A., and Mahuran, D. J. (1992a). Two small deletion mutations of the HEXB gene are present in DNA from a patient with infantile Sandhoff disease. *Biochim. Biophys. Acta* **1138,** 315–317.

McInnes, B., Potier, M., Wakamatsu, N., Melancon, S. B., Klavins, M. H., Tsuji, S., and Mahuran, D. J. (1992b). An unusual splicing mutation in the HEXB gene is associated with dramatically different phenotypes in patients from different racial backgrounds. *J. Clin. Invest.* **90,** 306–314.

Mitsuo, K., Nakano, T., Kobayashi, T., Goto, I., Taniike, M., and Suzuki, K. (1990). Juvenile Sandhoff disease: A Japanese patient carrying a mutation identical to that found earlier in a Canadian patient. *J. Neurol. Sci.* **98,** 277–286.

Mules, E. H., Dowling, C. E., Petersen, M. B., Kazazian, H. H. Jr., and Thomas, G. H. (1991). A novel mutation in the invariant AG of the acceptor splice site of intron 4 of the beta-hexosaminidase alpha-subunit gene in two unrelated American black GM2-gangliosidosis (Tay-Sachs disease) patients. *Am. J. Hum. Genet.* **48,** 1181–1185.

Mules, E. H., Hayflick, S., Dowling, C. E., Kelly, T. E., Akerman, B. R., Gravel, R. A., and Thomas, G. H. (1992a). Molecular basis of hexosaminidase A deficiency and pseudodeficiency in the Berks County Pennsylvania Dutch. *Hum. Mutat.* **1,** 298–302.

Mules, E. H., Hayflick, S., Miller, C. S., Reynolds, L. W., and Thomas, G. H. (1992b). Six novel deleterious and three neutral mutations in the gene encoding the alpha-subunit of hexosaminidase A in non-Jewish individuals. *Am. J. Hum. Genet.* **50,** 834–841.

Myerowitz, R. (1988). Splice junction mutation in some Ashkenazi Jews with Tay-Sachs disease: Evidence against a single defect within this ethnic group. *Proc. Natl. Acad. Sci. (USA)* **85,** 3955–3959.

Myerowitz, R., and Costigan, F. C. (1988). The major defect in Ashkenazi Jews with Tay-Sachs disease is an insertion in the gene for the alpha-chain of beta-hexosaminidase. *J. Biol. Chem.* **263,** 18587–18589.

Myerowitz, R., and Hogikyan, N. D. (1986). Different mutations in Ashkenazi Jewish and non-Jewish French Canadians with Tay-Sachs disease. *Science* **232,** 1646–1648.

Myerowitz, R., and Hogikyan, N. D. (1987). A deletion involving Alu sequences in the beta-hexosaminidase alpha-chain gene of French Canadians with Tay-Sachs disease. *J. Biol. Chem.* **262,** 15396–15399.

Myerowitz, R., and Proia, R. L. (1984). cDNA clone for the alpha-chain of human beta-hexosaminidase: Deficiency of alpha-chain mRNA in Ashkenazi Tay-Sachs fibroblasts. *Proc. Natl. Acad. Sci. (USA)* **81,** 5394–5398.

Myerowitz, R., Piekarz, R., Neufeld, E. F., Shows, T. B., and Suzuki, K. (1985). Human beta-hexosaminidase alpha chain: Coding sequence and homology with the beta chain. *Proc. Natl. Acad. Sci. (USA)* **82,** 7830–7834.

Nagarajan, S., Chen, H. C., Li, S. C., Li, Y. T., and Lockyer, J. M. (1992). Evidence for two cDNA clones encoding human GM2-activator protein. *Biochem. J.* **282**, 807–813.

Nakano, T., and Suzuki, K. (1989). Genetic cause of a juvenile form of Sandhoff disease. Abnormal splicing of beta-hexosaminidase beta chain gene transcript due to a point mutation within intron 12. *J. Biol. Chem.* **264**, 5155–5158.

Nakano, T., Muscillo, M., Ohno, K., Hoffman, A. J., and Suzuki, K. (1988). A point mutation in the coding sequence of the beta-hexosaminidase alpha gene results in defective processing of the enzyme protein in an unusual GM2-gangliosidosis variant. *J. Neurochem.* **51**, 984–987.

Nakano, T., Nanba, E., Tanaka, A., Ohno, K., Suzuki, Y., and Suzuki, K. (1990). A new point mutation within exon 5 of beta-hexosaminidase alpha gene in a Japanese infant with Tay-Sachs disease. *Ann. Neurol.* **27**, 465–473.

Navon, R., and Proia, R. L. (1989). The mutations in Ashkenazi Jews with adult GM2 gangliosidosis, the adult form of Tay-Sachs disease. *Science* **243**, 1471–1474.

Navon, R., and Proia, R. L. (1991). Tay-Sachs disease in Moroccan Jews: Deletion of a phenylalanine in the alpha-subunit of beta-hexosaminidase. *Am. J. Hum. Genet.* **48**, 412–419.

Navon, R., Argov, Z., and Frisch, A. (1986). Hexosaminidase A deficiency in adults. *Am. J. Med. Genet.* **24**, 179–196.

Navon, R., Kolodny, E. H., Mitsumoto, H., Thomas, G. H., and Proia, R. L. (1990). Ashkenazi-Jewish and non-Jewish adult GM2 gangliosidosis patients share a common genetic defect. *Am. J. Hum. Genet.* **46**, 817–821.

Navon, R., Khosravi, R., Korczyn, T., Masson, M., Sonnino, S., Fardeau, M., Eymard, B., *et al.* (1995). A new mutation in the HEXA gene associated with a spinal muscular atrophy phenotype. *Neurology* **45**, 539–543.

Neote, K., Bapat, B., Dumbrille-Ross, A., Troxel, C., Schuster, S. M., Mahuran, D. J., and Gravel, R. A. (1988). Characterization of the human HEXB gene encoding lysosomal beta-hexosaminidase. *Genomics* **3**, 279–286.

Neote, K., Brown, C. A., Mahuran, D. J., and Gravel, R. A. (1990a). Translation initiation in the HEXB gene encoding the beta-subunit of human beta-hexosaminidase. *J. Biol. Chem.* **265**, 20799–20806.

Neote, K., McInnes, B., Mahuran, D. J., and Gravel, R. A. (1990b). Structure and distribution of an Alu-type deletion mutation in Sandhoff disease. *J. Clin. Invest.* **86**, 1524–1531.

O'Brien, J. S., Tennant, L., Veath, M. L., Scott, C. R., and Bucknall, W. E. (1978). Characterization of unusual hexosaminidase A (HEX A) deficient human mutants. *Am. J. Hum. Genet.* **30**, 602–608.

O'Dowd, B. F., Quan, F., Willard, H. F., Lamhonwah, A. M., Korneluk, R. G., Lowden, J. A., Gravel, R. A., *et al.* (1985). Isolation of cDNA clones coding for the beta subunit of human beta-hexosaminidase. *Proc. Natl. Acad. Sci. (USA)* **82**, 1184–1188.

O'Dowd, B. F., Klavins, M. H., Willard, H. F., Gravel, R., Lowden, J. A., and Mahuran, D. J. (1986). Molecular heterogeneity in the infantile and juvenile forms of Sandhoff disease (O-variant GM2 gangliosidosis). *J. Biol. Chem.* **261**, 12680–12685.

Ohno, K., and Suzuki, K. (1988a). Mutation in GM2-gangliosidosis B1 variant. *J. Neurochem.* **50**, 316–318.

Ohno, K., and Suzuki, K. (1988b). Multiple abnormal beta-hexosaminidase alpha chain mRNAs in a compound-heterozygous Ashkenazi Jewish patient with Tay-Sachs disease. *J. Biol. Chem.* **263**, 18563–18567.

Oonk, J. G., van der Helm, H. J., and Martin, J. J. (1979). Spinocerebellar degeneration: Hexosaminidase A and B deficiency in two adult sisters. *Neurology* **29**, 380–384.

Ozkara, H. A., Akerman, B. R., Ciliv, G., Topcu, M., Renda, Y., and Gravel, R. A. (1995). Donor splice site mutation in intron 5 of the HEXA gene in a Turkish infant with Tay-Sachs disease. *Hum. Mutat.* **5**, 186–187.

Paw, B. H., and Neufeld, E. F. (1988). Normal transcription of the beta-hexosaminidase alpha-chain gene in the Ashkenazi Tay-Sachs mutation. *J. Biol. Chem.* **263**, 3012–3015.

Paw, B. H., Kaback, M. M., and Neufeld, E. F. (1989). Molecular basis of adult-onset and chronic GM2 gangliosidoses in patients of Ashkenazi Jewish origin: Substitution of serine for glycine at position 269 of the alpha-subunit of beta-hexosaminidase. *Proc. Natl. Acad. Sci. (USA)* **86,** 2413–2417.

Paw, B. H., Moskowitz, S. M., Uhrhammer, N., Wright, N., Kaback, M. M., and Neufeld, E. F. (1990a). Juvenile GM2 gangliosidosis caused by substitution of histidine for arginine at position 499 or 504 of the alpha-subunit of beta-hexosaminidase. *J. Biol. Chem.* **265,** 9452–9457.

Paw, B. H., Tieu, P. T., Kaback, M. M., Lim, J., and Neufeld, E. F. (1990b). Frequency of three Hex A mutant alleles among Jewish and non-Jewish carriers identified in a Tay-Sachs screening program. *Am. J. Hum. Genet.* **47,** 698–705.

Paw, B. H., Wood, L. C., and Neufeld, E. F. (1991). A third mutation at the CpG dinucleotide of codon 504 and a silent mutation at codon 506 of the HEX A gene. *Am. J. Hum. Genet.* **48,** 1139–1146.

Peleg, L., Meltzer, F., Karpati, M., and Goldman, B. (1995). GM2 gangliosidosis B1 variant: Biochemical and molecular characterization of hexosaminidase A. *Biochem. Mol. Med.* **54,** 126–132.

Poenaru, L., and Akli, S. (1994). Molecular epidemiology of Tay-Sachs disease in Europe. *Biomed. Pharmacother.* **48,** 341–346.

Proia, R. L. (1988). Gene encoding the human beta-hexosaminidase beta chain: Extensive homology of intron placement in the alpha- and beta-chain genes. *Proc. Natl. Acad. Sci. (USA)* **85,** 1883–1887.

Proia, R. L., and Neufeld, E. F. (1982). Synthesis of beta-hexosaminidase in cell-free translation and in intact fibroblasts: An insoluble precursor alpha chain in a rare form of Tay-Sachs disease. *Proc. Natl. Acad. Sci. (USA)*. **79,** 6360–6364.

Proia, R. L., and Soravia, E. (1987). Organization of the gene encoding the human beta-hexosaminidase alpha-chain. *J. Biol. Chem.* **262,** 5677–5681.

Raghavan, S., Krusell, A., Lyerla, T. A., Bremer, E. G., and Kolodny, E. H. (1985a). GM2-ganglioside metabolism in cultured human skin fibroblasts: Unambiguous diagnosis of GM2-gangliosidosis. *Biochim. Biophys. Acta.* **834,** 238–248.

Raghavan, S. S., Krusell, A., Krusell, J., Lyerla, T. A., and Kolodny, E. H. (1985b). GM2-ganglioside metabolism in hexosaminidase A deficiency states: Determination in situ using labeled GM2 added to fibroblast cultures. *Am. J. Hum. Genet.* **37,** 1071–1082.

Redonnet-Vernhet, I., Mahuran, D. J., Salvayre, R., Dubas, F., and Levade, T. (1996). Significance of two point mutations present in each HEXB allele of patients with adult GM2 gangliosidosis (Sandhoff disease): Homozygosity of the Ile207Val substitution is not associated with a clinical or biochemical phenotype. *Biochim. Biophys. Acta* **1317,** 127–133.

Ribeiro, M. G., Pinto, R., Miranda, M. C., and Suzuki, K. (1995). Tay-Sachs disease: Intron 7 splice junction mutation in two Portuguese patients. *Biochim. Biophys. Acta* **1270,** 44–51.

Ribeiro, M. G., Sonin, T., Pinto, R. A., Fontes, A., Ribeiro, H., Pinto, E., Palmeira, M. M., *et al.* (1996). Clinical, enzymatic, and molecular characterization of a Portuguese family with a chronic form of GM2-gangliosidosis B1 variant. *J. Med. Genet.* **33,** 341–343.

Richard, M. M., Erenberg, G., and Triggs-Raine, B. L. (1995). An A-to-G mutation at the +3 position of intron 8 of the HEXA gene is associated with exon 8 skipping and Tay-Sachs disease. *Biochem. Mol. Med.* **55,** 74–76.

Schepers, U., Glombitza, G., Hoffmann, A., Chabas, A., Ozand, P., and Sandhoff, K. (1996). Molecular analysis of a GM2-activator deficiency in two patients with GM2-gangliosidosis AB variant. *Am. J. Hum. Genet.* **59,** 1048–1056.

Schroder, M., Klima, H., Nakano, T., Kwon, H., Quintern, L. E., Gartner, S., Suzuki, K., *et al.* (1989). Isolation of a cDNA encoding the human GM2 activator protein. *FEBS Lett.* **251,** 197–200.

Schroder, M., Schnabel, D., Suzuki, K., and Sandhoff, K. (1991). A mutation in the gene of a glycolipid-binding protein (GM2 activator) that causes GM2-gangliosidosis variant AB. *FEBS Lett.* **290,** 1–3.

Schroder, M., Schnabel, D., Hurwitz, R., Young, E., Suzuki, K., and Sandhoff, K. (1993). Molecular genetics of GM2-gangliosidosis AB variant: A novel mutation and expression in BHK cells. *Hum. Genet.* **92,** 437–440.

Shore, S., Tomczak, J., Grebner, E. E., and Myerowitz, R. (1992). An unusual genotype in an Ashkenazi Jewish patient with Tay-Sachs disease. *Hum. Mutat.* **1,** 486–490.

Sonderfeld-Fresko, S., and Proia, R. L. (1988). Synthesis and assembly of a catalytically active lysosomal enzyme, beta-hexosaminidase B, in a cell-free system. *J. Biol. Chem.* **263,** 13463–13469.

Tanaka, A., Ohno, K., and Suzuki, K. (1988). GM2-gangliosidosis B1 variant: A wide geographic and ethnic distribution of the specific beta-hexosaminidase alpha chain mutation originally identified in a Puerto Rican patient. *Biochem. Biophys. Res. Commun.* **156,** 1015–1019.

Tanaka, A., Ohno, K., Sandhoff, K., Maire, I., Kolodny, E. H., Brown, A., and Suzuki, K. (1990a). GM2-gangliosidosis B1 variant: Analysis of beta-hexosaminidase alpha gene abnormalities in seven patients. *Am. J. Hum. Genet.* **46,** 329–339.

Tanaka, A., Punnett, H. H., and Suzuki, K. (1990b). A new point mutation in the beta-hexosaminidase alpha subunit gene responsible for infantile Tay-Sachs disease in a non-Jewish Caucasian patient (a Kpn mutant). *Am. J. Hum. Genet.* **47,** 568–574.

Tanaka, A., Sakuraba, H., Isshiki, G., and Suzuki, K. (1993). The major mutation among Japanese patients with infantile Tay-Sachs disease: A G-to-T transversion at the acceptor site of intron 5 of the beta-hexosaminidase alpha gene. *Biochem. Biophys. Res. Commun.* **192,** 539–546.

Tanaka, A., Sakazaki, H., Murakami, H., Isshiki, G., and Suzuki, K. (1994). Molecular genetics of Tay-Sachs disease in Japan. *J. Inher. Metab. Dis.* **17,** 593–600.

Tews, I., Perrakis, A., Oppenheim, A., Dauter, Z., Wilson, K. S., and Vorgias, C. E. (1996). Bacterial chiotobiase structure provides insight into catalytic mechanism and the basis of Tay-Sachs disease. *Nat. Struct. Biol.* **3,** 638–648.

Thomas, G. H., Raghavan, S., Kolodny, E. H., Frisch, A., Neufeld, E. F., O'Brien, J. S., Reynolds, L. W., *et al.* (1982). Nonuniform deficiency of hexosaminidase A in tissues and fluids of two unrelated individuals. *Ped. Res.* **16,** 232–237.

Tomczak, J., and Grebner, E. E. (1994). Three novel β-hexosaminidase A mutations in obligate carriers of Tay-Sachs disease. *Hum. Mutat.* **4,** 71–72.

Tomczak, J., Boogen, C., and Grebner, E. E. (1993). Distribution of a pseudodeficiency allele among Tay-Sachs carriers. *Am. J. Hum. Genet.* **53,** 537–539.

Triggs-Raine, B. L., Feigenbaum, A. S., Natowicz, M., Skomorowski, M. A., Schuster, S. M., Clarke, J. T., Mahuran, D. J., *et al.* (1990). Screening for carriers of Tay-Sachs disease among Ashkenazi Jews. A comparison of DNA-based and enzyme-based tests. *N. Engl. J. Med.* **323,** 6–12.

Triggs-Raine, B. L., Akerman, B. R., Clarke, J. T., and Gravel, R. A. (1991). Sequence of DNA flanking the exons of the HEXA gene, and identification of mutations in Tay-Sachs disease. *Am. J. Hum. Genet.* **49,** 1041–1054.

Triggs-Raine, B. L., Mules, E. H., Kaback, M. M., Lim-Steele, J. S., Dowling, C. E., Akerman, B. R., Natowicz, M. R., *et al.* (1992). A pseudodeficiency allele common in non-Jewish Tay-Sachs carriers: Implications for carrier screening. *Am. J. Hum. Genet.* **51,** 793–801.

Triggs-Raine, B., Richard, M., Wasel, N., Prence, E. M., and Natowicz, M. R. (1995). Mutational analyses of Tay-Sachs disease: Studies on Tay-Sachs carriers of French Canadian background living in New England. *Am. J. Hum. Genet.* **56,** 870–879.

Trop, I., Kaplan, F., Brown, C., Mahuran, D., and Hechtman, P. (1992). A glycine250→aspartate substitution in the alpha-subunit of hexosaminidase A causes juvenile-onset Tay-Sachs disease in a Lebanese-Canadian family. *Hum. Mutat.* **1,** 35–39.

Vecht, J., Zeigler, M., Segal, M., and Bach, G. (1983). Tay-Sachs disease among Moroccan Jews. *Isr. J. Med. Sci.* **19,** 67–69.

Vidgoff, J., Buist, N. R., and O'Brien, J. S. (1973). Absence of β-N-acetyl-D-hexosaminidase A activity in a healthy woman. *Am. J. Hum. Genet.* **25,** 372–381.

Wakamatsu, N., Kobayashi, H., Miyatake, T., and Tsuji, S. (1992). A novel exon mutation in the human beta-hexosaminidase beta subunit gene affects 3′ splice site selection. *J. Biol. Chem.* **267,** 2406–2413.

Xie, B., McInnes, B., Neote, K., Lamhonwah, A. M., and Mahuran, D. (1991). Isolation and expression of a full-length cDNA encoding the human GM2 activator protein. *Biochem. Biophys. Res. Commun.* **177,** 1217–1223.

Xie, B., Wang, W., and Mahuran, D. J. (1992). A Cys138-to-Arg substitution in the GM2 activator protein is associated with the AB variant form of GM2 gangliosidosis. *Am. J. Hum. Genet.* **50,** 1046–1052.

Yoo, H. W., Astrin, K. H., and Desnick, R. J. (1993). Comparison of enzyme and DNA analysis in a Tay-Sachs disease carrier screening program. *J. Kor. Med. Sci.* **8,** 84–91.

Zhang, Z. X., Wakamatsu, N., Mules, E. H., Thomas, G. H., and Gravel, R. A. (1994). Impact of premature stop codons on mRNA levels in infantile Sandhoff disease. *Hum. Mol. Genet.* **3,** 139–145.

Zhang, Z. X., Wakamatsu, N., Akerman, B. R., Mules, E. H., Thomas, G. H., and Gravel, R. A. (1995). A second, large deletion in the HEXB gene in a patient with infantile Sandhoff disease. *Hum. Mol. Genet.* **4,** 777–780.

Zokaeem, G., Bayleran, J., Kaplan, P., Hechtman, P., and Neufeld, E. F. (1987). A shortened beta-hexosaminidase alpha-chain in an Italian patient with infantile Tay-Sachs disease. *Am. J. Hum. Genet.* **40,** 537–547.

18

Targeting the Hexosaminidase Genes: Mouse Models of the G_{M2} Gangliosidoses

Richard L. Proia
Genetics of Development and Disease Branch
National Institute of Diabetes and Digestive and Kidney Diseases
National Institutes of Health
Bethesda, Maryland 20892

In 1983, I started my own laboratory while I was a tenure track scientist at the National Institutes of Health. My laboratory's primary effort was focused on the isolation and characterization of the human *HEX A* and *HEX B* genes (Proia and Soravia, 1987; Proia, 1988). The fundamental structural and sequence information that we ultimately acquired fueled the flurry of activity over the next few years that led to the identification of the genetic lesions in Tay-Sachs and Sandhoff diseases (Gravel *et al.*, 1995). When Mario Capecchi and Oliver Smithies devised a method for introducing mutations in the mouse genome by gene targeting, it became clear that this technique would be extremely powerful, both for producing mouse models of genetic disease and for analyzing gene function (Figure 18.1) (Capecchi, 1989; Koller and Smithies, 1992). I believed that animal models would be essential tools in advancing toward a therapy for the G_{M2} gangliosidoses in much the same way that the information on the structure of the human genes was central for the identification of disease mutations. After a few years of struggling with the techniques required to accomplish the project, we finally produced our first litter of mice homozygous for the disrupted *HEX A* gene (Yamanaka *et al.*, 1994).

The onset of neurological symptoms in Tay-Sachs disease patients normally occurs by 3 to 5 months after birth. Accordingly, we anxiously watched the

Gene Targeting In Embryonic Stem Cells

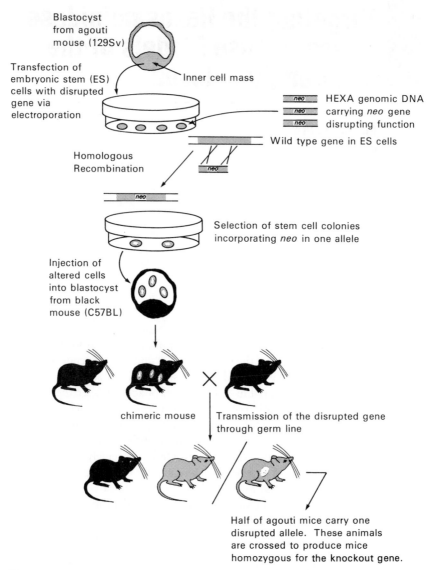

Blastocyst
from agouti
mouse (129Sv)

Transfection of
embryonic stem (ES)
cells with disrupted
gene via
electroporation

Inner cell mass

neo HEXA genomic DNA
neo carrying *neo* gene
neo disrupting function

Wild type gene in ES cells

Homologous
Recombination

neo

neo

Selection of stem cell colonies
incorporating *neo* in one allele

Injection of
altered cells
into blastocyst
from black
mouse (C57BL)

chimeric mouse │ Transmission of the disrupted gene
through germ line

Half of agouti mice carry one
disrupted allele. These animals
are crossed to produce mice
homozygous for the knockout gene.

Figure 18.1. Gene targeting to establish mouse models of human disease. Embryonic stem cells are isolated from the blastocyst of a mouse strain (129/Sv) with agouti colored fur. A targeting
vector is introduced into these cells, which can integrate into the Hex A gene through
homologous recombination. Since the targeting vector contains the antibiotic resistance
gene, neomycin (*neo*), only cells that incorporate the vector can survive in culture.

litter for what we were sure would be inevitable signs of the disease. We were quite astonished and somewhat disappointed when we finally realized that these mice could live happily without β-hexosaminidase A. Although these mice did not display any neurologic defects, they did exhibit the biochemical and pathologic hallmarks of the disease. As a result of their β-hexosaminidase A deficiency, G_{M2} ganglioside accumulated in the brains of the Tay-Sachs mice in an age-dependent fashion. This G_{M2} ganglioside accumulation resulted in storage neurons with membranous cytoplasmic bodies (MCBs) that were identical to affected neurons in Tay-Sachs disease patients. However, there was an important difference. Unlike Tay-Sachs disease patients in which, ultimately, all neurons are affected, certain regions of the nervous system in the *Hex A* −/− mice were spared or showed limited ganglioside storage.

Initially, the reason for the phenotypic difference between humans and mice with Tay-Sachs disease was a puzzle. The answer, the unexpected involvement of β-hexosaminidase B in the ganglioside degradation pathway, became clear when the Sandhoff disease model mouse was subsequently produced through the targeted disruption of the *Hex B* gene (Sango *et al.*, 1995). These mice displayed a deficiency of both β-hexosaminidase A and β-hexosaminidase B in contrast to the Tay-Sachs mice who were deficient only in β-hexosaminidase A.

In contrast to the asymptomatic Tay-Sachs mice, the Sandhoff mice showed severe, progressive neurologic manifestations beginning about 3 months after birth. The most prominent phenotypic features of the Sandhoff disease mice were abnormalities in motor function initially presenting as defects in balance and coordination at about 3 months of age and evolving, within a few weeks, into an almost complete loss of hind limb movement. Consistent with a severe phenotype was a diffuse accumulation of storage material throughout the central nervous system. Another striking difference between the two disease models was the significant accumulation of asialo-G_{M2} ganglioside (G_{A2}) in the Sandhoff disease mice.

Introduction of radiolabeled G_{M1} ganglioside into fibroblasts derived from the knockout mice revealed the existence of two independent catabolic pathways for the degradation of G_{M2} that explained the phenotypic difference (Figure 18.2). In human cells, β-hexosaminidase A degrades G_{M2} ganglioside to form G_{M3}. In addition to this pathway, mouse cells also have a detour pathway where G_{M2} is converted to G_{A2} by the action of sialidase. G_{A2} can now be

---

The *neo* containing ES cells are checked for homologous recombination at the locus of interest. The targeted cells are injected into the blastocyst of a black mouse strain (C57BL). Chimeric offspring are obtained with tissues derived from both the C57BL blastocyst and the targeted ES cells. The chimeric mice are crossed with C57BL mice. Since agouti is the dominant fur color, the agouti offspring must be derived from the mutant ES cells. These mice have a 50% chance of carrying the disrupted gene because the targeted ES cells had just one allele disrupted. The heterozygous mice are then bred together to obtain homozygous knockouts.

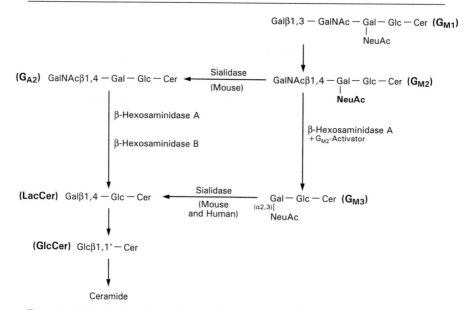

Figure 18.2. The ganglioside degradation pathway. The major degradative pathway in the mouse proceeds from G_{M2} to G_{A2} via removal of sialic acid (NeuNAc).

further degraded by β-hexosaminidase B (or A) and then by the action of other glycosidases to yield ceramide. In the Tay-Sachs mice the G_{M2} to G_{M3} conversion is blocked because of the absence of β-hexosaminidase A. The detour pathway in the mouse, via the conversion of G_{M2} to G_{A2}, remains intact. Thus β-hexosaminidase B in mice, but not in humans, can effectively participate in the G_{M2} ganglioside degradation because mouse sialidase has much greater activity against G_{M2} ganglioside than does the corresponding human enzyme. It is, therefore, the sialidase specificity difference between mice and humans that allows the degradation of G_{M2} by β-hexosaminidase B and explains the unexpectedly mild phenotype of the Tay-Sachs disease mice. Both degradation pathways are blocked in the Sandhoff disease mice where β-hexosaminidase A and B are missing. As a result both G_{A2} and G_{M2} ganglioside accumulate to high levels, causing a severe neurodegenerative disease.

Our next surprise came when we bred the Tay-Sachs and Sandhoff disease mice together to produce mice deficient in both the α- and β-subunits (Sango et al., 1996) These are so-called "double-knockout" mice. In addition to the expected gangliosidosis, mice with both hexosaminidase genes disrupted showed phenotypic, pathologic, and biochemical features of a second type of lysosomal storage disease—the mucopolysaccharidoses (Neufeld and Muenzer, 1995). These are a group of human-inherited diseases characterized by the inability to degrade glycosaminoglycans resulting in the storage of these substrates in tissues and their excretion in urine. Some of the major clinical features are short stature, facial dysmorphism, restricted joint mobility, shortened life span, and a constellation of skeletal abnormalities known as dysostosis multiplex. Like the patients, the double-knockout mice displayed these features. Analysis of the glycosaminoglycans in the urine of the double-knockout mice indicated the presence of dermatan sulfate and keratan sulfate. The absence of the characteristic clinical features of the mucopolysaccharidoses in the Tay-Sachs or Sandhoff mice showed that this phenotype was dependent on the deficiency of both genes.

Each of the human mucopolysaccharidoses involves the heritable absence of one of the enzymes that are required for the lysosomal degradation of glycosaminoglycans. β-hexosaminidase has long been suspected of being required for the degradation of glycosaminoglycans because of the presence of β-N-hexosaminyl residues in dermatan sulfate, keratan sulfate, and the chondroitin sulfates. It had been a mystery, however, why patients with Tay-Sachs and Sandhoff diseases showed none of the biochemical or phenotypic features associated with mucopolysaccharidosis. The severe mucopolysaccharidosis phenotype exhibited by double-knockout mice proves that β-hexosaminidase is an essential enzyme for the degradation of glycosaminoglycans. The lack of a mucopolysaccharidosis phenotype in humans and mice with one β-hexosaminidase gene mutated is, therefore, due to functional redundancy in the enzyme system. Although β-hexosaminidase A and S are absent in Tay-Sachs disease patients and mice, a large amount of the B isozyme is available for glycosaminoglycans degradation. In Sandhoff disease patients and mice where the major β-hexosaminidase isozymes, A and B, are absent, a low residual level of β-hexosaminidase S (an αα homodimer) is still present. This enzyme, previously thought to be useless, gives rise to just a few percent of normal levels of the total hexosaminidase activity. The lack of a mucopolysaccharidosis phenotype in the Tay-Sachs mice proves that β-hexosaminidase S functions to prevent a clinically significant accumulation of glycosaminoglycan and underscores how small amounts of residual enzyme activity can dramatically alter the course of a lysosomal storage disease (Conzelmann and Sandhoff, 1991).

A major reason for producing the G$_{M2}$ gangliosidosis mice was to establish disease models. The Sandhoff mice, with a neurologic phenotype comparable in severity to the human disease, have been extremely valuable for the identification of new therapeutic approaches for the G$_{M2}$ gangliosidoses, as well as for defining the molecular mechanisms underlying neurodegeneration (Norflus *et al.*, 1998; Liu *et al.*, 1999; Jeyakumar *et al.*, 1999; Wada *et al.*, 2000; Jeyakumar *et al.*, 2001). The Tay-Sachs and double-knockout mice, with their unexpected phenotypes, have also been valuable but for a different reason. They have provided new insights into the function of the hexosaminidase system by revealing an alternative degradative pathway in ganglioside catabolism and a new category of critical substrates for the enzyme system—the glycosaminoglycans.

Acknowledgments

I am grateful to my talented collaborators for their essential contributions to the work I have described. In particular I thank Shoji Yamanaka, Kazunori Sango, Masako Taniike, Kinuko Suzuki, Mark Johnson, Alex Grinberg, Heiner Westphal, Jacqueline Crawley, Michael McDonald, Chris Starr, Alexander Hoffmann, and Konrad Sandhoff.

References

Capecchi, M. R. (1989). Altering the genome by homologous recombination. *Science* **244**, 1288–1292.

Conzelmann, E., and Sandhoff, K. (1991). Biochemical basis of late-onset neurolipidoses. *Dev. Neurosci.* **13**, 197–204.

Gravel, R. A., Clarke, J. T. R., Kaback, M. M., Mahuran, D., Sandhoff, K., and Suzuki, K. (1995). The G_{M2} gangliosidoses. *In* C. R. Scriver, A. L., Beaudet, W. S., Sly and D. Valle, eds. "The Metabolic and Molecular Basis of Inherited Disease," 7th ed. pp. 2839–2879. McGraw-Hill, New York.

Jeyakumar, M., Butters, T. D., Cortina-Borja, M., Hunnam, V., Proia, R. L., Perry, H. V., Dwek, R. A., and Platt, F. M. (1999). Delayed symptom onset and increased life expectancy in Sandhoff disease mice treated with N-butyldeoxynojirimycin. *Proc. Natl. Acad. Sci. U.S.A.* **96**, 6388–6393.

Jeyakumar, M., Norflus, F., Tifft, C. J., Cortina-Borja, M., Butters, T. D., Proia, R. L., Perry, V. H., Dwek, R., and Platt, F. M. (2001). Enhanced survival in Sandhoff disease mice receiving a combination of substrate deprivation therapy and bone marrow transplantation. *Blood* **97**, 327–329.

Koller, B. H., and Smithies, O. (1992). Altering genes in animals by gene targeting. *Annu. Rev. Immunol.* **10**, 705–730.

Liu, Y., Wada, R., Kawai, H., Sango, K., Deng, C., Tai, T., McDonald, M. P., Araujo, K., Crawley, J. N., Bierfreund, U., Sandhoff, K., Suzuki, K., and Proia, R. L. (1999). A genetic model of substrate deprivation therapy for a glycosphingolipid storage disorder. *J. Clin. Invest.* **103**, 497–505.

Neufeld, E. F., and Muenzer, J. (1995). The mucopolysaccharidoses. *In* C. R., Scriver, A. L., Beaudet, W. S., Sly, and D. Valle eds. "The metabolic and molecular basis of inherited disease," 7th ed. pp. 2465–2494. McGraw-Hill, New York.

Norflus, F., Tifft, C. J., McDonald, M. P., Goldstein, G., Crawley, J. N., Hoffmann, A., Sandhoff, K., Suzuki, K., and Proia, R. L. (1998). Bone marrow transplantation prolongs life span and ameliorates neurologic manifestations in Sandhoff disease mice. *J. Clin. Invest.* **101**, 1881–1888.

Proia, R. L. (1998). The gene encoding the human β-hexosaminidase β-chain: Extensive homology of intron positions in the α- and β-chain genes. *Proc. Natl. Acad. Sci. U.S.A.* **85**, 1883–1887.

Proia, R. L., and Soravia, E. (1987). Organization of the gene encoding the human β-hexosaminidase α-chain. *J. Biol. Chem.* **262**, 5677–5681.

Sango, K., McDonald, M. P., Crawley, J. N., Mack, M. L., Tifft, C. J., Skop, E., Starr, C. M., Hoffmann, A., Sandhoff, K., Suzuki, K., and Proia, R. L. (1996). Mice lacking both subunits of lysosomal β-hexosaminidase exhibit mucopolysaccharidosis and gangliosidosis. *Nature Genet.* 14, 348–352.

Sango, K., Yamanaka, S., Hoffmann, A., Okuda, Y., Grinberg, A., Westphal, H., McDonald, M. P., Crawley, J. N., Sandhoff, K., Suzuki, K., and Proia, R. L. (1995). Mouse models of Tay-Sachs and Sandhoff diseases differ in neurologic phenotype and ganglioside metabolism. *Nature Genet.* **11**, 170–176.

Yamanaka, S., Johnson, M. D., Grinberg, A., Westphal, H., Crawley, J. N., Taniike, M., Suzuki, K., and Proia, R. L. (1994). Targeted disruption of the Hexa gene results in mice with biochemical and pathologic features of Tay-Sachs disease. *Proc. Natl. Acad. Sci. U.S.A.* **91**, 9975–9979.

Wada, R., Tifft, C. J., and Proia, R. L. (2000). Microglial activation precedes acute neurodegeneration in Sandhoff disease and is suppressed by bone marrow transplantation. *Proc. Natl. Acad. Sci. U.S.A.* **97**, 10954–10959.

19

Molecular Epidemiology of Tay-Sachs Disease

Neil Risch
Department of Genetics
Stanford University School of Medicine
Stanford, California 94305

I. Introduction
II. Mutations and Their Frequencies
   A. Tay-Sachs Disease
   B. Gaucher Disease
   C. Niemann-Pick Disease
   D. Mucolipidosis Type IV
   E. Other Ashkenazi Genetic Diseases
III. The Demographic History of the Ashkenazim
IV. Statistical Modeling
V. Conclusion
   References

I. INTRODUCTION

The Ashkenazi Jewish population is subject to a number of genetic diseases, perhaps the most prominent of which is Tay-Sachs disease (TSD). Although Tay-Sachs is not the most common genetic disease among the Ashkenazim, it has received a great deal of attention, probably because of its severity (lethality at a young age), its early recognition as a disease common among the Ashkenazim (Sachs, 1896), as well as the early identification of its biochemical defect (Okada and O'Brien, 1969; Sandhoff, 1969).

    The recognition that a disease occurs at a high frequency in a specific ethnic group provokes interest among population geneticists and evolutionists to

Advances in Genetics, Vol. 44

233

Table 19.1. Authors Arguing for Selection or Genetic Drift

Selection	Drift
Myrianthopoulos and Aronson, 1966	Chase and McKusick, 1972
Chakravarti and Chakraborty, 1978	Rao and Morton, 1973
Motulsky, 1979	Fraikor, 1977
Yokoyama, 1979	Wagener et al., 1978
Rotter and Diamond, 1987	Risch et al., 1995a
Zlotogora et al., 1988	
Jorde, 1992	
Beutler et al., 1993	
Diamond, 1994	

provide an explanation. The usual list of potential explanations includes a high mutation rate, stochastic or chance effects, typically manifest by founder effects, and selective advantage. Because nearly all genetic diseases with ethnic predilections are recessive, the selection argument usually takes the form of heterozygote advantage, either through increased reproductive fitness, or enhanced survival in the presence of a debilitating or lethal selective environmental agent.

Debate regarding the high frequency of TSD in the Ashkenazi population has a long history, dating back at least three decades. Elevated mutation rates in a particular population or ethnic group have not been given much credence on biological grounds. The two major contending hypotheses are natural selection (heterozygote advantage) and genetic drift (founder effect). Table 19.1 provides a partial list of the numerous commentators on this topic. Roughly, they fall into two groups, one favoring selection, the other drift. It should be noticed that while the debate was quite active through the end of the 1970s, after this time the pendulum appeared to swing toward the selectionist's point of view. This may have been due to increasing knowledge regarding other related genetic disorders occurring at high frequency among Jews, as well as a new understanding of the molecular basis for these disorders. However, with a detailed genetic analysis of another Ashkenazi genetic disease, idiopathic torsion dystonia, Risch et al. (1995a) have reopened the debate by once again arguing for the effect of genetic drift.

The arguments most often cited in this debate are listed in Table 19.2. From the point of view of those favoring drift, the demographic history of the Jewish population, with a small initial size and subsequently several bottlenecks and expansions, has provided several opportunities for the operation of stochastic effects. According to those favoring selection, mathematical analysis has shown that the probability of observing such a high TSD allele frequency ($q = .0133$) in the Ashkenazim, when the frequency in non-Jews is about 10-fold lower, is very small, on the order of 1% or less (Chakravarti and Chakraborty 1978; Yokoyama, 1979). By contrast, using similar mathematical arguments, and obtaining the

Table 19.2. Arguments for Selection versus Genetic Drift

Selection	Drift
Frequencies too high for drift alone	Small founder population
Statistical analysis shows low probability of such high gene frequencies	Several historical bottlenecks
Four different lysosomal storage enzyme deficiencies unlikely by chance alone	Observed diseases are not random, but extreme outliers
Multiple mutations at elevated frequency at same locus	No selective agent demonstrated, only speculated
	No apparent comparable selection on neighboring non-Jews

same probabilities, others have argued that such probabilities do not apply as they would in a standard statistical test, because we are observing only the most visible extreme tail of a distribution (a lethal recessive disease), and thus the low probability is to be expected (Rao and Morton, 1973; Wagener et al., 1979).

If selection has operated in the past, what is the selective agent? Most authors favor epidemics due to bacterial infections, such as tuberculosis, which had a major impact on the populations of Europe after urbanization (Myrianthopoulos et al., 1972). In fact, tuberculosis appears to be most often cited as a likely cause. However, others argue that no such demonstration of survival advantage for heterozygote carriers against tuberculosis, or for that matter any postulated infectious agent, has been made. Furthermore, they argue that tuberculosis and other infectious diseases struck non-Jews as well as Jews living in close proximity, yet the elevated gene frequencies appear only among the Jews; selective effects should have operated on non-Jews as well (Wagener et al., 1978).

As additional support for the selection hypothesis, it has been noted that four different diseases, all involving lysosomal storage enzymes [in addition to TSD, there are Gaucher disease (GD), Niemann-Pick disease (NPD), and mucolipidosis type IV (MLIV)], occur at elevated frequencies among Ashkenazi Jews. It appears beyond coincidence that four such diseases involving related pathways should be found.

A significant advance in our understanding of the population genetics of TSD and other Jewish genetic disease has occurred through the discovery over the past decade of the actual mutations underlying them. This has provided the opportunity, through random population screening as well as molecular analysis of patients and/or their carrier parents, of determining the actual allele frequencies for the various mutations. It has been argued that, under genetic drift, one would only expect to observe a single mutant allele at a given locus at elevated frequency; the existence of multiple, distinct mutations at elevated frequency provides support for selective advantage (e.g., Rao and Morton, 1973). As it has turned out, molecular analysis has revealed significant mutational heterogeneity in the Jewish population underlying TSD, as well as GD and NPD. Some have interpreted this observation as significant support for selection—for example, arguing that "Jewish lysosomes" have been hit by at least 10 distinct mutations (Diamond, 1994; also Bach et al., 1992). For TSD, there are three different mutations found commonly in the Ashkenazi population, two for the infantile form of the disease, and one for the adult, chronic form. Using a mathematical analysis similar to that of Chakravarti and Chakraborty (1978) and Wagener et al. (1978), Jorde (1992) has shown that even the second most common mutation (splice site mutation in exon 12) has a frequency too high to be readily explained by chance, supporting the selectionist argument.

In the following, I will review what is currently known about the frequency of the various mutations underlying TSD and the related disorders GD, NPD, and MLIV, the frequency of mutations in the same genes in other, non-Jewish populations, as well as in non-Ashkenazi Jews, the demographic history of the Ashkenazim in Europe, and what conclusions can be drawn from mathematical analysis of these data. Further, I suggest that additional molecular studies using haplotype analysis with linked genetic markers can help resolve the population genetic questions regarding these lysosomal storage diseases in the Jewish population by providing likely dates for their origins and relationship to other populations, as has been done for another Jewish genetic disease, idiopathic torsion dystonia (Risch et al., 1995a).

II. MUTATIONS AND THEIR FREQUENCIES

A. Tay-Sachs disease

TSD is caused by mutations in the *HEXA* gene locus, which encodes the alpha subunit of the enzyme hexosaminidase A. The most common mutation in the Ashkenazi population is a 4-bp insertion (TATC) in exon 11, which leads to a frameshift and subsequent downstream stop codon (Myerowitz and Costigan, 1988). The second most common mutation is a G→C transversion in the donor splice site of intron 12 (IVS12), which leads to misspliced and unstable mRNA (Arpaia et al., 1988; Myerowitz, 1988). These two mutations lead to the infantile severe form of the disease. A third mutation occurs in the Ashkenazi population; this is a missense mutation (G805A) which has residual but reduced enzyme activity and leads to the adult chronic form of the disease. The frequencies of these three mutations in the Ashkenazi population, based on enzyme deficiency-detected carriers from three random population screens, have been combined into a single estimate and are given in Table 19.3. In addition, the frequencies obtained from a large screen (n = 13,349) of an Orthodox community in Brooklyn by direct DNA analysis of all individuals are given (Ekstein, personal communication). There are some differences in the derived frequencies, which may represent actual

Table 19.3. Frequencies of HEXA Mutations

Mutation	Ashkenazi Jews		Non-Ashkenazi Jews		Non-Jews[e]
	3 Studies[a]	Dor-Yeshorim[b]	Proximate[c]	Distant[d]	
(1278TATC)	.0128	.0184	.0008	.0001	.0004
(+1 IVS12)	.0027	.0014	.0003	.0002	0
(G805A)	.0005	.0006	.0001	0	.0001
(+1 IVS9)	0	0	0	0	.0003
Other	.0008	.0009	.0008	.0016	.0012

[a]Based on an assumed total mutation frequency of .0168; relative mutation frequencies taken from Triggs-Raine *et al.* (1990); Paw *et al.* (1990); and Grebner and Tomczak (1991).
[b]Based on DNA analysis of random population screen; Ekstein (personal communication).
[c]Jews from Bulgaria, Turkey, Azerbeijan, Georgia; assumed total mutation frequency of .002. Peleg *et al.* (1994).
[d]Jews from Libya, Iraq, Tunisia, Yemen, Morrocco; assumed total mutation frequency of .002; Peleg *et al.* (1992).
[e]Assumed total mutation frequency of .002; Gravel *et al.* (1995).

differences in the populations studied, or just sampling variation. The frequency of the 1278 TATC mutation appears higher in the Brooklyn data, but the +1 IVS 12 mutation appears to have a lower frequency. Taking approximate averages, the estimated allele frequency for 1278 TATC would be .0156, and for +1 IVS12 it would be .0020. The frequency of the less common adult mutation, G805A, is about .0005. Other mutations have also been found in this population, none with a substantial frequency; cumulatively, the frequency is .0008 for these remaining mutations.

The frequency of these mutations, and other mutations, in both non-Jewish populations as well as non-Ashkenazi Jewish groups, is of considerable interest because it can shed light on potential mechanisms for their existence and frequency in the Ashkenazi population. Peleg *et al.* (1994) examined the frequency of the major HEXA mutations in enzyme deficiency-detected carriers of non-Ashkenazi Jewish extraction living in Israel. They separated these subjects into two groups based on country of origin: (1) those living in a proximate location to Eastern and Central Europe, whence most Ashkenazi Jews derive—this group includes the countries of Bulgaria, Turkey, Azerbeijan, and Georgia; and (2) those living at a greater distance from Europe, including the countries of Libya, Iraq, Tunisia, Yemen, and Morocco. Based on the relative frequencies of the various mutations among carriers, I have estimated their absolute population frequencies assuming a total allele frequency of .002 (Gravel *et al.*, 1995) (Table 19.3). The results are quite revealing. The Ashkenazi mutations are present but occur at a much lower frequency in these other Jewish groups. Notably, while the frequency ratio of the TATC mutation to the IVS-12 mutation in Ashkenazim is quite high, around 7.8, the ratio in non-Ashkenazi Jews is much less, about 2.7 for those living in proximity and 0.5 for those living distally. The milder G805A mutation also has a much lower frequency in the Jewish non-Ashkenazi groups. The estimated frequencies in non-Jews, again assuming a total allele frequency of .002, and using the observed relative frequencies of the various mutations in enzyme-detected carriers, are also given in Table 19.3. Here, the common Ashkenazi TATC mutation is also the most frequent; the IVS-12 mutation appears to be absent, while another mutation, not found in Jews, a splice site mutation in intron 9, is the second most common (Gravel *et al.*, 1995). The G805A mutation is also found at low frequency.

It is tempting to speculate about the numbers in Table 19.3. The dramatically higher frequency of the 1278 TATC mutation in the Ashkenazim compared to other populations suggests that the mutation may have originated in this population; then its frequency in other populations would be the result of recent admixture. This hypothesis is also consistent with the higher frequency (.0008) observed in Jews deriving from countries bordering Eastern and Central Europe than in those from farther away (.0001) or in European non-Jews (.0004). If so, we can estimate the admixture rate from the Ashkenazim into these other populations.

If we assume that the allele frequency among non-Ashkenazim represents exclusively admixture from the Ashkenazim, then the admixture proportion $\alpha = p_1/p_2$, where p_1 is the allele frequency in the population under consideration and p_2 is the frequency in the Ashkenazim. Then, according to the frequencies in Table 19.3, we would estimate α to be 5% for Jews from Bulgaria, Turkey, Azerbeijan, and Georgia, 0.7% for Jews from the MidEast and North Africa, and 2.6% for European non-Jews. These values appear plausible. This model can also be tested, in theory, by haplotype analysis using linked markers (described later).

Using the same admixture rate, we would predict a smaller proportion of the alleles of the second most common mutation, IVS-12, in the non-Ashkenazi populations to be due to Ashkenazi admixture. For the proximate non-Ashkenazi Jews, we would expect a frequency of $(.002)(.05) = .0001$ due to admixture; for the distant Jews, the frequency would be $(.002)(.007) = .000014$, and in non-Jews $(.002)(.026) = .00005$. The observed frequencies of .0003 for proximate Jews, .0002 for distant Jews, and 0 for non-Jews suggests that this mutation, despite being less frequent in the Ashkenazim, may actually be older, being present in various Jewish groups predating their separation, and thus having a Middle Eastern origin (Peleg *et al.*, 1994). This would also be consistent with its relative absence in non-Jews.

The G805A adult mutation is quite infrequent in the non-Ashkenazi groups, but also only of modest frequency in the Ashkenazim (.0006). Its presence in the non-Ashkenazi groups may also represent recent Ashkenazi admixture.

Finally, a distinct mutation, another splice site mutation in intron 9 (+1 IVS-9), occurs relatively frequently in non-Jews (.0003), but appears to be absent in the Jewish groups.

The geographic origins of carriers of disease mutations is also of some interest, as it can provide potential clues regarding the site and time of origin of their foundings. Although TSD carriers were originally thought to aggregate in Eastern Europe, especially Lithuania (Goodman, 1979), a later, larger systematic study of enzyme-detected carriers showed that the highest carrier frequencies occurred in Jews from Central Europe, such as Austria, Hungary, and Czechoslovakia (Petersen et al., 1983). No such similar analysis has been done based on DNA-characterized mutations. Because most carriers have the 1278 TATC mutation, we can assume that the geographic distribution observed by Peterson et al. (1983) most likely represents the 1278 TATC mutation. The high frequency in Central Europe is of interest, because it might suggest the presence of this mutation prior to migration of the Jews eastward into Poland and Russia. However, a large portion of the gene pool of Jews from Central Europe actually derives from migrations back from the east to the west. A finer analysis of geographic origin of carriers separated by the specific mutations would be of considerable interest. Such an analysis has been done for the French Canadians, who also have two prominent TSD mutations, revealing two distinct sites of origin.

It is also important to comment on the elevated frequency of HEXA mutations in other select non-Jewish populations. Perhaps the best studied are the French Canadians, who have a TSD prevalence approaching that of the Ashkenazim, and who also harbor two common mutations. The most common mutation, and the one first discovered, is a 7.6-kb deletion covering the first exon and flanking sequences (Myerowitz and Hogikyan, 1986). Using the excellent genealogical data in the French Canadian population, DeBraekeleer et al. (1992) were able to identify the likely source of this founder effect to be the south shore of the St. Lawrence River in the Gaspé region of Quebec. This mutation appears to account for approximately 80% of HEXA mutations found in this population. A second mutation, a G→A transition leading to a splice site mutation in intron 7, is the second most common mutation in French Canadians. It has been found to derive from founders in the Saguenay-Lac St. Jean region of Quebec. Overall, the mutational frequency distribution of HEXA mutations in French Canadians parallels that in Ashkenazim. One mutation predominates, accounting for approximately 80% of all mutations (Hechtman et al., 1992). A second mutation accounts for the majority of the remainder.

At least five other population isolates have been discovered to harbor a high frequency of HEXA mutations, including the Acadians of Louisiana, a Pennsylvania Dutch religious isolate, Moroccan Jews, and groups in Switzerland and Japan (Gravel et al., 1995). To the extent studied, the mutational pattern in these various isolates are distinct. In the Acadians, the Ashkenazi 1278 TATC and non-Jewish IVS-9 mutations appear; in the Pennsylvania group, the IVS-9 mutation predominates, but a second pseudo-deficiency allele is also common. In the Japanese, a unique splice site mutation in intron 5 is most common. Interestingly, in the Moroccan Jews, a unique mutation, a 3-bp deletion (ΔTTC) in exon 8 predominates.

It is also worth commenting briefly on the *HEXB* gene locus, mutations in which lead to the clinically related Sandhoff disease. Population screens based on enzyme activity have revealed a carrier frequency of about 1/500 in Ashkenazi Jews, and 1/278 in non-Jews. Thus, it appears that the Ashkenazim have a reduced frequency of mutations in the *HEXB* gene compared to non-Jews.

B. Gaucher disease

The second lysosomal lipid storage disease we consider is Gaucher disease (GD), which is the most common genetic disease in Ashkenazi Jews. GD is clinically subgrouped into three forms: type I, the mildest form, without nervous system involvement (or nonneuronopathic); type II, which is the most severe, involving neuronic pathology and death by age 2; and type III, which is also neuronopathic but less severe than type II.

It is type I disease that is markedly increased in the Ashkenazim. GD is caused by mutations in the gene for the enzyme glucocerebrosidase, which catabolizes β-glucosides. The most common mutation in the Ashkenazi population is the missense mutation 1226G (or, in amino acid notation, N370S). This mutation has been estimated to have a frequency of 3.4% in the Ashkenazim (Table 19.4). In addition, there are numerous other mutations that occur in this and the non-Jewish population. The five most prominent are the 84GG mutation (indicating an insertion of a G at nucleotide 84), with a frequency of .0053, the missense mutation 1604A, with a frequency of .0029, the missense mutation 1448C with

Table 19.4. Frequencies of Glucocerebrosidase Mutations

Mutation	Ashkenazi Jews	
	Beutler *et al*[a]	Dor-Yeshorim[b]
1226G	.0321	.0340
84GG	.0022	.0053
1604A	—	.0029
1488C	<.001	.0010
1297T	—	.0006
+1 IVS2	<.001	.0002

[a]DNA analysis of random population screen; Beutler *et al*. (1993).
[b]DNA analysis of random population screen; Ekstein *et al*. (1993).

a frequency of .0010, the missense mutation 1297T with a frequency of .0006, and the splice site mutation with a frequency of .0002. These gene frequency estimates are obtained from a large population study ($n = 21,369$) of Ashkenazim. No such comparable studies have been performed in non-Ashkenazi Jews or non-Jews. Rather, the frequencies of these various mutations have been determined in affected individuals. The latter may not give an accurate representation of their general population distribution because the various mutations are associated with differing clinical severity (Sibille *et al.*, 1993). Homozygotes for the 1226G mutation appear to have the mildest manifestation, those heterozygous for 1226G and allele 84GG are the most severe, while those with a 1226G/1448C genotype are intermediate (Sibille *et al.*, 1993). Indeed, asymptomatic homozygotes for 1226G have been detected, and many affected such individuals have a late age of onset. Thus, one might expect a reduced relative frequency of the 1226G allele in a patient sample compared to an unselected population sample. In fact, such an observation was made by Beutler *et al.* (1993), who noted a ratio of 14.8 for the frequency of 1226G relative to 84GG in a sample of 2000 normal Ashkenazim, but only 6.4 in a sample of 121 Jewish patients. They attributed this to a deficit of 1226G homozygotes among the patients.

However, results from the much larger population sample ($n = 21,369$, Table 19.4) tell a somewhat different story. Here, the frequency of the 84GG allele appears to be nearly 2.5 times higher than previously estimated; the frequency ratio of the 1226G and 84GG alleles is now .0340/.0053 $= 6.4$, identical to what is observed in patients (Beutler *et al.*, 1993; Sibille *et al.*, 1993; Horowitz *et al.*, 1993) On the other hand, the allele 1448C does appear to occur more frequently in patients (3.4% of mutant alleles; Horowitz *et al.*, 1993) than in normal carriers (2.3%; Ekstein, personal communication), although this difference may not be significant.

In any event, accurate estimates of mutant allele frequencies in non-Ashkenazi Jews or non-Jews are not readily obtained. However, based on the mutation distribution among patients, it is possible to draw some conclusions. The 84GG mutation appears to be absent or extremely uncommon, suggesting that this mutation is unique to the Ashkenazim and probably of recent origin. As in the Jews, the 1226G mutation is the most frequent, but accounts for a much smaller proportion of all mutations, while the 1448C mutation, relatively uncommon in Jews, is the second most common. The ratio of 1226G to 1448C alleles in non-Jews is 3.2 (Horowitz *et al.*, 1993), far less than the corresponding ratio of 34 observed in Ashkenazim (Ekstein, 1997). Without accurate estimates of frequency in non-Jews, it is impossible to draw conclusions regarding the origins of these other alleles. It is likely that some of the 1226G alleles found in non-Jews or non-Ashkenazi Jews is due to Ashkenazi admixture. However, its high frequency (.0045) in the general Portuguese population and even higher frequency in a northern Portuguese isolate (Lacerda

Table 19.5. Frequencies of Acid Sphingomyelinase Mutations

Mutation	Ashkenazi Jews[a]
R469L	.0045
L302P	<.001
fsP330	.001

[a]DNA analysis of random population screen; Caggana et al. (1994).

et al., 1994) suggests that it may be older and predate the isolation of the Ashkenazim in Eastern Europe. A similar argument applies to the less common 1448C mutation.

C. Niemann-Pick disease

Niemann-Pick disease is a third lysosomal storage disorder found at increased frequency in Ashkenazi Jews. This disease is subdivided into two variants, labeled A and B. Type A NPD is characterized by early onset (infancy) and a rapidly progressing course involving significant neurodegeneration and early fatality. Type B NPD is a milder form, with onset in childhood and little or no neurologic involvement. In both forms, patients have an accumulation of sphingomyelin due to a deficiency in the enzyme acid sphyngomyelinase (ASM) (Schuchman and Desnick, 1995). Type A patients have little or no ASM activity, while type B patients have about 5–10% normal activity.

Both forms are due to mutations in the ASM gene, with clinical manifestation depending on the mutations present. It is type A disease that occurs at an increased frequency in the Ashkenazim. Three mutations have been identified in Ashkenazi patients, all of which appear to eliminate enzyme activity (Schuchman and Desnick, 1995): two point mutations leading to amino acid (missense) substitutions, R496L and L302P, and a single-base deletion leading to a frameshift and early protein termination (fsP330). These three mutations are found in approximately equal proportions among type A patients (36%, 24%, and 32%, respectively). Assuming a prevalence of 1/40,000 of type A NPD in Ashkenazim (Goodman, 1979), or a total allele frequency of .0050, gives population frequencies for these three alleles of .0018, .0012, and .0016, respectively. However, in a population screen of 1000 normal Ashkenazi subjects, 9 carriers of R469L were detected, 2 carriers of fsP330, and no L302P carriers (Table 19.5). These results are highly discordant from expectations based on the genotypes of patients, for reasons as yet unclear.

For these three alleles, only the R496L mutant has been found in non-Jews, and it accounts for 4.7% of all mutations (Levran, 1992). If, in fact, the frequency of this mutation is elevated in the Ashkenazim, these few alleles in non-Jews may represent Ashkenazi admixture; alternatively, they could represent an older mutation predating Ashkenazi isolation. The absence of the other two mutations in non-Jews suggests that these are likely to be recent and unique to the Ashkenazim.

Only two type B NPD patients have been reported to date; one carried the R496L allele, while both carried another mutation, ΔR608, a 3-bp deletion leading to the deletion of an arginine at position 608. This mutation has also been deleted in non-Jewish type B patients, suggesting an older origin.

D. Mucolipidosis type IV

The fourth lipid storage disease is mucolipidosis type IV (MLIV). Most patients identified to date with this disorder are Ashkenazi Jews. The deficient enzyme in this disease has yet to be discovered, so no population carrier screens have been undertaken. However, based on the disease incidence, Bach et al. (1992) have estimated the allele frequency to be .01. Significant clinical variation in this disorder exists, including an age of onset ranging from 1 year to mid-thirties (Bach et al., 1992), which may relate to underlying mutational heterogeneity.

E. Other Ashkenazi genetic diseases

While much attention has been given to the elevated frequency of four lysosomal storage diseases in the Ashkenazim, in fact numerous other unrelated disorders also occur at increased frequency in this population. Many mutations underlying these disorders have achieved frequencies comparable to or exceeding those observed for the lipid storage disorders.

The following is a list of those currently identified; undoubtedly more will be discovered in the future: Bloom syndrome, BRCA1 (two mutations), BRCA2, Canavan disease (two mutations), cystic fibrosis, cystinuria, factor XI deficiency (two mutations), familial dysautonomia, familial hypercholesterolemia, familial hyperinsulinisin (two mutations), Fanconi anemia, glycogenesis type VII (two mutations), hereditary non-polyporis colon cancer, idiopathic torsion dystonia, and nonclassical 21-hydroxylase deficiency. Notable among these are the 1281L mutation at 21-hydroxylase, which is reported to have a frequency of .20, the HNPCC mutation with a frequency of .030, the two factor XI mutations (type II and type III) with frequencies of .022 and .025, respectively, the E285A mutation of aspartoacylase (Canavan disease) with a frequency of .011, the Bloom syndrome mutation with a frequency of .009, the W1282X mutation for cystic fibrosis, with a frequency of .011, the cystinuria mutation with a frequency of .008, the Fanconi anemia C mutation with a frequency of .006, and the two breast cancer mutations 185 del AG at BRCA1 and 6174 del T at BRCA2, each with a frequency around .006. These mutations all represent significant founder effects in the Ashkenazi population.

III. THE DEMOGRAPHIC HISTORY OF THE ASHKENAZIM

In evaluating the distribution of mutations for the various lipid storage disorders, as well as the numerous other disorders found at elevated frequency in the Ashkenazim, it is important to consider the demographic history of this population. Its origins are highly debated, but it is known that a substantial number of Jews were living in Europe, in particular in the Rhine region of Central Europe, and in northern France, in the eleventh century. The number of Jews living there at the time is unknown, but may have numbered as many as several hundred thousand (Engelman, 1960). Events at the end of this century occurring in Christian Europe were to have a major impact on the future demographics of Jews from this region. A burgeoning religious fanaticism, which may also have been fueled by a resentment of Jewish economic success, led to discrimination and ultimately violent attacks on Jews of Western Europe. The series of Crusades, which began in 1096, led to devastation of communities in the Rhineland and northern France. Such events continued unabated for nearly two centuries, at which point the remaining Jews were formally expelled from England, France, and Germany. In contrast to events occurring in Western Europe, new opportunities were presenting themselves in Poland and Lithuania. In fact, Jews living in the Polish kingdom were protected from persecution, and were granted privileges. They were encouraged to settle and create livelihoods. These protections and privileges, originally granted by Prince Boleslas in 1264 and extended by King Casimir, survived for nearly five centuries. It was this tolerant attitude toward Jews in Poland that shaped their destiny in Europe. It allowed their numbers to increase steadily so that they ultimately became the dominant Jewish population in the world.

The evidence that Polish Jews originated primarily from Western Europe, especially Germany, is quite compelling, although a minority may also have arrived from Italy or the east (Weinryb, 1972). This is also reflected in their name Ashkenazi, which is a Hebrew form referring to German, as well as in the language Yiddish, which has primarily a basis in German.

It is also generally accepted that the Jewish population of Europe reached a nadir around the end of the fifteenth century (Engleman, 1960; Fraikor, 1977). At this time, it has been estimated that between 10,000 and 20,000 Jews were living in the territories of Poland and Lithuania. This population was to grow into one of 6 million over the next four centuries, despite some setbacks in between (Weinryb, 1972).

Although the history of European Jews prior to the migrations to the east is unclear, it seems most likely that events in the Polish empire from the four centuries between 1500 and 1900 had an impact on the genetic uniqueness of this population, based simply on numbers. The gene pool of current-day Ashkenazim is dominated by the descendants of the early settlers in this region.

Probably the best demographic history of the Jews of Poland has been given by Weinryb (1972), who gives a scholarly account of the likely number of

Jews living in the territory at various points in history. The first reliable census in the region was not until 1765, so numbers for prior to this time are necessarily speculative. These numbers are often estimated from tax rolls and texts of early commentators, although Weinryb cautions on the reliability of these sources. Weinryb's estimate for the number of Jews dwelling in Eastern Europe in 1500 is 10,000. Although some have argued for a higher number, such as 24,000, Weinryb effectively refutes these claims as distortions. He further estimates the growth of these numbers to approximately 200,000 by 1648, the eve of the Cossack massacres. The next 20 years, until 1667, were a time of unprecedented destruction and devastation of the burgeoning Jewish communities of Eastern Europe, particularly in the regions of the Ukraine, but spreading further north as well. The loss of life during this time has been the subject of much debate, with some arguing for the loss of over 90% of the population (Fraikor, 1977). However, a more realistic estimate, given by Weinryb, is 25%, which would have brought the numbers down to 160,000 by 1667. The diminishing fanaticism after this time point once again allowed the Jewish population to expand rapidly, and it has been estimated that at the time of the 1765 census (with adjustment for underreporting) there were a total of 750,000 Jews in Poland and Lithuania. The next census, in 1897, put the number at 6 million, including all those residing in the Russian empire as well as the Galician part of Austria-Hungary. A rough estimate of the number of Jews at various time points is summarized in Table 19.6 (taken from Risch *et al.*, 1995b). Examination of the numbers in the table suggests a reasonably consistent growth pattern of approximately 50% per 25-year generation during this time period. While some have argued that growth prior to the nineteenth century was much less, due to high mortality rates from disease epidemics (Zoosmann-Diskin, 1995; Ettinger, 1969), the numbers in Table 19.6 suggest otherwise, as also argued by Weinryb (1972).

Two other factors require additional comment. Early on, the Jews of this region led a seminomadic existence, living in small groups in relative isolation (Fraikor, 1977). Over time, as the numbers increased, they became organized into larger communities, primarily in the larger cities such as Lublin, Krakow, and Poznan. However, it is probably true that for a substantial period of time, the population was subdivided into relatively isolated subgroups with reduced

Table 19.6. Estimated Number of Ashkenazim in Eastern Europe at Various Time Points

Year	Number
1500	10,000
1648	200,000
1667	160,000
1765	750,000
1897	6,000,000

From Weinryb (1972).

migration and matings between them. This would have had the effect of reducing the effective population size for the entire population as a whole.

The second important point is that once Jews had settled in the major cities and their population growth had begun, the reproductive patterns were not uniform across all segments of society. In fact, it is likely that large differences in the number of surviving offspring existed based on social stratification. As pointed out by Weinryb (1972) numerous times, the wealthier had larger and more sanitary accommodations, and better and healthier food, allowing them to support a large number of children through adulthood; by contrast, the poorer individuals could not support such numbers. He indicates that the socially prominent often had 8–10 children, while the poor could only support 1–2.

Another essential point made by Weinryb (discussed further below) was that in the time of epidemics, the wealthier Jews were able to escape the cities and live in the country, where the impact was much less; the poor had no such option. Thus, the demographic fact was that much of the population growth was produced by those from the upper social classes, while the poorer classes were reproducing only at replacement level. Such a pattern would serve to further reduce the effective population size (Risch et al., 1995a, 1995b).

IV. STATISTICAL MODELING

An important feature of the debate regarding the role of selection versus drift in explaining the high rates of certain genetic diseases in the Ashkenazim relates to statistical modeling of the observed frequencies in Jews versus non-Jews. This modeling has primarily addressed the frequency of Tay-Sachs disease. The question posed is: What is the probability of observing an allele frequency as high as that observed in the Ashkenazim given the much lower frequency in non-Jews? Originally, Rao and Morton (1973) calculated this probability assuming selective neutrality for the TSD allele; the calculation was subsequently modified to allow for the allele being a recessive lethal by Chakravarti and Chakraborty (1978) and by Wagener et al. (1978). The probability is calculated from a formula of Sewall Wright (1937) based on several assumptions: (1) that the allele frequency represents a steady-state frequency (i.e., equilibrium) resulting from a balance among mutation, migration, and drift; (2) that the effective population size is a given constant value. Using an assumed TSD allele frequency of .0133 in Ashkenazi Jews and a frequency of .0015 in non-Jews, Chakravarti and Chakraborty (1978) and Wagener et al. (1978) calculated a probability of .0012 for an effective population size N of 10,000, .007 for an N of 5000 and .031 for an N of 1000. Naturally, the smaller the effective population size, the greater the probability of a high allele frequency by chance.

Interestingly, Chakravarti and Chakraborty (1978) interpreted these low probabilities as supportive of selection operating, while Wagener et al. (1978)

reached the opposite conclusion. While they agreed regarding the relatively small probability for such a high allele frequency, Wagener *et al.* (1978) argued that it is inappropriate to evaluate this probability in a conventional statistical setting, as in formal hypothesis testing. They argued that TSD was not a randomly selected locus, but was picked precisely because of its unusually high frequency, and therefore needs to be considered as in the tail of a distribution. From this perspective, then, the low probabilities are not unexpected.

However, the identification of other lysosomal storage diseases at high frequency in the Ashkenazim and the recent molecular characterization of them have provided new fuel for the debate. It is now known that there are four such diseases at elevated frequency in the Ashkenazim (GD, NPD-A, and MLIV, in addition to TSD), and that multiple different mutations characterize each of them (except MLIV, which has yet to be cloned). The sizable array of mutations affecting these enzymes found in the Ashkenazim has been taken by some as conclusive evidence for selection; for example, Diamond (1994) has characterized the situation in terms of "Jewish lysosomes" historically subjected to selection in Eastern Europe due to crowding and disease.

From the statistical perspective, Jorde (1992) asked the question: What is the probability for such a high allele frequency for the *second* most common TSD mutation in the Ashkenazim? Using the same Wright (1937) formula as used by Chakravarti and Chakraborty (1978) and Wagener *et al.* (1978), he calculated that for an effective population size N of 10,000, and a frequency of 1/300 for the intron 12 splice site mutation in Ashkenazim versus 1/1,200 in non-Jews, the probability is only .03 for such a high allele frequency in Ashkenazim. Thus, he argued that, in the absence of selection, it is unlikely that the *second* most common mutation would also be statistically increased in frequency.

As mentioned above, these arguments are based on several assumptions. The most important of these is that the observed allele frequency represents an equilibrium state between the opposing forces of mutation and migration and selection. In fact, this is unlikely to be the case, because of the remarkable demographic history of the Jews, with a substantial contraction after their expulsion from Palestine to the sixteenth century, when the numbers began to rapidly increase. In searching for explanations for allele frequencies in the Ashkenazim, it is critical to examine their demographic history.

It is also possible to apply some statistical arguments to the selection hypothesis. First, it is important to characterize the time and location over which the selection operated. Those arguing for selection usually assume that the selective agent was an infectious disease (or diseases) leading to epidemics in Europe. Also, because the diseases are unique to the Ashkenazim among the Jewish groups, the selection must have been operating after their relative isolation, and so must be confined to the past 1000 years, at most, or 40 generations, assuming 25 years per generation. How much could an allele frequency increase during that time due to heterozygote advantage? Suppose that the selective advantage of the heterozygote is s, that is, the relative survival/reproductive success of the heterozygote Aa is $1 + s$ compared to the normal homozygote AA. Then, to a close approximation, the frequency of the allele a will increase by a factor $1 + s$ per generation (this is true whether the abnormal homozygote aa is genetically lethal or normal, as this will have little effect when q is small). Thus, after n generations, the allele frequency q_n will increase from an initial value of q_0 to $q_0(1 + s)^n$. Suppose q_0 is the observed allele frequency in non-Jews. Then $(1 + s)^n = q_n/q_0$, so that $s = e^{1/n \log(q_n/q_0)} - 1$. For the exon 11 TATC mutation, the estimated allele frequency in the Ashkenazim is .015, while in non-Jews an estimated frequency is .004, giving $q_n/q = 37.5$. Assuming $n = 40$, the above equation for s gives $s = .095$; if the selection were acting over only half that time period, say 500 years ($n = 20$), the selective advantage $s = .199$. Thus, to explain the high frequency of the TATC mutation based on selection alone requires very high selection coefficients; it appears that such a large advantage, as seen, for example, with hemoglobin variants and malaria, should be readily demonstrable. To date, the selective agent and the advantage due to lysosomal enzyme deficiency have been speculated about, without confirmatory support.

Alternatively, Jewish demographic history needs to be considered as an explanation for the numerous founder effects observed in this population. Thompson and Neel (1978), through statistical analysis, showed that for a population experiencing rapid expansion from a small original pool, the probability of elevated allele frequencies at some loci, or "private polymorphisms," is quite high. They applied this theory to explain the sizable number of private polymorphisms found in Amerindian groups. More recently, Thompson and Neel (1997) showed that rapid population expansion leads to sizable probabilities of founder effects, and that when considering a particular (disease) locus, it is not unexpected to find several different mutations at elevated frequencies. Because many populations have undergone significant expansions in recent history, they argue that selective advantage is not required to explain population-specific genetic diseases.

It appears that the Ashkenazim provide a remarkably comparable demographic history to the Amerinds. The tribal analogy is actually quite accurate, as in the Middle Ages the population had a nomadic-type existence, living in small subpopulations or bands with limited mobility and intermarriage (Fraikor, 1977). In fact, Risch et al. (1995a, 1995b) noted that the enormous expansion that occurred over many centuries (up to 1900) could readily explain the observed distribution, and indeed the large region of linkage disequilibrium observed on chromosomes carrying the idiopathic torsion dystonia mutation in this population. In effect, the statistical modeling previously applied, assuming equilibrium with a fixed population size, was inappropriate; the correct modeling strategy is one determining probabilities of founder effects in a rapidly expanding population.

V. CONCLUSION

The last decade has added considerable new data regarding the identity and frequency of mutations found at high frequency in the Ashkenazi population, as well as their distribution and frequency in non-Ashkenazi Jewish and non-Jewish populations. This is true for both lysosomal storage and other diseases, several of which were previously unrecognized. These new data enable a fresh view and insight regarding the likely explanation for the high frequency of TSD and the other lysosomal storage diseases found in the Ashkenazim.

The essential observations are as follows: four lysosomal storage diseases are found at high frequency in the Ashkenazim. Multiple mutations have been identified for three of the four, while the fourth gene has yet to be identified. For TSD, the three most common mutations have frequencies of about .015. (1278 TATC), .0014 (+1 IVS12), and .0006 (G805A); for GD, the five most common mutations have frequencies of about .033 (1226G), .004 (84GG), .003 (1604A), .001 (1488C), and .0006 (1297T); for NPD-A the two most common alleles have frequencies of .0045 (R469L) and < .001 (2302P). On the surface, the numerous mutations and their frequencies present a picture consistent with selective advantage of carriers, especially the two most common mutations (1278TATC for TSD and 1226G for GD).

However, this pattern appears striking only in isolation. There are at least a dozen other diseases found at high frequency in the Ashkenazim, many with mutation frequencies comparable to or exceeding those for TSD and GD. These include Bloom syndrome (mutation frequency ≈ .01), BRCA1 (2 mutations, frequencies ≈ .005 and .001), BRCA2 (mutation frequency ≈ .006), Canavan disease (2 mutations, frequencies ≈ .011 and .002), cystinuria (frequency ≈ .008), factor XI deficiency (2 mutations, frequencies = .025 and .021), familial dysautonomia [mutation(s) undiscovered, probable frequency ≈ .01], familial hypercholesterolemia (mutation frequency .0025), familial hyperinsulinism (2 mutations, frequencies unknown), Fanconi anemia C (mutation frequency ≈ .006), phosphofructotinase deficiency (mutation frequency unknown), and nonclassical 21-hydroxylase deficiency (mutation frequency ≈ .20) These mutations and their associated diseases cover a broad spectrum of clinical medicine, ranging from metabolism to cancer to neurology to cardiovascular disease. It seems beyond plausibility to suggest that all these mutations have increased frequency due to previous carrier advantage, especially the dominant ones (BRCA1, BRCA2, FH, ITD). On the other hand, the observed distribution is consistent with genetic drift and random founder effects. Among the lysosomal storage diseases, only two mutations have strikingly high frequencies (1226G for GD and 1278 TATC for TSD). However, numerous other mutations are also in this range or greater (NC210HD, two factor XI mutations, Canavan disease).

Another major question is why mutations in these loci are found exclusively, or nearly so, in the Ashkenazim. If lysomomal storage enzyme deficiencies offer a selective advantage for heterozygotes, why do they not appear among non-Jews who were living in proximity to Jews during the same period of time that selection was presumably operating? This is especially true if the selective factor was disease epidemics, because these diseases affected non-Jews as well as Jews.

Furthermore, it is not clear the extent to which epidemics would have been a strong selective agent in this population historically. The fact is that rapid population expansion occurred between 1500 and 1900 despite epidemics. This was most likely due to the fact that wealthier Jews, the ones who had large families, were able to escape from the cities to the country in times of epidemics and survive; also, their children received better nutrition and housing conditions in general (Weinryb, 1972). It seems more likely that the disease mutations were fortuitously carried by those having many children, rather than providing them with a protection against disease (Risch et al., 1995a).

The pattern of disease mutations found in the Ashkenazim is also comparable to that found in other populations with small founder groups which have undergone recent population expansions. Such populations include the French Canadians, Finns, Amish, Amerindians, Afrikaners, and others.

It is likely that additional molecular work will provide further useful data. Specifically, haplotypes at genetic markers linked to individual disease mutations can be constructed; their history can then be characterized by examining the extent of chromosome preserved around the mutation (which is a reflection of the time since its founding in the population), as well as explanations for the presence of the same mutation in multiple populations (Risch et al., 1995a). For example, if two populations share a common ancient mutation but separated long ago, we would expect more haplotype similarity within the populations than between populations. In contrast, recent gene flow from one population to the other should reveal very similar haplotype patterns between the two populations.

Whereas predicting the future is usually left to psychics and seers, predicting the past may also be problematic. Although we gather high-tech molecular data on current-day individuals in an attempt to infer past events, definitive proof is difficult to obtain, and arguments remain in the realm of speculation. Thus, although we have mounting evidence for the importance of chance, or "bad luck" in the existence of numerous genetic diseases in the Ashkenazi population, it may also be difficult to disprove a prior role of selection as well.

References

Arpaia, E., Dumbrille-Ross, A., Maler, T., et al. (1988). Identification of an altered splice site in Ashkenazi Tay-Sachs disease. Nature **333**, 85–86.

Bach, G., Zlotogora, J., and Ziegler, M. (1992). Lysosomal storage disorders among jews. In "Genetic Diversity among Jews" (B. Bonne-Tamir and A. Adam, eds.), pp. 301–304. Oxford University Press, New York.

Beutler, E., Nguyen, N. J., Henneberger, M. W., *et al.* (1993). Gaucher disease: Gene frequencies in the Ashkenazi Jewish population. *Am. J. Hum. Genet.* **52,** 85–88.

Caggana, M., Eng, C. M., Desnick, R. J., and Schuchman, E. H. (1994). Molecular population studies of Niemann-Pick disease type A. *Am. J. Hum. Genet. Suppl.* **55,** A147.

Chakravarti, A., and Chakraborty, R. (1978). Elevated frequency of Tay-Sachs disease among Ashkenazic Jews unlikely by genetic drift alone. *Am. J. Hum. Genet.* **30,** 256–261.

Chase, G. A., and McKusick, V. A. (1972). Controversy in human genetics: Founder effect in Tay-Sachs disease. *Am. J. Hum. Genet.* **24,** 339–340.

DeBraekeleer, M., Hechtman, P., and Andermann, E., *et al.* (1992). The French Canadian Tay Sachs disease deletion mutation: identification of probable founders. *Hum. Genet.* **89,** 83–87.

Diamond, J. M. (1994). Jewish lysosomes. *Nature* **368,** 291–292.

Engelman, U. Z. (1960). Sources of Jewish statistics. *In* "The Jews: Their History Culture and Religion" (L. Finkelstein, ed.), Vol. 2, pp. 1172–1197. Harper, New York.

Ettinger, S. (1969). "History of the Jewish People in Modern Times." Dvir, Tel-Aviv.

Fraikor, A. L. (1977). Tay Sachs disease: Genetic drift among Ashkenazim Jews. *Soc. Biol.* **24,** 117–134.

Goodman, R. M. (1979). "Genetic Disorders among the Jewish People." Johns Hopkins University Press, Baltimore.

Gravel, R. A., Clarke, J. T. R., Kaback, M. M., Mahuran, D., Sandhoff, K., and Suzuki, K. (1995). The GM2 gangliosidoses. *In* "The Metabolic and Molecular Bases of Inherited Diseases" (C. R. Scriver, A. L. Beaudet, W. S. Sly, and D. Valle, eds.), Vol. II, pp. 2839–2879. McGraw-Hill, New York.

Grebner, E. E., and Tomczak, J. (1991). Distribution of three α-chain β hexosaminidase A mutations among Tay-Sachs carriers. *Am. J. Hum. Genet.* **48,** 604–607.

Hechtman, P., Boulay, B., DeBraekeleer, M., *et al.* (1992). The intron 7 donor splice site transition: A second Tay-Sachs disease mutation in French Canada. *Hum. Genet.* **90,** 402–406.

Horowitz, M., Tzuri, G., Eyal, N., Berebi, A., Kolodny, E. H., Brady, R. O., Barton, N. W., Abrahamov, A., and Zimran, A. (1993). Prevalence of nine mutations among Jewish and non-Jewish Gaucher disease patients. *Am. J. Hum. Genet.* **53,** 921–930.

Jorde, L. B. (1992). Genetic diseases in the Ashkenazi population: Evolutionary considerations. *In* "Genetic Diversity among Jews" (B. Bonne-Tamir, and A. Adam, eds.), pp. 305–318. Oxford University Press, New York.

Lacerda, L., Amoral, O., Pinto, R., *et al.* (1994). Gaucher disease: N370S glucocerebrosidase gene frequency in the Portuguese population. *Clin. Genet.* **45,** 298–300.

Levran, O., Desnick, R. J., and Schuchman, G. H. (1992). A common missense mutation (L302P) in Ashkenazi Jewish type A Niemann-Pick disease patients. Transient expression studies demonstrate the causative nature of the two common Ashkenazi Jewish Niemann-Pick disease mutations. *Blood* **80,** 2081.

Motulsky, A. G. (1979). Possible selective effects of urbanization on Ashkenazi Jewish populations. *In* "Genetic Disease among Ashkenazi Jews" (R. M. Goodman and A. G. Motulsky, eds.), pp. 301–312. Raven Press, New York.

Myerowitz, R. (1988). Splice junction mutation in some Ashkenazi Jews with Tay-Sachs disease: Evidence against a single defect within this ethnic group. *Proc. Natl. Acad. Sci. (USA)* **85,** 3955–3959.

Myerowitz, R., and Costigan, F. C. (1988). The major defect in Ashkenazi Jews with Tay-Sachs disease is an insertion in the gene for the α-chain of β-hexosaminidase. *J. Biol. Chem.* **253,** 18587–18589.

Myerowitz, R., and Hogikyan, N. D. (1986). Different mutations in Ashkenazi Jewish and non-Jewish French Canadians with Tay Sachs disease. *Science* **232,** 1646–1648.

Myrianthopoulos, N. C., and Aronson, S. M. (1966). Population dynamics of Tay-Sachs disease. I. Reproductive fitness and selection. *Am. J. Hum. Genet.* **18,** 313–327.

Myrianthopoulos, N. C., Naylor, A. F., and Aronson, S. M. (1972). Founder effect in Tay-Sachs disease unlikely. *Am. J. Hum. Genet.* **24,** 341–342.

Okada, S., and O'Brien, J. S. (1969). Tay-Sachs disease: Generalized absence of a beta-D-N-acetylhexosaminidase component. *Science* **165,** 698.

Paw, B. H., Kaback, M. M., and Neufeld, E. F. (1989). Molecular basis of adult-onset and chronic GM2 gangliosidoses in patients of Ashkenazi Jewish origin: Substitution of serine for glycine at position 269 of the α-subunit of β-hexosaminidase. *Proc. Natl. Acad. Sci. (USA)* **86**, 2413–2417.

Paw, B. H., Tieu, P. T., Kaback, M. M., *et al.* (1990). Frequency of three HexA mutant alleles among Jewish and non-Jewish carriers identified in a Tay-Sachs screening program. *Am. J. Hum. Genet.* **47**, 698–705.

Peleg, L., Karpati, M., Gazit, E., *et al.* (1994). Mutations of the hexosanunidose A gene in Ashkenazi and non-Ashkenazi Jews. *Biochem. Med. Metab. Biol.* **52**, 22–26.

Petersen, G. M., Rotter, J. I., Cantor, R. M., *et al.* (1983). The Tay-Sachs disease gene in North American Jewish populations: Geographic variations and origin. *Am. J. Hum. Genet.* **35**, 1258–1269.

Rao, D. C., and Morton, N. E. (1973). Large deviations in the distribution of rare genes. *Am. J. Hum. Genet.* **25**, 594–597.

Risch, N., deLeon, D., Ozelius, L., *et al.* (1995a). Genetic analysis of idiopathic torsion dystonia in Ashkenazi Jews and their recent descent from a small founder population. *Nat. Genet.* **9**, 152–159.

Risch, N., deLeon, D., Fahn, S., *et al.* (1995b). ITD in Ashkenazi Jews—Genetic drift or selection? *Nat. Genet.* **11**, 14–15.

Rotter, J. I., and Diamond, J. M. (1987). What maintains the frequencies of human genetic diseases? *Nature* **329**, 289–290.

Ruppin, A. (1972). "The Jewish Fate and Future." Greenwood Press, Westport, CT.

Sachs, B. (1896). A family form of idiocy, generally fatal, associated with early blindness. *J. Nerv. Ment. Dis.* **21**, 475.

Sandhoff, K. (1969). Variation of beta-N-acetylhexosaminidase pattern in Tay-Sachs disease. *FEBS Lett.* **4**, 351.

Schuchman, E. H., and Desnick, R. J. (1995). Niemann-Pick disease types A and B: Acid sphingomyelinase deficiencies. *In* "The Metabolic and Molecular Bases of Inherited Disease" (C. R. Scriver, A. L. Beaudet, W. S. Sly, and D. Valle, eds.), Vol. II, pp. 2839–2879. McGraw-Hill, New York.

Sibille, A., Eng, C. M., Kim, S.-J., Pastores, G., and Grabowski, G. A. (1993). Phenotype/genotype correlations in Gaucher disease type 1: Clinical and therapeutic implications. *Am. J. Hum. Genet.* **52**, 1094–1101.

Thompson, E. A., and Neel, J. V. (1978). Probability of founder effect in a tribal population. *Proc. Natl. Acad. Sci. (USA)* **75**, 1442–1445.

Thompson, E. A., and Neel, J. V. (1997). Allelic disequilibrium and allele frequency distribution as a function of social and demographic history. *Am. J. Hum. Genet.* **60**, 197–204.

Triggs-Raine, B., Feigenbaum, M. B., and Natowicz, C. B. M. (1990). Screening for carriers of Tay-Sachs disease among Ashkenazi Jews: A comparison of DNA-based and enzyme-based tests. *N. Engl. J. Med.* **323**, 6–12.

Wagener, D., Cavalli-Sforza, L. L., and Barakat, R. (1978). Ethnic variation of genetic disease: Roles of drift for recessive lethal genes. *Am. J. Hum. Genet.* **30**, 262–270.

Weinryb, B. D. (1972). "The Jews of Poland. A Social Economic History of the Jewish Community in Poland from 1100–1800." The Jewish Publication Society of America, Philadelphia.

Wright, S. (1937). The distribution of gene frequencies in populations. *Proc. Natl. Acad. Sci. (USA)* **23**, 307–320.

Yokoyama, S. (1979). Role of genetic drift in the high frequency of Tay-Sachs disease among Ashkenazic Jews. *Ann. Hum. Genet.* **43**, 133–136.

Zlotogora, J., Zieglier, M., and Bach, G. (1988). Selection in favor of lysosomal storage disorders? *Am. J. Hum. Genet.* **42**, 271–273.

Zoosmann-Diskin, A. (1995). ITD in Ashkenazi Jews—Genetic drift or selection? *Nat. Genet.* **11**, 13–14.

20

Screening and Prevention in Tay-Sachs Disease: Origins, Update, and Impact

Michael M. Kaback
Departments of Pediatrics and Reproductive Medicine
University of California, San Diego School of Medicine
San Diego, California 92123

I. Program Origins: The Place
II. The Events and the People
III. The Program is Conceived
IV. From Baltimore to Jerusalem
V. Results and Update
VI. Impact and Conclusion
   Acknowledgments
   References
   Appendix I

I. PROGRAM ORIGINS: THE PLACE

Many individuals have asked how the Tay-Sachs program began. What was behind it? Why did it happen? Like many other innovations, the education, screening, and counseling program for the prevention of Tay-Sachs disease (TSD) occurred, not as the result of a single individual, but rather because of a host of factors: timing, location, events, and most important, people.

During residency training at The Johns Hopkins Hospital, I had been involved in the care of three or four children with TSD. The severity and hopelessness of their condition made a deep and lasting impression. After joining the faculty at Hopkins in 1968, my laboratory interests were focused on the study of human genetic disorders, particularly Down syndrome and lysosomal storage disorders, in somatic cells (skin fibroblasts, amniocytes, etc.) cultivated *in vitro*.

Advances in Genetics, Vol. 44

Much of this work was done in the laboratory of Rod Howell, my friend and mentor, during my fellowship at Hopkins. During the summer of 1968, John O'Brien came to Hopkins to carry out some studies on several patients with Victor McKusick. John worked in Rod's lab, and there our long-standing friendship began. John and Shintaro Okada at the University of California, San Diego, already had begun their landmark studies on hexosaminidase isozyme profiles in TSD. As Rod and I were studying metachromatic leukodystrophy and other lysosomal storage disorders, John and Shintaro's work was of great interest and excitement to us.

II. THE EVENTS AND THE PEOPLE

In the fall of 1969, I first met the Gershowitz family, Bayla, Harold, and their children Amy and Stephen. Stephen, age 16 months, had TSD. He was being cared for at home, as there were no adequate facilities in the family's area (suburban Washington, D.C.) where the parents could be satisfied with the care provided for Stephen. I began to appreciate the far-reaching impact of such a tragic disorder. It was not simply a tragedy that the child was doomed, but the enormity for the parents, relatives, friends, and siblings became obvious. Fortunately, we were able to place Stephen for several months in the John F. Kennedy Institute at Hopkins, where a superb level of care could be provided while awaiting the opening of a new state facility in the region where the family resided.

Not long after my encounter with the Gershowitz family, I was called by a pediatric neurologist at Hopkins who was seeing the 10-month-old son of one of our most outstanding pediatric interns. The child seemed to be regressing and the parents were duly concerned about him, their first child. A strikingly abnormal neurologic exam, bilateral macular cherry-red spots, and both parents being of Ashkenazi Jewish background made the possibility of TSD highly probable. The results of serum hexosaminidase assays several hours later confirmed our suspicions.

Perhaps the most powerful aspect of that diagnosis was that Karen, the infant's mother, was 7 months pregnant with their second child at that time. Needless to say, the anxiety and sadness that prevailed were beyond description. Many hours of discussion and deliberation ensued over the following few days. Although clearly beyond the point where any intervention in the pregnancy could be carried out legally, some advocated amniocentesis on the 75% probability that the results would obviate the notion of a second affected child. On the other hand, a positive diagnosis of TSD would unquestionably be devastating.

After extensive consultations, discussions, soul searching, and prayers, a decision was made. No fetal testing would be done, fearing the psychological impact of completing the remainder of pregnancy, knowing that their next child would deteriorate and die while watching that process accelerate in their beautiful son. Rather, it was decided to carry the pregnancy to term, with the plan that the parents would not see the newborn until it could be determined that the infant was unaffected. If the child did have TSD, then it would be placed in foster care until such time as medical needs required more long-term placement, without the parents ever seeing it.

At that juncture, TSD had never been diagnosed in a newborn infant. Prenatal diagnosis with cultured amniotic fluid cells had been achieved, and serum, white blood cell, and other tissues had been utilized successfully for the diagnosis in older children. Certainly the probability was great that hexosaminidase profiles in serum, leukocytes, placental tissues, or cultured skin cells would reflect the genotype, vis-a-vis TSD, in the newborn as well.

Contacts were made with John O'Brien in San Diego and Ed Kolodny at the National Institutes of Health (NIH). Both agreed to participate with my laboratory in making this critical diagnosis at the time of the child's birth. We began by preparing serum and leukocyte samples from cord blood specimens obtained from a large cohort of normal newborns at Hopkins. I divided each preparation into three fractions, one for each lab. These were shipped frozen at several intervals over the months before the critical birth. Clearly, all three labs agreed that hexosaminidase, and particularly Hex A, were readily detectable and quantifiable in such materials. The control data for the critical determination were therefore at hand.

In May 1970, Karen delivered a full-term female infant. Her husband and I were both with her in the delivery room. Local anesthesia was employed, and both parents were fully involved. The scene in the delivery room, where I obtained the cord sample, is one I shall never forget. Two wonderful young people covering their eyes, afraid to look at a beautiful crying baby girl, for fear that they would never see her again.

She was born in the early afternoon. I quickly returned to my laboratory to begin preparing the samples. A call to Friendship Airport confirmed that if I could get the package there by 4:00 p.m., it could reach O'Brien in San Diego on the last flight out that day. I called John; he would meet the plane and begin right away (thanks to the time difference). Ed, also, said he would wait at his laboratory until I drove there from Baltimore, and would start his assays immediately.

After giving the package to a stewardess and telling her that she would be met by a handsome young doctor named O'Brien in San Diego, I drove on to the NIH and left the material with Ed. I got back to Baltimore and began working at about 6:00 p.m. We had guessed that it might take a couple of days before we could be certain where we stood. We agreed that no information would be released until we all had completed our studies. Obviously, we were hopeful there would be no disagreement as to the findings!

At midnight that evening, I heard some movement in the hallway, outside my laboratory. It was our young intern, the baby's father, getting a chair to begin the long vigil. When it became clear that he was not going to leave, I told him to come in and try to relax. We chatted as I worked till about 4:00 a.m. that morning. I had spent many, many hours with him and his wife in the months since this incredible drama began. They were both extraordinary young people. We had become bound together as lifelong friends, and they were nearing the end of the nightmare.

By 4:00 a.m., my results were in and they were unequivocal—clearly, approximately half of the total leukocyte hexosaminidase was of the Hex A form— heat labile, and also obviously evident on cellulose acetate gel electrophoresis. Cord serum revealed similar findings. A call to San Diego (about 1:00 a.m. there) reached John. He had just completed his column and heat profile studies and his results were the same—an unaffected infant. Next, we reached Ed—the same conclusion! Clearly nothing borderline—the baby was *not affected with Tay-Sachs disease*. No need for further analysis—it couldn't be clearer!

At about 5:00 a.m., Bob and I went to see Karen. She was awake, unable to sleep. Together, the three of us walked to the pediatric floor, and they held their new baby for the first time. The joy, relief, and sheer ecstasy for this couple was of a magnitude like nothing I had ever witnessed.

III. THE PROGRAM IS CONCEIVED

It was in the next few hours that the impact of this experience really began to sink in. What had happened to this young family need never be replicated. It was, at least in theory, possible to prevent such a devastating experience from recurring. Clearly, this disease was seen, at least in the great majority of instances, among children of a defined ethnic origin—Ashkenazi Jews. Second, based on O'Brien's work, serum quantification of Hex A could identify healthy carriers of the trait for TSD, and third, fetal diagnosis was possible in early pregnancy through amniocentesis. In this way, it should be possible to screen young Jewish adults to identify carriers of the TSD gene (and, most critically, to identify couples in which both man and woman are carriers). Such at-risk couples, once identified, could, through genetic counseling, choose to monitor each pregnancy and carry to term only those fetuses shown to be unaffected. Although not a perfect solution, involving the complex (but legal) alternative of abortion of the fatally afflicted pregnancies, such an approach could assure all tested couples, regardless of carrier status, an option by which they could have children of their own without fear of TSD.

Thus, the idea of community-based heterozygote screening was conceived. The labors, plans, and initial efforts have been presented elsewhere (Kaback, 1977; Kaback et al., 1974a). Suffice it to say that 12 months of planning and organization transpired, heavily contributed to by the two aforementioned couples, who played such important roles in sensitizing this young physician to the critical importance of disease prevention. From technical planning (Kaback, 1972) and assay automation (Lowden *et al.*, 1973), to developing volunteer educators in the target community—from the rabbinate to the media—from physicians to synagogues—a broad-based planning effort unfolded (Kaback *et al.*, 1974b).

IV. FROM BALTIMORE TO JERUSALEM

On a rainy Sunday afternoon, in May 1971, at Congregation Beth El in Bethesda, Maryland, more than 1800 young Jewish men and women presented themselves for the first community-based screening of this type ever conducted (Kaback and Zeiger, 1972). The extraordinary efforts of many individuals and organizations underscored this flawless first effort. Similar testings followed over the ensuing months in both the Baltimore and Greater Washington, D.C., areas. Within 2 years, similar efforts were initiated in Toronto, Montreal, Philadelphia, Milwaukee, Miami, Minneapolis, Cleveland, Boston, and New York. The first statewide program began in California in 1973.

To date, more than 100 major cities throughout the world have established testing programs. Such community-based education, screening, and counseling efforts, modeled to varying degrees on the Baltimore/Washington prototype, now exist in 15 countries on six continents (North and South America, Europe, Middle East, Africa, and Australia) (Kaback *et al.*, 1993).

V. RESULTS AND UPDATE

With the efforts and support of the National Tay-Sachs Disease and Allied Disorders Association, an International Center has been established for the annual survey of worldwide programs involved in TSD screening and prevention. This center, based in San Diego, also conducts annual quality control assessments of all laboratories involved in TSD carrier screening and provides reference standards for interlaboratory comparisons. This aspect of the program, blinded proficiency testing and accuracy evaluation, have been described previously (Kaback *et al.*, 1977, 1993). Through this mechanism, an extremely high level of testing accuracy has been maintained.

Table 20.1 summarizes the worldwide experience with TSD heterozygote detection and prenatal diagnosis from 1969 through June 1998. In this 30-year interval, since the discovery of the underlying defect in lysomal hexsaminidase A activity, more than 1,331,000 young adults have volunteered for the TSD carrier detection test throughout the world. The international experience, by country or region, is provided in Table 20.2. The individual experiences of each of the worldwide testing centers are tabulated in the Appendix.

Table 20.1. Tay-Sachs Disease Carrier Screening and
Prenatal Diagnosis in 102 International
Centers, 1969–1998

Total persons screened	1,332,047
Total carriers identified	48,864
At-risk couples identified	1,350
Pregnancies monitored	3,146
Tay-Sachs fetuses identified	604
Unaffected offspring born	2,466

Employing either serum and/or leukocyte Hex A determinations (and, in recent years, specific mutation analysis in Ashkenazi Jewish individuals), nearly 49,000 TSD heterozygotes have been identified in this interval. Most critically, 1350 couples, none of whom had prior offspring with TSD, have been identified as being at risk for TSD in their children (both partners: carriers). Approximately 3150 pregnancies, either in couples with prior offspring with TSD or in those identified to be at risk through screening, have been monitored electively by amniocentesis or, more recently, by chorionic villus sampling. This experience is summarized in Table 20.3. More than half of these pregnancies have been to screening-identified, at-risk couples.

Of the 604 monitored pregnancies in which fetuses were identified with TSD, 583 were electively terminated. The 21 which continued to term did so either because diagnosis was made too late in preganancy for intervention to be conducted legally (all occurred in the early years of TSD prevention), or because rabbinical proscriptions against abortion prohibited couples from pursuing this option (among certain ultraorthodox Jewish families). One of these 21 resulted in an offspring with adult-onset TSD (as diagnosed by mutation analysis *in utero*),

Table 20.2. Tay-Sachs Disease Heterozygote Screening,
1971–1998

Country	Total # tested	Total # carriers	At-risk couples
United States	925,876	35,372	795
Israel	302,395	7,277	380
Canada	65,813	3,301	62
South Africa	15,138	1,582	52
Europe	17,725	1,127	37
Brazil	1,027	72	20
Mexico	655	26	0
Australia	3,334	102	4
Argentina	84	5	0
	1,332,047	48,864	1,350

Table 20.3. Prenatal Diagnosis of Tay-Sachs Disease,
Summary Experience, 1969–1998

Total at-risk pregnancies monitored		3,146
# TSD fetuses identified	604*	
Fetuses with TSD missed	3	
Spontaneous abortions	54	
Stillbirths	7	
Perinatal deaths	10	
Other	2	
Subtotal	680	

and the remaining 20 developed classical infantile TSD as predicted. Three fetuses with infantile TSD were incorrectly diagnosed as being unaffected.

Most critically, nearly 2500 infants unaffected by TSD have been born to these at-risk families. Clearly, data developed during the years before the alternative of prenatal diagnosis existed indicate that many of these healthy children might never have been conceived or brought to term had not this critical genetic counseling option been available (Aronson, 1964). It is these healthy children and their families that speak most dearly about the importance of the screening and prevention effort.

VI. IMPACT AND CONCLUSION

In 1970, when the prevention program strategies were first implemented, there were estimated to be about 50–60 infants newly diagnosed with TSD each year in the United States and Canada (see Figure 20.1). Of these, 40–45 infants were of Jewish ancestry and 10–15 were of non-Jewish heritage. From 1983 through the most recent compilation in 1998, the average total number of new cases per year in the United States and Canada was 12, with Jewish cases comprising 4 per year (range 2–6). In the same interval, TSD among non-Jewish infants averaged 8 per year (range 5–12). Thus, a 90% or greater reduction in the incidence of TSD in the high-risk Jewish population has occurred, undoubtedly contributed to, in large part at least, by the education, screening, and counseling programs targeted to Jewish adults of child-bearing age.

This prototype, developed for TSD prevention, has also served as a model for "disease control" in other areas as well. Similar efforts directed to community education, voluntary heterozygote screening, and genetic counseling (with the option of prenatal diagnosis) for the prevention of β-thalassemia have been initiated effectively in several parts of the world. Most particularly, such efforts in Sardinia, Southern Italy, Greece, Cyprus, and among the Cypriot communities in

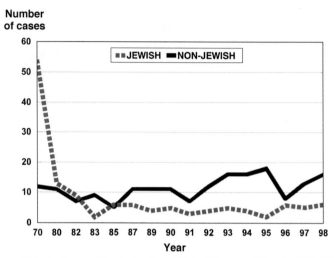

Figure 20.1. Incidence of Tay-Sachs disease, United States and Canada, 1970–1998.

the United Kingdom have had major impact in reducing the incidence of these severe hemoglobinopathies (Modell *et al.*, 1980; Cao *et al.*, 1989). With recent advances in gene isolation and molecular diagnosis, it is likely that comparable efforts targeted to the "control" of other serious hereditary disorders will be mounted in the near future. It is hopeful that many of the technical and social issues that have been addressed and are ongoing in the worldwide TSD screening and prevention programs will facilitate comparable efforts directed to different genetic disorders in other populations.

Acknowledgments

The contributions of the many parents, volunteers, organizations, and colleagues throughout the world who participated in these efforts are most gratefully acknowledged. This work was supported in part by a contract from the Maternal and Child Health Section, Genetic Disease Branch, Department of Health Services, State of California, and a grant from the National Tay-Sachs Disease and Allied Disorders Association.

References

Aronson, S. M. (1964). Epidemiology. *In* Volk, B. W., ed. "Tay-Sachs Disease" (B. W. Volk, ed.). Grune & Stratton, New York.

Cao, A., Rosatelli, C., and Galanello, R. (1989). The prevention of thalassemia in Sardinia. *Clin. Genet.* **36**, 277–285.

Kaback, M. M. (1972). Thermal fractionation of serum hexosaminidases: Applications to heterozygote detection and diagnosis of Tay-Sachs disease. *Meth. Enzymol.* **28**, 862–867.

Kaback, M. M., ed. (1977). Tay-Sachs disease: Screening and prevention. *Prog. Clin. Biol. Res.* **18.**

Kaback, M. M., and Zeiger, R. S. (1972). Heterozygote detection in Tay-Sachs disease: A prototype community screening program for the prevention of recessive genetic disorders. *In* "Sphingolipids, Sphingolipidosis, and Allied Disorders: Advances in Experimental Medicine and Biology," Vol. 10, (B. W. Volk and S. M. Aronson, eds.), pp. 613–632. Plenum, New York.

Kaback, M. M., Zeiger, R. S., Reynolds, L. W., and Sonneborn, M. (1974a). Tay-Sachs disease: A model for the control of recessive genetic disorders. *In* "Birth Defects, Proceedings of the Fourth International Conference" (A. Motulsky and W. Lenz, eds.), pp. 248–262. Excerpta Medica, Amsterdam.

Kaback, M. M., Zeiger, R. S., Reynolds, L. W., and Sonneborn, M. (1974b). Approaches to the control and prevention of Tay-Sachs disease. *In* "Progress in Medical Genetics," Vol. 10 (A. G. Steinberg and A. Bearn, eds.), pp. 103–134. Grune & Stratton, New York.

Kaback, M. M., Shapiro, L. J., Kirsch, P., and Roy, C. (1977). Tay-Sachs disease heterozygote detection: A quality control study. *Prog. Clin. Biol. Res.* **18,** 267–281.

Kaback, M., Lim-Steele, J., Dabholkar, D., Brown, D., Levy, N., and Zeiger, K. (1993). Tay-Sachs Disease–Carrier screening, prenatal diagnosis, and the molecular era. An International Perspective, 1970 to 1993. *J. Am. Med. Assoc.* **270.**

Lowden, J. A., Skomorowski, M. A., Henderson, F., and Kaback, M. M. (1973). The automated assay of hexosaminidases in serum. *Clin. Chem.* **19,** 1345–1349.

Modell, B., Ward, R., and Fairweather, D. (1980). Effect of introducing antenatal diagnosis on reproductive behavior of families at-risk for thalassemia major. *Br. Med. J.* **1,** 1347–1355.

Myrianthopoulos, N. C., and Aronson, S. M. (1966). Population dynamics of Tay-Sachs disease, I: Reproductive fitness and selection. *Am. J. Hum. Genet.* **18,** 313–325.

APPENDIX I

Summary of Individual Programs, 1971–1998.

Program		Total # tested	Total # carriers	At-risk couples
Arizona				
	Scottsdale	3,125	101	2
	Tucson	895	54	1
California				
	Los Angeles	122,239	3,696	92
	Sacramento	7,488	117	3
	San Diego	29,154	1,026	15
	San Francisco	48,274	1,146	28
	Santa Barbara	12,446	233	2
	Misc. Cal.	353	2	0
Colorado				
	Denver	5,424	313	4
Connecticut				
	Hartford	14,134	437	2
Delaware				
	Newark	575	35	0

continues

continued

Program		Total # tested	Total # carriers	At-risk couples
Florida				
	Gainesville	948	41	2
	Miami (A)	16,418	615	37
	Miami (B)	631	66	8
	Orlando	342	11	0
	Tampa	4,757	252	14
	West Palm Beach	38	2	0
Georgia				
	Atlanta	6,498	300	7
Hawaii				
	Honolulu	18	1	0
Iowa				
	Iowa City	23	0	0
Illinois				
	Chicago (A)	5,635	245	9
	Chicago (B)	1,313	76	3
	Chicago (C)	7,590	349	12
	Chicago (D)	608	66	5
	Chicago (E)	300	4	0
	Parkridge	3,660	150	4
	Rockford	45	1	0
Indiana				
	Indianapolis	662	50	0
Louisiana				
	New Orleans	2,147	149	12
Maryland				
	Baltimore (A)	28,670	1,356	27
	Baltimore (B)	6,332	233	6
	Bethesda	508	59	3
Massachusetts				
	Boston	532	8	1
	Waltham	45,983	1,721	68
Michigan				
	Ann Arbor	283	16	1
	Detroit	12,385	618	15
Minnesota				
	Minneapolis	4,289	159	3
	Rochester	63,613	2,476	13
Mississippi				
	Jackson	1	0	0
Missouri				
	St. Louis	1,962	74	3
Nebraska				
	Omaha	890	45	0
Nevada				
	Las Vegas	14	0	0
	Reno	2	0	0

continued

Program	Total # tested	Total # carriers	At-risk couples
New Hampshire			
Portsmith	1	0	0
New Jersey			
Elizabeth	106	2	0
Hackensack	1,344	64	1
Newark	14,678	665	18
Teterboro	14,278	595	2
New Mexico			
Albuquerque	42	2	0
Santa Fe	5,054	162	1
New York			
Beth Israel	1,106	41	1
Bronx	65,571	2,826	40
Brooklyn	147,945	5,182	80
Buffalo	3,356	189	3
Cornell	2,975	139	1
Manhasset	10,188	424	22
Mt. Sinai	45,930	2,117	101
New Hyde Park	9,548	493	19
NYU	2,985	130	3
Port Jefferson	757	33	1
Staten Island	1,132	53	0
Valhalla	4,123	158	4
North Carolina			
Chapel Hill	407	24	2
Research Triangle Park	7,067	524	7
Winston Salem	1,068	6	0
Oklahoma			
Oklahoma City	648	32	1
Ohio			
Cleveland	5,958	153	4
Columbus	140	4	0
Toledo	20	0	0
Oregon			
Portland	2,185	86	2
Pennsylvania			
Harrisburg	293	8	0
Philadelphia	79,996	3,545	41
Pittsburgh	749	43	2
Rhode Island			
Providence	6,712	250	9
South Carolina			
Charleston	192	8	0
Greenwood	1,622	75	0

continues

continued

Program		Total # tested	Total # carriers	At-risk couples
Tennessee				
	Knoxville	379	20	2
	Memphis	707	24	0
	Nashville	1,773	36	0
Texas				
	Dallas	1,526	39	1
	Denton	1,773	86	3
	Galveston	215	0	0
	Houston	6,278	218	9
Utah				
	Salt Lake City	137	4	1
Virginia				
	Charlottesville	2,646	141	9
	Richmond	3,829	89	0
Washington				
	Seattle	583	47	0
Washington, D.C.		2,510	130	1
Wisconsin				
	Madison	2,024	101	0
	Milwaukee	2,116	101	2
USA totals		925,876	35,372	795
Canada				
Alberta				
	Calgary	747	62	0
British Columbia				
	Vancouver	2,735	87	1
Manitoba				
	Winnipeg	2,403	83	2
Nova Scotia				
	Halifax	708	47	2
Ontario				
	Toronto	30,711	1,554	33
Quebec				
	Montreal	28,509	1,468	24
Canada totals		65,813	3,301	62
Israel				
	Jerusalem	24,109	908	33
	Tel-Aviv	278,286	6,369	347
Israel totals		302,395	7,277	380
Europe				
Belgium		306	22	1
Denmark		30	6	1
England		14,017	918	19

continued

Program	Total # tested	Total # carriers	At-risk couples
France	938	102	11
Germany	1,535	27	0
Italy	328	5	1
Norway	4	3	0
Sweden	459	33	4
Switzerland	108	11	0
Europe totals	17,725	1,127	37
South Africa	15,138	1,582	52
Mexico	655	26	0
Brazil	1,027	72	20
Australia	3,334	102	4
Argentina	84	5	0
Total	1,332,047	48,864	1,350

21

Not Preventing—Yet, Just Avoiding Tay-Sachs Disease

Charles R. Scriver*

Department of Human Genetics, McGill University
and the deBelle Laboratory for Biochemical Genetics
McGill University—Montreal Children's Hospital Research Institute
Montreal, Quebec, Canada H3H 1P3.

I. Introduction
II. Context
III. The Patient with the Disease
   A. To Understand the Disease the Patient Has
IV. Strategies to Avoid Tay-Sachs Disease
V. Tay-Sachs Disease Carrier Testing: An Illustration
   of "Community Genetics"
VI. Conclusion
   References

I. INTRODUCTION

This particular chapter has four origins. The first is a Conference held in 1988 to celebrate the anniversary of Dr. Bernard Sachs' description of the disease that came to be known as "Tay-Sachs disease" (TSD), and now recognized as a disorder of G_{M2} ganglioside catabolism. Mutations in the HEXA gene causing

*Eva Andermann, Annie Capua, Lola Cartier, Edgard Delvin, Carol Clow, Reynold Gold, Peter Hechtman, Feige Kaplan, John Mitchell, Keo Phommarinh, Leon Wolfe, Sina Yak, and Susan Zeesman, our students Janet Bayleran, Ellen Beck, Shirley Blaichman, and many unnamed colleagues, belong as "co-authors" and certainly as co-participants in the Montreal TSD project.
   Address for correspondence: E-mail: cseriv@po-box.mcgill.ca. Fax (514) 934-4329. Telephone: (514) 934-4417.

Advances in Genetics, Vol. 44   267

TSD were beginning to be described in 1988; the number of alleles recorded at the HEXA locus now approaches 100 (see the locus-specific mutation database at http://data.mch.mcgill.ca/hexadb). The second origin is a report on the worldwide impact of TSD carrier screening (Kaback *et al.*, 1993). The third is an analysis of process and outcome variables in an unusual carrier screening program that focuses on high school students and has been in operation for more than 20 years in Quebec (Mitchell *et al.*, 1996). Whereas its approach is "conventional" in the locus of the community being served, it has been viewed with some skepticism by outsiders (Scriver, 1995), and from this experience comes the fourth origin for the thoughts expressed here. If genetics is about inherited variation, perhaps it is natural that there are various ways in which individuals, families, and communities will deal with the consequences of genetic variation. High school screening is one way; carrier testing with matchmaking in the Dor Yeshorim program can be another. Such variety of viewpoints colors this volume.

II. CONTEXT

Dr. Bernard Sachs (the neurologist) and G_{M2} gangliosidosis (TSD) were the dual subjects of the Centennial Symposium held at Mount Sinai Hospital and School of Medicine in 1988. On that occasion, someone suggested that the symposium might also be dedicated to the patients, their families, and their communities because, without them, we would not be gathered for the stated purpose. Although knowledge of Tay-Sachs disease has grown enormously in a century, and very much more so since the 1988 symposium, the best we can offer still is its *avoidance*; we may have to wait for the next centennial, perhaps earlier, to report its *prevention* through treatment. Meantime, we avoid the disease by a combination of carrier screening and testing,* and reproductive counseling with fetal diagnosis.

III. THE PATIENT WITH THE DISEASE

Among the many things accomplished in his long career, Dr. Bernard Sachs achieved two of particular interest here: he identified the eponymous disease (along with Dr. Warren Tay), and he concerned himself with its patients who came to his clinic because of their disease. To say the first was a form of clinical investigation and the second was simply good medical practice would be the conventional "medical" view.

*Genetic *screening* implies a population approach; *testing* involves family members at known or presumably elevated risk

Another and somewhat different point of view sees Dr. Sachs acting both as a conventional physician when he dealt primarily with the disease his patient had (which is different from helping the patient who has the disease), and also acting like a modern medical geneticist if he had asked the question: Why does my patient have this (Tay-Sachs) disease now? The latter question is all about "the patient with the disease." Perhaps the Mount Sinai archives will reveal that Dr. Sachs actually asked such a question; it was a matter of some interest to physicians practicing medicine long before him, such as Caleb Parry of Bath (King, 1988).

After Sachs came Garrod who conceptualized the inborn error of metabolism, of which Tay-Sachs disease is an example. Garrod's larger theme was chemical individuality (Bearn, 1993) and how its inheritance conferred susceptibility to oneself or offspring. To avoid Tay-Sachs disease, and we can do that now, we use the phenomenon of its chemical individuality to identify carriers of the gene. Circumstances of his times and contemporary deficits in knowledge did not permit Garrod to apply his concepts of individuality to prevent or avoid Tay-Sachs or any other "genetic" disease. But a century after the discovery of Tay-Sachs disease, the concepts of one great man (Garrod) are being applied around the world, in the form of carrier screening and testing (Kaback et al., 1993) to reduce the burden of the disease described by two others (Tay and Sachs).

A. To understand the disease the patient has

The medical model of disease states that manifestations are the result of a pathologic process, which has a cause. We learned about Tay-Sachs disease slowly (Gravel et al., 1995), beginning with its manifestations, then discovering its pathogenesis, and then its cause. Now, we can approach the disease in reverse. The cause of Tay-Sachs disease is a set of mutant disease-causing alleles at a locus on chromosome 15q encoding the alpha-polypeptide of heteropolymeric hexosaminidase A; a database (http://data.mch.mcgill.ca) records alleles by systematic and trivial names, the type of nucleotide alterations and the effects on mRNA and protein. Patients of different ethnic background are likely to inherit different forms of the mutant allele. The "French-Canadian" mutation, the first to be formally described, is a large deletion (Myerowitz and Hogikyan, 1986); it is different from any inherited by Ashkenazi Jews. From this, one learns that the former is not a "Jewish" allele; it has a possible founder effect in French Canada (DeBraekeleer et al., 1992), and it was not an assimilated allele. The finding, of considerable interest to historians and geneticists, continues a story begun not so long ago by another participant at this celebration (Andermann et al., 1977).

Specific DNA reagents can be used to diagnose most carriers of the Tay-Sachs alleles when they have a particular demographic or geographical association.

Thus, knowing that there is chemical individuality (heterogeneity) at the level of DNA in the infantile form of the disease, might we anticipate subtle yet significant clinical heterogeneity in manifestations associated with the different alleles? If there is clinical heterogeneity in the infantile form of the disease, it is clearly not as striking as in the Tay-Sachs allelic variants that affect patients with different ages at onset in other age groups. Accordingly, there is every reason to learn about the nucleotide–polypeptide relationships in Hex A; unit protein expression analysis of mutation is one way to learn; transgenic and knockout animal models are another (Igdura et al., 1999; Jeyakumar et al., 1999).

It is of some theoretical interest that Tay-Sachs alleles show a clear gene dosage effect in Hex A enzyme activity, yet have no overt metabolic phenotype and do no harm to health in the heterozygote; perhaps just the opposite is the case. Some people speculate that the high frequency of heterozygosity (2–4% in high risk populations and outside obvious founder lineages) reflects a balanced polymorphism with compensating advantage against tuberculosis. TSD allele frequencies are also compatible with the effects of founders, population bottlenecks, and genetic drift (see elsewhere, this volume). Even so, the fact that a "bad" allele might have been "good" under certain conditions is a helpful concept to take aboard in the high risk TSD communities.

IV. STRATEGIES TO AVOID TAY-SACHS DISEASE

Careful measurement of serum hexosaminidase A activity with the 4MUG substrate detects carriers of the Tay-Sachs gene. It is an easy procedure in principle but a finicky one in practice. The challenge is to measure the gene dosage effect against a background of non-hexosaminidase A activities, with sufficient accuracy to classify the heterozygote at maximum sensitivity and specificity. We developed an approach that does this as well as any (Delvin et al., 1974; Gold et al., 1974) by taking the a priori risk into account in the development of a density discriminant function. A complementary approach, even more accurate, is now available. The artificial sulfated substrate (4MUGS) binds preferentially to the alpha-subunit of Hex A (Bayleran et al., 1984), and Hex A activity can be discriminated from non-Hex A activity in a single assay step. Thus assay of enzyme activity with the sulfated substrate (properly purified of contaminants), and mutation detection by DNA analysis together will permit accurate classification of any heterozygote and will raise population screening and carrier testing to levels of specificity and sensitivity heretofore not possible.

Carrier screening and testing precede reproductive (genetic) counseling. In the majority of societies where Tay-Sachs screening and testing are practiced today, carriers have the option to choose their spouse.* Carrier couples have

---

*The Dor Yeshorin program is an exception but it provides an alternative (see elsewhere in this volume).

the option of prenatal diagnosis and termination of pregnancy when the fetus is affected. At present, they are the best options we have. They constitute *avoidance* of Tay-Sachs disease. They are used because they are perceived by those who accept them as better than nothing, and because we do not yet know how to prevent the disease through an effective treatment. [When one reflects on the nature of Tay-Sachs disease, the organelle and cells it involves, and the stage of development at which it begins its course, one is inclined to believe it may remain, for a long time, in that apparently untreatable category of Mendelian disease (Hayes *et al.*, 1985; Treacy *et al.*, 1995). Yet from transgenic studies in knockout mice with a Tay-Sachs enzyme deficiency it is possible to see that the absence of the Tay-Sachs disease phenotype in these mice is explained by compensating brain sialidase activity (Igdoura *et al.*, 1999). Could the human disease be treated by manipulating brain sialidase?]

Persons at risk for having an offspring with Tay-Sachs disease do indeed participate voluntarily in screening and avoidance programs, imperfect procedures that they are, even high school students (Beck *et al.*, 1974; Clow and Scriver, 1977; Zeesman *et al.*, 1984; Mitchell *et al.*, 1996). The student carriers are not abused of or significantly diminished in self image by the screening program and there is no substantive evidence to the contrary. In the TSD carrier screening experience, citizens learn that chemical individuality is just another form of what we all have—a diathesis in one aspect or another (Garrod, 1931; Williams, 1956).

V. TAY-SACHS DISEASE CARRIER TESTING: AN ILLUSTRATION OF "COMMUNITY GENETICS"

The Montreal program has existed since 1972 (Beck *et al.*, 1974); a 20-year analysis of its process and outcome variables has now been completed (Mitchell *et al.*, 1996). The program, which is centered in the senior grades of high schools, has 3 separate components: one for education, another for screening, and a third for counseling after the results of the screening test are distributed. The program, which focuses on schools in regions of the city where the majority of Ashkenazi Jews live, was initiated by the community itself, followed by pilot studies, and approved by school boards, principals, teachers, parents, and community leaders, both religious and secular. This long-term program has been supported by the health care system of Quebec and by community funds. The program continues to be operated for, by, and in the community it serves.

Three process and outcome variables have been measured (Mitchell *et al.*, 1996): 1) the voluntary participation rate in the high school program, and the uptake rate for the screening test; 2) the origin of carrier couples seeking the prenatal diagnosis option; and 3) change in incidence of Tay-Sachs disease over the past 2 decades. By 1992 we had screened 14,844 Ashkenazi Jewish students and identified 521 carriers of a Tay-Sachs disease mutation (carrier frequency, 1 in

28); carriers are identified by a semi-automated heat-denaturation serum hexosaminidase assay using the conventional 4MUG substrate (Delvin *et al.*, 1974) and a statistical classification of Hex A and Hex B values (Gold *et al.*, 1974). The educational program reaches 89% of the demographic cohort. Participation in the screening phase was accomplished in the first decade with the students' own consent under Quebec law. In the second decade, signed consents by parent and student were required and added. Voluntary participation rate in the screening phase is 67% (average for 2 decades).

From the demographic data we learned that virtually all carriers identified in the high school screening program, are taking up the options of reproductive counseling and prenatal diagnosis. This finding implies that "student" carriers remember test results, learn the genotype of their partner, and are not deterred by it in their choice. Nor does the possibility of being a carrier couple deter them from having children later in life. In the two decades covered by our analysis, 16 couples sought prenatal diagnosis, 10 coming from the screening program, the rest from families affected earlier by Tay-Sachs disease. These 16 couples had 32 pregnancies, 17 belonging to the screened couples. There were eight affected fetuses: four carried by couples who had a prior experience with Tay-Sachs disease and 4 carried by naïve couples. All eight affected fetuses were voluntarily terminated. In this same cohort of couples, there were also 24 unaffected live births—13 born to the naïve couples. At the present time, all couples in Montreal using prenatal diagnosis for Tay-Sachs disease trace their knowledge to the high school screening program.

The idea that Tay-Sachs disease *can* be avoided has permeated the Montreal community, which is now well aware of the potential benefits in participation in the program. Accordingly, the incidence of Tay-Sachs disease in Quebec has fallen by 95%. The rare cases born now occur outside the region of the screening program and always to an unscreened, usually non-Ashkenazi, couple. The validity of "community genetics" (Modell and Kuliev, 1991) has advocacy as strong in Montreal as anywhere today.

VI. CONCLUSION

The reality that one—meaning myself—might have a "private," potentially deleterious, phenotype is a new awareness for most of us and it takes some getting used to. In my part of the world, citizens have experienced genetic screening, found out about their chemical individualities, and have found it satisfactory. Their experience of the present, derived from knowledge gained in the past, shows us a bit of the future. It may help to dispel some of the fears expressed about genetic screening and testing, and about the way we carry out these practices (Scriver, 1995). And it may help us to recognize that social structures and values can influence (and mitigate or aggravate) these concerns (Scriver, 1995). A high school program successful in Montreal may fail elsewhere for social, not scientific, reasons. Meanwhile, the enlightened uptake of processes that help people to avoid Tay-Sachs by whatever means, are developments that I imagine would have delighted both Archibald Garrod and Bernard Sachs.

References

Andermann, E., Scriver, C. R., Wolfe, L. S., Dansky, L., and Andermann, F. (1977). Genetic variants of Tay-Sachs disease: Tay-Sachs disease and Sandhoff's disease in French Canadians, juvenile Tay-Sachs disease in Lebanese Canadians, and a Tay-Sachs screening program in the French-Canadian population. *Prog. Clin. Biol. Res.* **18**, 161–188.

Bayleran, J., Hechtman, P., and Saray, W. (1984). Synthesis of 4 methylumbelliferyl–D-N-acetylglu-cosamine–sulfate and its use in classification of GM2 gangliosidosis genotypes. *Clin. Chim. Acta* **143**, 73–89.

Bearn, A. G. (1993). *Archibald Garrod and the Individuality of man.* Oxford, Clarendon Press.

Beck, E., Blaichman, S., Scriver, C. R., and Clow, C. L. (1974). Advocacy and compliance in genetic screening. Behaviour of physicians and clients in a voluntary pro-tram of testing for the Tay-Sachs gene. *New. Engl. J. Med.* **291**, 1166–1170.

Clow, C. L., and Scriver, C. R. (1977). Knowledge about and attitudes toward genetic screening among high-school students: The Tay-Sachs experience. *Pediatrics* **59**, 86–91.

De Braekeleer, M., Hechtman, P., Andermann, E., and Kaplan, F. (1992). The French Canadian Tay-Sachs disease deletion mutation: Identification of probable founders. *Hum. Genet.* **89**, 83–87.

Delvin, E., Pottier, A., Scriver, C. R., and Gold, R. J. M. (1974). The application of an automated hexosaminidase assay to genetic screening. *Clin. Chim. Acta* **53**, 135–142.

Garrod, A. E. (1931). *The Inborn Factors of Disease.* Oxford Univ. Press.

Gold, R. J. M., Maag, U. R., Neal, J. L., and Scriver, C. R. (1974). The use of biochemical data in screening for mutant alleles and in genetic counselling. *Ann. Hum. Genetics* **37**, 315–326.

Gravel, R. A., Clarke, J. T. R., Kaback, M. M., Mahuran, D., Sandhoff, K., and Susuki, K. (1995). The G$_{M2}$ gangliosidoses. *In* Scriver, C. R., Beaudet, A. L., Sly, W. S., Valle, D. (eds). *The Metabolic and Molecular Bases of Inherited Disease,* 7e. McGraw Hill Book Co. New York, p. 2839–2879.

Hayes, A., Costa, T., Scriver, C. R., and Childs, B. (1985). The effect of Mendelian disease on human health II: Response to treatment. *Amer. J. Med. Genet.* **21**, 243–255.

Igdoura, S. A., Mertineit, C., Trasier, J. M., and Gravel, P. A. (1999). Sialidase-mediated depletion of GM2 ganglioside in Tay-Sachs neuroglia cells. *Hum. Molec. Genet.* **8**, 1111–1116.

Jeyakumar, M., Butters, T. D., Cortina-Borja, M., Hunnam, V., Proia, R. L., Perry, V. H., Dwek, R. A., and Platt, F. M. (1999). Delayed symptom onset and increased life expectancy in Sandhoff Disease mice treated with N-butyldeoxynojirimycin. *Proc. Natl. Acad. Sci. USA* **96**, 6388–6393.

Kaback, M., Lim-Steele, J., Dabholkar, D., Brown, D., Levy, N., and Zeiger, K. (1993). Tay-Sachs disease—Carrier screening, prenatal diagnosis, and the molecular era. *JAMA* **270**, 2307–2315.

King, L. S. (1982). *Medical Thinking. A Historical Preface.* Princeton Univ. Press, Princeton, N. J. pp. 175–177.

Mitchell, J. J., Capua, A., Clow, C. L., and Scriver, C. R. (1996). Twenty-year outcome analysis of genetic screening programs for Tay-Sachs and β-thalassemia disease carriers in high schools. *Am. J. Hum. Genet.* **59**, 793–798.

Modell, B., and Kuliev, A. M. (1991). Services for Thalassemia as a model for cost-benefit analysis of genetics services. *J. Inher. Metab. Dis.* **14**, 640–651.

Myerowitz, R., and Hogikyan, N. D. (1986). Different mutations in Ashkenazi Jewish and non-Jewish French Canadians with Tay-Sachs disease. *Science* **232,** 1646–1648.

Scriver, C. R. (1995). Book review: *Assessing Genetic Risks. Am. J. Hum. Genet.* **56,** 814–816.

Treacy, E., Childs, B., and Scriver, C. R. (1995). Response to treatment in hereditary metabolic disease: 1993 survey and 10-year comparison. *Am. J. Hum. Genet.* **56,** 359–367.

Williams, F. J. (1956). *Biochemical Individuality. The Basis for the Genetotrophic Concept.* John Wiley and Sons, New York.

Zeesman, S., Clow, C. L., Cartier, L., and Scriver, C. R. (1984). A private view of heterozygosity: Eight year follow-up study on carriers on Tay-Sachs gene detected by high school screening in Montreal. *Amer. J. Med. Genet.* **18,** 769–778.

22 Experiences in Molecular-Based Prenatal Screening for Ashkenazi Jewish Genetic Diseases

Christine M. Eng* and Robert J. Desnick†
Departments of Human Genetics and Pediatrics
Mount Sinai School of Medicine of New York University
New York, New York 10029

I. Introduction
II. Common Recessive Diseases in the Ashkenazim
   A. Tay-Sachs Disease
   B. Cystic Fibrosis
   C. Type 1 Gaucher Disease
   D. Canavan Disease
   E. Types A and B Niemann-Pick Disease
   F. Bloom Syndrome
   G. Fanconi Anemia Complementation Group C
   H. Familial Dysautonomia
III. Sensitivity of Enzymatic and DNA-Based Carrier Screening
IV. Experience with Multiple-Option Prenatal Carrier Screening
V. Rationale for Multiple-Option Carrier Screening
VI. Strategy for Multiple-Option Carrier Screening
VII. Enzyme and DNA Testing
VIII. Demographics and Test Acceptance
IX. Frequency of Detected Carriers

*Current affiliation: Department of Molecular and Human Genetics, Baylor College of Medicine, Houston, Texas 77030
†Address Correspondence to Robert J. Desnick, Ph.D., M.D., Professor and Chairman, Department of Human Genetics, Mount Sinai School of Medicine, 5th Avenue at 100th Street, Box 1498, New York, NY 10029. E-mail: rjdesnick@vaaxa.crc.mssm.edu.

X. Detected Carrier Couples Choose Prenatal Diagnosis
XI. Importance of Educational Intervention
XII. Group Counseling Preferred
XIII. Couple Screening Reduces Anxiety
XIV. Acceptance and Selection of Prenatal Screening Tests
XV. Confidentiality Issues
XVI. Lessons Learned and Future Prospects
XVII. Type A Niemann-Pick Disease Detectability and Carrier
       Frequency in the Ashkenazi Population
XVIII. Canavan Disease Detectability and Carrier Frequency
       in the Ashkenazi Population
XIX. Multiple-Option Carrier Screening for Five Disorders
XX. Summary
       Acknowledgments
       References

I. INTRODUCTION

The remarkable effectiveness of prenatal heterozygote screening for the prevention of severe recessive genetic diseases has been documented by over almost 25 years of experience with carrier screening for Tay-Sachs disease (TSD) in the Ashkenazi Jewish community (Kaback et al., 1993). Initiated in the early 1970s by Kaback and colleagues (Kaback and Zeigler, 1972; Kaback et al., 1977), the paradigm of voluntary carrier screening, genetic counseling of at-risk "carrier couples," and the availability of accurate prenatal diagnosis has resulted in the marked reduction and near elimination of this devastating disease in the Ashkenazi Jewish population. Previously, between 40 and 60 Tay-Sachs babies were born each year in this group, in which 1 in 25 is a disease gene carrier and 1 in every 625 couples has a 1-in-4 risk for an affected child. Prenatal genetic screening for TSD has dramatically reduced the number of Jewish babies born with this devastating disease, and this has become the prototype for the prevention of recessive genetic diseases, a paradigm that has been effectively replicated for other recessive disorders that occur at high frequency in defined populations, such as α-thalassemia and Sickle Cell disease (Rucknagel, 1983; Scriver et al., 1984).

Over the past decade, there has been increasing interest in expanding prenatal screening to other severe and debilitating recessive disorders so that at-risk couples can make informed reproductive choices. However, efforts to expand population-based prenatal screening have been limited either by the fact that the disease gene was unknown or by the lack of an accurate carrier detection test. As a result of the Human Genome Project, the specific gene defects are being identified in a large number of disorders whose causative genes were previously unknown (cystic fibrosis, Fragile X syndrome, Fanconi anemia, Bloom syndrome, etc.). These advances are providing precise DNA-based diagnostic methods for accurate diagnosis and carrier detection in an ever-increasing number of disorders. Thus, prenatal carrier screening in populations where the incidence of certain recessive genetic diseases is high, such as cystic fibrosis, in which 1 in 25 Caucasians is a carrier, is now possible. In the future, it is likely that DNA chip technologies will be developed that permit screening of individuals for numerous genes causing severe recessive diseases and malignancies, allowing at-risk couples in the general population to make better-informed reproductive decisions. Thus, the lessons learned from TSD screening have become increasingly important as the number of identified and characterized disease genes continues to increase at an accelerating rate. The current challenge is to design and evaluate delivery programs for prenatal carrier testing of multiple diseases so that the benefits associated with predictive medicine are optimized and the potential risks are minimized.

II. COMMON RECESSIVE DISEASES IN THE ASHKENAZIM

Among ethnic groups, the Ashkenazi Jewish population is unique, being at risk for at least eight recessive disorders for which genetic testing is currently feasible. These include TSD, cystic fibrosis (CF), type 1 Gaucher disease (GD), Canavan disease (CD), types A and B Niemann-Pick disease (NPD), Bloom syndrome (BS), Fanconi anemia complementation group C (FACC), and Familial Dysautonomia (FD). As indicated in Table 22.1, these seven disorders have a broad range of clinical severities (severe neurodegenerative diseases of infancy to mild disabilities in adulthood), variable levels of carrier detectabilty by DNA-based testing (94–98%), and are currently untreatable, have nonspecific supportive care, or specific, albeit expensive therapy. It is estimated that about 1 in 6 to 7 Ashkenazi

Table 22.1. Comparison of Eight Recessive Diseases Frequent Among Ashkenazi Jewish Individuals

Disease	Severity	Life expectancy	Treatment	Carrier frequency
Tay-Sachs (Infantile)	Severe	3–5 years	None	~1 : 25
Canavan	Severe	3–5 years	None	~1 : 40
Niemann–Pick Type A	Severe	3–5 years	None	~1 : 80
Cystic Fibrosis	Moderate	26 years	Symptomatic	~1 : 29
Familial Dysautonomia	Moderate	adulthood	Symptomatic	~1 : 30
Bloom Syndrome	Moderate	early adulthood	Symptomatic	~1 : 107
Fanconi Anemia (C)	Moderate	early adulthood	Symptomatic	~1 : 90
Gaucher Disease Type 1	Severe–Mild	adulthood	Enzyme Replacement	~1 : 19

Jewish individuals is a carrier for one of these eight disorders. Although each of these diseases is discussed in detail elsewhere in this book, the molecular genetics and epidemiology of each are presented briefly here.

A. Tay-Sachs disease

Tay-Sachs disease is an autosomal recessive disorder caused by the defective activity of the lysosomal enzyme, β-hexosaminidase A, and the progressive accumulation of G_{M2} ganglioside and related metabolites in the central nervous system. There are two major forms, an infantile-onset neurodegenerative disease that is uniformly fatal in early childhood, and a much less frequent, later-onset disorder characterized by clinical onset in adolescence, progressive muscle weakness, ataxia, dysarthyria, and mild intellectual impairment (Gravel et al., 1995). There is no treatment for either form. Infantile TSD occurs predominantly in the Ashkenazi Jewish population, with a frequency of about 1 in every 2500 births (carrier frequency of about 1 in 25). Using a combination of enzymatic and DNA-based testing, nearly all carriers of this disorder can be identified. Molecular testing for the common Jewish infantile TSD mutations, 1278insTATC and IVS12^{+1}, and the later-onset disease mutation, G269S, permits identification of over 99% of TSD in the Ashkenazi population (Navon and Proia, 1989; Triggs-Raine et al., 1990). It should be noted that the enzyme assay will detect both Jewish and non-Jewish carriers, and that molecular testing will identify two pseudo-deficient alleles (R247W, R249W) that are common in non-Jews (Triggs-Raine et al., 1992; Cao et al., 1993).

B. Cystic fibrosis

Cystic Fibrosis is the most common autosomal recessive disorder affecting Caucasians, with a carrier frequency among both Ashkenazi Jewish and non-Jewish individuals of about 1 in 29. Clinical manifestations in affected individuals result from defects in the cystic fibrosis transmembrane regulatory (CFTR) gene that cause the abnormal regulation of chloride ion channels and the generalized dysfunction of exocrine glands (Welsh et al., 1995). Affected individuals develop chronic pulmonary disease, pancreatic insufficiency with steatorrhea, and abnormally high sweat electrolytes. To date, therapy for these patients remains supportive and expensive. Even with optimal care, the median life expectancy of affected individuals is 31 years (Cystic Fibrosis Foundation Patient Registry, 1997). To date, over 900 mutations in the CFTR gene have been identified, which poses a challenge to the development of effective carrier screening programs. However, common mutations occur in certain populations. For example, the ΔF508 mutation is present in about 70–80% of Caucasian CF patients of northern European ancestry (Cutting et al., 1992). The frequencies of various mutations in different populations have been reviewed (Cutting and Rosenstein, 1998). For prenatal carrier screening, a panel of about 30 CF mutations would detect about 80%

of Caucasian carriers of non-Jewish descent, and about 69% of Black American carriers. Among Ashkenazi Jewish individuals, five mutations (ΔF508, W1282X, G542X, N1303K, 3849 + 10kbCΔT) will detect approximately about 96% of carriers (Cutting *et al.*, 1992; Eng *et al.*, 1995, 1997), the highest detectability among any ethnic, racial, or demographic group to date.

C. Type 1 Gaucher disease

Type 1 Gaucher disease is an autosomal recessive lysosomal storage disease that is one of the most prevalent disorders among Ashkenazi Jewish individuals, with a carrier frequency of about 1 in 18 (Beutler and Grabowski, 1995; Eng *et al.*, 1997). GD results from the deficient activity of the lysosomal enzyme, acid β-glucosidase, and the resultant accumulation of the glycolipid, glucocerebroside, primarily in cells of the macrophage-monocyte system. The disorder is clinically heterogeneous, ranging from early onset of severe disease resulting in major disability or death in childhood or adolescence, to a mild disease course compatible with a relatively normal and productive life (Beutler and Grabowski, 1995). In fact, mildly affected or presymptomatic homozygous individuals have been diagnosed during prenatal carrier screening (Eng *et al.*, 1997). The major acid β-glucosidase mutations in Ashkenazi Jewish patients have been identified (N370S, 84GG, L444P, and IVS2^{+1}) (Beutler *et al.*, 1991, 1992; Sibille *et al.*, 1993; Beutler and Grabowski, 1995), permitting a carrier detectability of about 95%. Based on our recent study of 200 symptomatic type 1 GD patients, certain genotype/phenotype correlations are evident (Sibille *et al.*, 1993). Type 1 patients who are homozygous for the common N370S mutation tend to be less severely affected than patients who are heteroallelic for the N370S allele and an 84GG, IVS2, or L444P mutation. Notably, enzyme replacement therapy is both biochemically and clinically effective, particularly when initiated prior to the development of bone disease.

D. Canavan disease

Canavan disease, also known as infantile spongy degeneration of the brain, is an autosomal recessive neurodegenerative leukodystrophy resulting from the deficiency of the enzyme aspartoacylase (Matalon *et al.*, 1995). Typically, affected infants develop normally during the first month of life, and then are noted to have poor head control and generalized hypotonia by 2–4 months of age. Affected infants have an increased head circumference, develop generalized seizures, and experience a marked loss or regression of early milestones. Affected children usually expire in the first decade of life. In the Ashkenazi Jewish population the carrier frequency is about 1 in 40 (Kaul *et al.*, 1994). Carrier detection is based on the identification of the two common Ashkenazi mutations, E285A and, Y231X, which detect about 97% of Ashkenazi carriers. The A305E lesion accounts for approximately 50% of non-Jewish mutant alleles.

E. Types A and B Niemann-Pick disease

Types A and B Niemann-Pick disease are autosomal recessive disorders resulting from the deficient activity of the lysosomal enzyme, acid sphingomyelinase (Schuchman and Desnick, 1995). Type A disease is a rapidly progressive neurodegenerative disease of infancy, and affected infants usually die by 3–5 years of age. There is no specific treatment available. In contrast, type B NPD is a milder form with primarily hematologic and pulmonary symptomatology, and little, if any, neurologic involvement. The life span of an individual affected with type B disease is variable, with most surviving into adulthood. Three common acid sphingomyelinase mutations (R496L, L302P, fsP330) account for approximately 94% of the mutations in Ashkenazi type A patients (Levran et al., 1991a, 1992), whereas a single common mutation (ΔR608) is frequent in type B patients (Levran, 1991b). Screening of 1900 unrelated Ashkenazi Jewish individuals with no known family history of NPD indicated that the carrier frequency was approximately 1 in 82 for the type A mutations (Li et al., 1997).

F. Bloom syndrome

Bloom syndrome is an autosomal recessive disease that results from mutations in the BLM gene, a RecQ helicase (Ellis et al., 1995), and is characterized clinically by severe pre- and postnatal growth deficiency, a sun-sensitive telangiectatic rash, and a predisposition for various malignant and benign tumors that develop in adolescence or early adulthood (German, 1995). There is no specific treatment available. BS is common in the Ashkenazi Jewish population, with a heterozygote frequency of 1 : 107 for a single causative mutation (a 6-bp deletion and 7-bp insertion, delATCTGA 2281insTAGATTC), and a detectability among Ashkenazi Jewish patients of 99% (Ellis and German, 1996; Li et al., 1998). By DNA testing, carrier detection and prenatal diagnosis are now feasible (Li et al., 1998).

G. Fanconi anemia complementation group C

FACC is an autosomal recessive disorder resulting from mutations in the FACC gene (Auerbach, 1997). The disease is characterized by progressive pancytopenia, variable congenital abnormalities, and a predisposition to acute myelogenous leukemia. The only treatment is bone marrow transplantation (BMT) (Guardiola et al., 1998). A single common mutation found in nearly all Ashkenazi Jewish patients has been identified ($1VS4^{+4A \to T}$), which has a carrier frequency of about 1 : 89 (Verlander et al., 1995), and a detectability among Ashkenazi Jewish patients of 99% (A. Auerbach, personal communication, 1998).

H. Familial dysautonomia

Familial dysautonomia is an autosomal recessive neuropathy that almost exclusively affects infants of Ashkenazi Jewish descent (Axelrod, 1990). The incidence of the disease is about 1 in 3700 births, with a carrier frequency of about 1 in 30. This disorder of sensory and autonomic function is characterized by episodes of protracted vomiting, decreased discrimination to pain and temperature, and cardiovascular instability. The autonomic crises are life-threatening, but with improved pharmacologic management, many patients survive into adulthood. By linkage analysis, the gene was mapped to chromosome 9q31–33 (Blumenfeld *et al.*, 1993), and prenatal diagnosis and carrier testing in known pedigrees was accomplished by family studies (Eng *et al.*, 1995). Recently the gene underlying FD was identified as IKBKAP, which encodes a protein IKAP that is a transcriptional regulator (Slaugenhaupt *et al.*, 2001; Anderson *et al.*, 2001). However, linkage analysis has assigned the gene to chromosome 9q31-33, and analysis of several closely linked markers permits prenatal diagnosis and carrier testing in families with a previously affected child (Blumenfeld *et al.*, 1993; Eng *et al.*, 1995). Although the gene has not been identified, the genetic map is sufficiently well resolved that linkage disequilibrium testing for carrier detection can be offered to Ashkenazi partners of known carriers. It is likely that the gene causing FD will be isolated imminently, making carrier detection feasible in the Ashkenazi Jewish population.

III. SENSITIVITY OF ENZYMATIC AND DNA-BASED CARRIER SCREENING

When the enzymatic defect causing TSD was identified as the deficient activity of the enzyme β-hexosaminidase A (Hex A) in 1969 (Okada and O'Brien, 1969), precise diagnosis of affected infants became possible. Studies immediately determined that their carrier parents had half-normal Hex A activities, and it was quickly realized that most carriers could be distinguished from noncarriers by the enzyme. Enzymatic diagnosis could be performed using a commercially available fluorogenic substrate, and nongenetic factors (such as pregnancy, birth control pills, etc.) which artificially lowered the enzymatic activity in plasma and serum were found not to alter the activity in isolated peripheral leukocytes or cultured cells. Recognizing the availability of highly accurate serum and leukocyte assays for carrier detection, Kaback and colleagues initiated mass prenatal TSD screening in the Ashkenazi community of Baltimore/Washington in 1971 (Kaback and Zeigler, 1972).

In the late 1980s, the Hex A (α-chain) gene was isolated and the common infantile (1278insTATC and IVS12^{+1}) and adult (G269S) mutations were identified (Arpaia *et al.*, 1988; Myerowitz and Costigan, 1988; Navon and Proia, 1989). Since DNA testing is expensive and mutation-specific, whereas enzyme-based testing is comparatively inexpensive and detects all mutations, enzyme-based testing has remained the mainstay for TSD carrier screening. In addition, enzymatic testing effectively identifies carriers of any ethnic group.

In contrast, efforts to screen for type 1 GD or types A and B NPD by enzyme assay were not feasible, due to the lack of a sensitive assay that could distinguish between carriers and normal individuals. Thus, carrier screening for these disorders awaited the isolation of the genes encoding their respective enzymes, and the identification of the major mutations causing each disease in Ashkenazi Jewish patients. These advances occurred over the last 12 years, permitting DNA-based diagnosis of carriers and affected individuals, as well as accurate prenatal diagnosis. However, the sensitivity (or detectability) of the gene-based tests was limited by the fact that testing for the common mutations would not detect all carriers, i.e., carriers of other rare mutations. Fortunately for GD, NPD, CF, BS, FACC, and FD carriers, the detectability of the common mutations in the Ashkenazi population is at least 94% or greater, making these DNA-based carrier detection tests reasonably sensitive and inexpensive. The ability to detect all mutations in the genome awaits the development of novel chip-based methods that would essentially sequence each gene and detect all mutations.

The sensitivity of detection, however, is currently a major feasibility issue in expanding carrier screening to the general population. For example, there are over 900 known mutations causing CF. Although ΔF508 is the major mutation among non-Jewish Caucasians, testing for about 20–30 mutations is required to detect approximately 80% of the mutations in non-Jewish affected Caucasian patients. Thus, mass prenatal screening in the non-Jewish Caucasian population would not detect 20% of carriers with rarer mutations. Couples in which one spouse was found to be a carrier and the other tested negative for the panel of over 20 mutations would still have a residual risk of 1 : 560 of having an affected child with each pregnancy. This concept is often difficult to understand, and referral to geneticists or genetic counselors should be an essential component of screening programs for diseases where the carrier detectability is not 100%.

Relevant to genetic diseases among the Ashkenazim, Table 22.2 indicates the detectability of carriers, and the risk for an affected child if both partners test negative or if one tests positive and the other negative. Clearly, the remarkably high detectabilities for these diseases among Ashkenazi Jewish individuals provides the rationale for pilot programs designed to evaluate prenatal carrier screening for multiple diseases in this population. Furthermore, the preference to screen couples decreases the risks associated with testing only a single spouse, as testing both partners will permit determination of the residual risk if one or both partners tests negative for a given disorder.

IV. EXPERIENCE WITH MULTIPLE-OPTION PRENATAL CARRIER SCREENING

A pilot program at the Center for Jewish Genetic Diseases of the Mount Sinai School of Medicine was conducted to develop an effective educational, carrier

Table 22.2. Detectability of Carriers for the Most Common Jewish Genetic Diseases and Residual Risk of a Negative Result

Disease	Carrier risk	% Detection	p(carrier) if −	p (affected fetus) if −/−[*]	p (affected fetus) if +/−[**]
TSD	1 : 25	98%[+]	1 : 1201	$1 : 5.7 \times 10^6$	1 : 4804
GD	1 : 18	95%	1 : 344	$1 : 4.7 \times 10^5$	1 : 1376
CF	1 : 25	96%	1 : 601	$1 : 1.4 \times 10^6$	1 : 2404

[*]Both partners have negative test results.
[**]One partner tests positive and one partner test negative.

testing, and counseling program for multiple diseases in the Ashkenazi Jewish community of Greater New York City (Eng *et al.*, 1997). Initially, a pilot multiple-option prenatal diagnosis screening program was conducted for three disorders, TSD, CF, and type 1 GD, since heterozygotes could be identified by enzyme and/or DNA-based assays with 99% detectability for TSD, 96% for CF, and 95% for Type 1 GD (Eng *et al.*, 1997; Table 22.2). This study was designed to identify the educational, laboratory, counseling, and psychological issues associated with genetic screening for CF and type 1 GD in a population that already accepted TSD screening and was highly motivated and health-oriented. Thus, they provided the unique opportunity to carry out a pilot study to assess the risks and benefits of multiple-option prenatal carrier screening. This program provided data to compare the attitudes and responses, including the psychological sequelae of screening for CF and type 1 GD with those for TSD, and provided an initial experience with the simultaneous screening for carriers of three genetic diseases with a spectrum of clinical severities in a defined population. When testing became feasible for CD and then types A and B NPD, screening was increased to include these diseases. The rationale, implementation issues, and our experience with multiple-option prenatal genetic screening in the Ashkenazi Jewish population are described below.

V. RATIONALE FOR MULTIPLE-OPTION CARRIER SCREENING

Although the rationale for TSD screening programs was based on the lethality of the disorder, the carrier frequency in the Ashkenazi Jewish population and the availability of inexpensive and accurate testing for carriers and affected fetuses, screening for disorders such as CF and type 1 GD raised additional issues related to disease severity and treatability that needed to be addressed in a research setting. With optimal care, the median life expectancy of an individual affected with CF today is about 31 years, although a significant proportion of patients die in infancy or childhood and others survive into their 40s. There is also optimism that gene therapy may provide definitive treatment in the near future.

Like CF, type 1 GD among Ashkenazi Jewish individuals can be clinically heterogeneous, ranging from early onset with significant involvement resulting in major disability or death in childhood or adolescence, to a mild disease course compatible with a relatively normal and productive life (Sibille *et al.*, 1993). Unlike CF or TSD, type 1 GD has effective therapy. Experience with enzyme replacement therapy in type 1 GD has indicated that both biochemical and clinical improvement can be achieved (Mistry and Abrahamov, 1997). However, the life-long therapy requires continued monitoring of its clinical effectiveness. The educational and genetic counseling issues as well as the psychological support programs for type 1 GD are similar to those required for CF in the Northern European population. Moreover, the issues raised by the clinical heterogeneity in type 1 GD and the availability of effective treatment provided corollaries with similar situations for CF in the acceptance of genetic screening for each. The comparison of attitudes and outcomes for genetic screening for CF and type 1 GD in the Ashkenazi population was particularly worthy of study as an initial experience in screening for carriers of nonlethal disorders. Thus, comparison of test acceptance following patient education and genetic counseling helped define the threshold for acceptance of prenatal carrier testing over a range of disease severities, detectabilities, and treatment options (Table 22.1), delineated patient preferences for genetic screening within this group, and provided experience for the development and evaluation of future testing programs for multiple disease options in this and other populations.

VI. STRATEGY FOR MULTIPLE-OPTION CARRIER SCREENING

The structure of the study was modeled on the Tay-Sachs Carrier Screening Program at Mount Sinai. Patients were usually referred by their obstetricians, preferably prior to conception, and usually only couples were accepted for screening. Screening couples, as opposed to only one spouse or single individuals, provided the most reliable means to ensure that a couple was not at risk. Moreover, screening couples avoided the anxiety that resulted when one member of the couple was found to be positive (1 in 7 screenees when testing for three diseases) and the other had not been tested. Our experience with screening only couples for over 25 years has reinforced our appreciation of the psychological benefits of this approach.

Prior to obtaining the appropriate specimens, educational brochures were distributed by mail and in-person group counseling sessions, conducted by a board-certified genetic counselor, discussed the major clinical symptoms and mode of inheritance of each disease and the risks, benefits, and limitations of carrier screening. Patients were given the opportunity to ask questions, and informed consent was obtained. Identified carriers were subsequently telephoned by a counselor and informed of their status. Testing of other at-risk family members was also recommended.

Carrier couples were contacted directly and an appointment was made for genetic counseling by a clinical geneticist. Prenatal diagnosis was offered to all couples at risk, and when accepted, was performed at our center on fetal tissue obtained by chorionic villus sampling at 9–11 menstrual weeks or by amniocentesis at 15–18 menstrual weeks.

VII. ENZYME AND DNA TESTING

Carrier identification for TSD was based on the determination of the percent Hex A activity. However, this method has been limited by the fact that approximately 3–5% of individuals fall into an inconclusive range between high carrier and low noncarrier values. In addition, pregnant women and women taking oral contraceptives have false positive results if only the serum assay is performed. Therefore, we and other investigators developed enzyme assays using leukocytes or other sources in which the enzyme levels were not affected by pregnancy or hormones (Goldberg et al., 1977). However, even with these assays, about 2–3% of values fall into the inconclusive range. Thus, more accurate testing was needed for the detection of carriers and for the resolution of inconclusive results generated by the enzyme assay. DNA testing was helpful in resolving inconclusive values and for confirming all enzyme-identified carriers. For such individuals, testing for the three major mutations in the β-hexosaminidase α-chain gene that cause TSD in Ashkenazi Jews (1278insTATC, IVS12^{+1}, and G269S) increased the accuracy of carrier detection in this population (Triggs-Raine et al., 1990).

For CF, carrier testing was performed by analysis of five mutations (ΔF508, W1282X, G542X, N1303K, 3849 + 10kbC$\rightarrow$T) that detected approximately 96% of Ashkenazi Jewish carriers (Eng et al., 1997), and for type 1 GD, four mutations (N370S, 84GG, L444P, IVS2 + 1) detected approximately 95% of Ashkenazi Jewish carriers (Eng et al., 1997). Molecular analyses for mutations causing CF and type 1 GD were performed by PCR amplification followed by either allele-specific oligonucleotide hybridization or restriction fragment analysis. Based on the assumed frequency of carriers in the Ashkenazi Jewish population and the sensitivity of each test to detect carriers, the residual risk of being a carrier who was not detected by the mutations analyzed for each disease was calculated and reported to screenees (Table 22.2).

VIII. DEMOGRAPHICS AND TEST ACCEPTANCE

Individuals enrolled as couples and testing was performed simultaneously on both partners. The mean age of participants was 33 years, and most (89%) had college and/or postgraduate education. Most participants (~80%) were informed of the program by their physician, while most of the others were recruited during visits to our genetics center for other indications. Among the 2824 Ashkenazi Jewish individuals who were not previously tested for any of the three disorders, 100% chose to be tested for TSD, 97% chose CF testing, and 95% chose type 1 GD testing. Others chose only type 1 GD and/or CF, as they had been previously tested for TSD or CF; 1053 tests for CF and 1082 tests for type 1 GD were performed for individuals previously tested for TSD and/or CF.

IX. FREQUENCY OF DETECTED CARRIERS

The frequency of detected carriers was 1:21 for the 2824 participants tested for TSD, 1:25 for the 3792 tested for CF, and 1:18 for the 3764 tested for Type 1 GD (Table 22.3). Of the 134 TSD carriers identified by enzymatic assay of β-hexosaminidase A, 113 were found to carry one of the three common mutations (DNA confirmed carrier frequency of 1:25). The distribution of mutations found in TSD carriers was 88%, 10%, and 2% for 1278insTATC, IVS12^{+1}, and G269S mutations, respectively. The 21 Ashkenazi Jewish individuals who were identified as carriers by enzymatic analysis but did not have one of the three common mutations were tested further for two pseudo-deficiency alleles and were found to be negative. These individuals were either carriers who had a mutation that was not among those analyzed, since only about 98–99% of TSD alleles are identified by analysis of the three mutations, or more likely, were noncarriers whose enzymatic activity fell in the high carrier range. For CF carriers, the distribution was 47%, 39%, 8%, 5%, and 1% for the ΔF508, W1282X, G542X, N1303K, and 3849+10kbC→T mutations, respectively. For type 1 GD, the distribution of mutations was 94% and 6% for N370S and 84GG, respectively. Three type 1 GD carriers were detected for the L444P mutation, while the IVS2^{+1} mutation was not detected. Of note, 1 in every 7.4 screenees was a carrier for one of three diseases, 11 screenees were carriers of mutations for two diseases, and one individual was found to be a carrier of all three disorders.

Table 22.3. Results of Carrier Screening

Disease	# Tested[*]	Carrier frequency	Carrier couples
TSD	2824	1:21	8
CF	3792	1:25	6
GD	3765	1:18	7

[*] = all individuals were of Ashkenazi Jewish descent.

X. DETECTED CARRIER COUPLES CHOOSE PRENATAL DIAGNOSIS

Twenty-one carrier couples (eight TSD, six CF, and seven type 1 GD) were iden-tified and counseled, and the 20 couples who had a current pregnancy all opted for prenatal diagnosis. Of the eight at-risk TSD pregnancies, three were predicted to be affected and these couples elected termination. Of the six at-risk CF pregnan-cies, three were predicted to be affected and these couples elected termination; two were predicted to be carriers and one was predicted to be unaffected, and all three continued. One CF carrier couple who had previously terminated an affected fetus had a second pregnancy monitored that was predicted to be unaf-fected. Of the seven type 1 GD carrier couples, five were at risk for offspring with the phenotypically milder genotype N370S/N370S, one was at risk for offspring with the more severe genotype N370S/L444P, and one was at risk for the appar-ently nonviable genotype 84GG/84GG. Four couples at risk for N370S/N370S fetuses were pregnant and pursued prenatal diagnosis. The analyses predicted two affected and two unaffected fetuses, and all four continued to term. The pregnancy at risk for an N370S/L444P fetus was predicted to be unaffected. The couple at risk for an 84GG/84GG fetus had three separate pregnancies, two affected that were terminated and one pregnancy that was predicted to be unaffected and continued. Follow-up testing by biochemical and molecular methods confirmed the results of all prenatal diagnoses. The number of carrier couples identified for TSD and CF exceeded the number predicted based on carrier frequencies of 1 : 25 for each disorder; however, the detected carrier frequencies fell within the 95% confidence intervals based on the relatively small number of identified carrier couples.

XI. IMPORTANCE OF EDUCATIONAL INTERVENTION

With respect to background knowledge, this highly educated population had an unexpectedly good understanding of recessive inheritance that was not sig-nificantly improved by the additional education or genetic counseling session. However, their initial limited knowledge of the genetic disorders common in their ethnic group was markedly increased by the educational brochure and ge-netic counseling session. Our follow-up studies at 3 and 12 months posttesting did not demonstrate a significant decline in knowledge. Of note, participants who achieved a higher degree of knowledge were less likely to report negative attitudes toward screening, were less anxious when one member of the couple received positive results, and were well informed regarding reproductive options if both members were found to be carriers of a particular disease. These findings emphasized the importance of pretest counseling and education. Importantly, in-dividuals understood the concept of test detectability and residual risk, thereby providing initial experience in the education of patients about a major risk of genetic testing. Thus, educational intervention was important for informed consent, test preferences, reproductive decision making, and for minimizing the possible risks and negative effects of carrier testing. However, it is likely that efforts to educate screenees about an increasing number of diseases will not be as successful, as there will probably be a limit to the number of disease descriptions that even a highly educated and motivated screenee can recall after a single educational session.

XII. GROUP COUNSELING PREFERRED

Based on follow-up studies at 3 and 12 months after testing, over 90% of participants indicated that the group genetic counseling session was preferred to the written materials for education. Clearly, the time and genetic counseling personnel required to adequately educate participants about carrier testing for three or more diseases would be a major limiting factor in delivering this or similar genetic testing in the generalist setting. As the number of disease options increases, educational methods that will augment in-person counseling will be needed. It is likely that educational intervention for the future mass screening of multiple diseases will have to focus on the genetic principles of inheritance and risk, and identify disease groups with different severities and therapies (e.g., severe untreatable neurodegenerative diseases of infancy such as TSD versus milder treatable disorders such as type 1 GD). Thus, it will be necessary to devise and evaluate other forms of education that are comparable in effectiveness, but less labor-intensive. Video education has been found to be comparable to written materials in other studies (Clayton et al., 1995), and computer-assisted education may be a valuable tool (Consoli et al., 1995), but it is still to be determined whether they can be as effective as in-person counseling.

XIII. COUPLE SCREENING REDUCES ANXIETY

In order to reduce potential risks, only couples were tested, since about 1 in 7 screenees would be a carrier of one of the three diseases. By screening couples, the positive screenee and his or her mate can be informed simultaneously of their risks. Since one spouse tested negative, the couple would have a minimal residual risk for an affected child. Thus, couple testing avoided the distress that would have been provoked while awaiting the results from testing the other spouse, if "two-step" testing were performed, i.e., testing of one partner, followed by testing only the partners of carriers. Couple screening becomes even more relevant when expanding this program to the seven currently screenable diseases which afflict Ashkenazi Jewish individuals, which will result in 1 in 6 to 7 being a carrier of one disease. Thus, it may be desirable that couples be tested in future screening programs for multiple disorders, particularly as advances in diagnostic technology make the testing of both partners technologically and economically feasible.

XIV. ACCEPTANCE AND SELECTION OF PRENATAL SCREENING TESTS

The high level of test acceptance in our program (95% of participants accepted testing for all three diseases) represents a unique finding as compared to the uptake of CF testing alone in other settings (Bekker *et al.*, 1993; Tambor *et al.*, 1994; Grody *et al.*, 1998). The high level of acceptance presumably was influenced by several factors. First, this ethnic group already accepted TSD testing and termination of affected pregnancies. Second, the patient population was predominantly prenatal, which supports recent observations in different ethnic and demographic groups that postconception individuals demonstrate a higher level of acceptance of genetic testing than other groups (Grody *et al.*, 1998).

The hypothesis that participants would discriminate among the three disorders based on clinical severity or availability of treatment was not borne out in their selection of tests. Preliminary experience with offering a choice of five tests by the addition of Canavan and Niemann-Pick diseases (which are less frequent than TSD, CF, or type 1 GD) indicated that testing for these less frequent but severe diseases was readily accepted as well (see below). It is likely that prenatal couples in the general population who would accept prevention by termination of prenatally diagnosed affected fetuses, also will broadly accept carrier testing for multiple severe diseases, with less regard to disease frequency, when such testing is technologically and economically feasible.

Although participants readily accepted testing for all three diseases, we found that they did discriminate among the diseases when asked about reproductive options if both partners were found to be carriers and a fetus were predicted to be affected. The difference in attitudes toward termination was based on disease severity and treatability. A high percentage of individuals would terminate a fetus diagnosed with TSD or CF, whereas participants were more likely to continue a pregnancy to term with type 1 GD, a non-neurologic and treatable disorder. As demonstrated by their attitudes toward type 1 GD, these results suggested that screenees sought information about their genetic makeup, but made reproductive decisions based on their understanding of disease severity and treatability, and that factors other than reproductive considerations influenced the choice of testing. Interestingly, individuals with lower knowledge scores were less likely to accept type 1 GD testing. Those who refused the type 1 GD test indicated that if they or their spouse were identified as a carrier of type 1 GD, they would have lower self-esteem or a lower opinion of their spouse. Further studies are needed to explore the reasons for these responses. Perhaps the greater uncertainty about the prognosis and treatment options for type 1 GD made choices more difficult than for disorders such as TSD and CF, which were perceived as more severe.

XV. CONFIDENTIALITY ISSUES

Several interesting and potentially concerning issues were raised about confidentiality. The majority of participants indicated that if identified as a carrier, they would share the information with their physician and immediate family. A majority also indicated that they would not tell their friends of their carrier status, suggesting that this information would be closely guarded by the individual and perhaps may be somewhat stigmatizing (Massarik and Kaback, 1981). In the future, efforts should be directed to broadly educate the general population that everyone is a carrier of five to seven lethal recessive disorders, and that carriers do not manifest symptoms or other adverse effects of a single recessive gene. These educational efforts should be expanded to include insurance carriers and employers, in order to prevent misuse and misunderstanding of genetic information.

XVI. LESSONS LEARNED AND FUTURE PROSPECTS

Since it is estimated that all individuals are carriers of five to seven lethal recessive genes, preventive strategies for multiple inherited diseases through preconception or prenatal carrier screening are likely to be introduced into the general population, particularly as advances in diagnostic technology permit testing for selected or virtually all serious debilitating diseases. Already, a National Institutes of Health Consensus Development Conference concluded that all prenatal patients should be informed of the availability of CF carrier testing (NIH, 1997 and guidelines for the implementation of broad prenatal screening have been developed (Grody et al., 2001). Thus, it is of paramount importance that safe and effective methods for multiplex genetic screening be developed and that the risks and benefits of such approaches be fully evaluated to avoid the potential anxiety and stigmatization that could result (Massarik and Kaback, 1981; Eng et al., 1997). With this perspective, the experience with multiple-option prenatal carrier testing in the Ashkenazi Jewish population is instructive for the future delivery of prenatal genetic services for population subgroups and/or the general population.

Recent advances in molecular genetics have resulted in the identification of the genes and common mutations in several other disorders that occur less frequently in this population, including FD, CD, NPD, FCCA, and BSOR, which have carrier frequencies estimated to range from ~1 : 30 to 1 : 107 (Table 22.1). Patient attitudes and preferences for testing this group of rarer disorders has yet to be fully determined. For example, it will be of interest to examine patient preferences for prenatal carrier screening if the disease choices include five to eight disorders, including CD and NPD, which are neurologically devastating diseases resulting in demise in early childhood, FCCA and BS, which occur rarely and result in death in early adulthood, usually from malignancies, and FD, which has chronic neurologic disability but normal intelligence and survival into adulthood.

Therefore, to assess the feasibility of offering carrier screening for these additional disorders and to obtain data for appropriate genetic counseling of the carrier risks and residual risks of negative results, we undertook studies to document the carrier frequency and detectability of carriers in this population for NPD and CD and to determine patient attitudes and acceptance of prenatal screening for these rarer, but severe neurodegenerative disorders.

XVII. TYPE A NIEMANN-PICK DISEASE DETECTABILITY AND CARRIER FREQUENCY IN THE ASHKENAZI POPULATION

To date, over 20 mutations in the acid sphingomyelinase gene have been identified in type A or B patients (Schuchman and Desnick, 1995). Three mutations, L302P, R496L, and fsP330, are recurrent in Ashkenazi Jewish type A patients (Levran et al., 1991a, 1992, 1993). Another common mutation, ΔR608, was found in both Ashkenazi and non-Ashkenazi type B patients (Levran et al., 1991b; Vanier et al., 1993). To facilitate large-scale molecular diagnostic studies for NPD, a robotics-assisted semiautomated DNA testing protocol was developed in our laboratory. Multiplex PCR was performed to amplify the two regions of genomic DNA that contain the three type A mutations. Using an automated workstation, replicate filters were produced for detection of the mutations by allele-specific hybridization.

In order to determine the sensitivity of these assays to detect carriers of NPD, studies were undertaken to identify the mutations in a sample of 28 Ashkenazi Jewish NPD type A patients. By screening for the three common mutations, 52 of 56 (94%) of mutant alleles were identified. The distribution of mutations was R496L, 53%; fsP330, 29%; and L302P, 12%. To determine the frequency of type A carriers, a population study was carried out to screen for the three common mutations in an anonymous sample of 1224 random Ashkenazi Jewish individuals with no family history of NPD. Fifteen heterozygotes were identified, giving a type A NPD carrier frequency of 1 in 82 in the general Ashkenazi population (Li et al., 1997).

XVIII. CANAVAN DISEASE DETECTABILITY AND CARRIER FREQUENCY IN THE ASHKENAZI POPULATION

Previous studies had determined that the frequency of carriers of the E285A and Y231X mutations in the aspartoacylase gene that cause Canavan disease was approximately 1 : 38 in the Ashkenazi population (Kaul et al., 1994). In addition, the detectability of these two mutations was found to be approximately 97% (Kaul et al., 1994). In order to confirm the carrier frequency of CD in our patient population, we tested 2385 anonymous DNAs from Ashkenazi Jewish individuals and determined that the frequency of these two mutations was 1 : 48, somewhat lower than the 1 : 38 carrrier frequency that was previously reported (Kaul et al., 1994). Thus, these studies for both NPD and CD generated important information that could be used in genetic counseling in order to accurately define risks and interpret results of carrier testing and prenatal testing.

XIX. MULTIPLE-OPTION CARRIER SCREENING FOR FIVE DISORDERS

Using the results of these studies, a pilot multiple-option carrier screening program for five disorders—TSD, CF, GD, CD, and NPD—was developed and evaluated. Brochures describing the five disorders were sent to all prospective participants prior to their attending a 40-minute group genetic counseling session. Participants were then asked to choose the diseases that they wished to be tested for, and informed consent was obtained. Testing was performed for couples in parallel, and results were reported to both partners simultaneously. The effectiveness of the education for five diseases as well as attitudes toward genetic testing were evaluated.

In the first 6 months of the pilot program, 497 individuals participated. Ninety-eight percent of participants accepted testing for NPD and 99% for CD, while the acceptance of the other three disorders was retained (see above). Six type A NPD carriers were identified and counseled; no carrier couples have been identified as yet. The carrier frequency in this group of 487 patients was 1 : 83 for NPD, which was in agreement with the results obtained in our pre-clinical studies, giving an overall carrier frequency of 1 : 82 in a total sample of 1711 individuals. The carrier frequency of CD in this group was 1 : 45 (11 carriers identified).

We evaluated participants attitudes toward genetic testing for NPD and CD as compared to other disorders in the test panel. In response to the question, "Suppose you and your spouse were pregnant and prenatal diagnosis indicated the fetus was affected with either TSD, CF, GD, NPD, or CD. How strongly would you feel about terminating the pregnancy?" For NPD and CD, 86% and 85% of participants indicated that they would favor termination of an affected pregnancy, respectively, compared to 89% who would favor termination of a fetus affected with TSD, 77% for CF, and 48% for GD. Thus, these results indicate that acceptance for these two additional severe, neurodegenative disorders will be high in this population. Like TSD, the acceptance of testing for NPD and CD was correlated with attitudes regarding reproductive decisions. It did not appear that the relatively low frequency of the genes is a negative factor in influencing acceptance. It will be interesting to monitor patient attitudes to determine if the acceptance of non-neurologic disorders such as GD and CF will decline in conjunction with the offering of testing for these neurodegenative disorders of infancy, and as well as for disorders such as FCCA, BS, and FD.

Recently, prenatal screening for FCCA and BS was added to the test panel, bringing the number of recessive diseases screened to seven. Despite the increase in the number of diseases, the screening program continues to offer pretest education consisting of brochures distributed prior to an in-person group genetic counseling session, thereby facilitating the informed choice of screenees. Among couples presenting for the first time for genetic screening, uptake has been over 95% for all seven disorders, closely reflecting previous experience with three and five diseases. For couples who were previously tested for TSD, CF, and GD and returned for counseling regarding a new pregnancy, acceptance has been over 95% as well. This continued participation in genetic testing, particularly when the added diseases are rare, suggests that their previous experience with prenatal genetic screening was positive and helpful in decision-making. Prenatal screening for FD has very recently been added to the panel for a total of eight disorders.

XX. SUMMARY

Multiple-option prenatal carrier testing in the Ashkenazi Jewish community for three and now eight disorders has been readily accepted in this prenatal, health-oriented and knowledgeable population. Counseling of screenees concerning the nature (severity, treatability, etc.), inheritance, and frequencies of each disorder was essential for informed test choices and future reproductive decision making. The value of couple testing for a group of disorders when 1 in about 6 would be found to be a carrier of at least one disease was emphasized. These studies identified issues of education, confidentiality, posttest anxiety, and self-esteem that must be continuously addressed in the Ashkenazi population. However, an important value of these studies is that they provide a framework for the development of mass carrier screening programs in the general population or in specific segments of the population with similar demographic characteristics and in more diverse prenatal populations.

Acknowledgments

This work was supported in part by the National Institutes of Health, including a research grant (1 R01 HG00644), a grant from the National Center for Research Resources for the Mount Sinai General Clinical Research Center (5 M01 RR00071), and a grant (5 P01 HD 28822) for the Mount Sinai Child Health Research Center.

References

Anderson, S. L., Coli, R., Daly, I. W., Kichula, E. A., Rork, M. J., Volpi, S. A., Ekstein, J., and Rubin, B. Y. (2001). Familial dysautonomia is caused by mutations of the IKAP gene. *Am. J. Hum. Genet.* **68,** 753–758.

Arpaia, E., Dumbrille-Ross, A., Maler, T., Neote, K., Tropek, M., Troxd, C., Stirling, J. L., Puts, J. S., Bapat, B., Lamhonwah, A. M., Mahuran, D. J., Shuster, S. M., Clarke, J. T. R., Lowden, J. A., and Gravel, R. A. (1988). Identification of an altered splice site in Ashkenazi Tay-Sachs disease. *Nature* **333,** 85–86.

Auerbach, A. D. (1997). Disorders of DNA replication and repair. *Curr. Opin. Pediatr.* **9,** 600–616.

Axelrod, F. B. (1990). Autonomic and sensory neuropathies. *In* "Principles and Practice of Medical Genetics," 2nd ed. (A. E. Emery and D. L. Rimoin, eds.), p. 465. Churchill-Livingston, Edinburgh.

Bekker, H., Modell, M., Deniss, G., Silver, A., Mathew, C., Bobrow, M., and Marteau, T. (1993). Uptake of cystic fibrosis testing in primary care: Supply push or demand pull. *Br. J. Med.* **306,** 1584–1589.

Beutler, E., and Grabowski, G. A. (1995). Gaucher disease. *In* "The Metabolic and Molecular Bases of Inherited Disease," 7th ed. (C. R. Scriver, A. L. Beaudet, W. S. Sly, and D. Valle, eds.), pp. 2641–2670. McGraw-Hill, New York.

Beutler, E., Gelbart, T., Kuhl, W., Sorge, J., and West, C. (1991). Identification of the second common Jewish Gaucher disease mutation makes possible population-based screening for the heterozygous state. *Proc. Natl. Acad. Sci. (USA)* **88,** 10544–10547.

Beutler, E., Gelbart, T., Kuhl, W., Zimran, A., and West, C. (1992). Mutations in Jewish patients with Gaucher disease. *Blood* **79,** 1662–1666.

Blumenfeld, A., Slaugenhaupt, S. A., Axelrod, F. B., Lucente, D. E., Maayan, C., Liebert, C. B., Ozelius, L. J., Trofatter, J. A., Haines, J. L., Breakefield, X. O., and Gusella, J. F. (1993). Localization of the gene for familial dysautonomia on chromosome 9 and definition of DNA markers for genetic diagnosis. *Nat. Genet.* **4,** 160–163.

Cao, Z., Natowicz, M. R., Kaback, M. M., Lin-Steele, J. S., Prence, E. M., Brown, D., Chabot, T., and Triggs-Raine, B. L. (1993). A second mutation associated with apparent beta-hexosaminidase A pseudodeficiency: Identification and frequency estimation. *Am. J. Hum. Genet.* **53,** 1198–1205.

Clayton, E. W., Hannig, V. L., Pfotenhauer, J. P., Parker, R. A., Campbell, P. W. III, and Phillips, J. A. (1995). Teaching about cystic fibrosis carrier screening by using written and video information. *Am. J. Hum. Genet.* **7,** 171–181.

Consoli, S. M., BenSaid, M., Jean, J., Menard, J., Plouin, P. F., and Chantellier, G. (1995). Benefits of a computer-assisted educational program for hypertensive patients compared with standard educational tools. *Patient Educ. Counsel.* **26,** 343–347.

Cutting, G. R., and Rosenstein, B. J. (1998). The diagnosis of cystic fibrosis: A consensus statement. Cystic Fibrosis Foundation Consensus Panel. *J. Pediatr.* **132,** 589–595.

Cutting, G. R., Curristin, S. M., Nash, E., Rosenstein, B. J., Lerer, I., Abeliovich, D., Hill, A., and Graham, C. (1992). Analysis of four diverse population groups indicates that a subset of cystic fibrosis mutations occur in common among Caucasians. *Am. J. Hum. Genet.* **50,** 1185–1194.

Cystic Fibrosis Foundation Patient Registry (1997). Annual Data Report.

Ellis, N. A., and German, J. (1996). Molecular genetics of Bloom's syndrome. *Hum. Mol. Genet.* **5,** 1457–1463.

Ellis, N. A., Groden, J., Ye, T. Z., Straughten, J., Lennon, D. J., Ciocci, S., Proyteneva, M., and German, J. (1995). The Bloom's syndrome gene product is homologous to RecQ helicases. *Cell* **83,** 655–666.

Eng, C. M., Burgert, T. S., Schecter, C., Robinowitz, J., Zinberg, R., Fulop, G., and Desnick, R. J. (1995a). Multiplex genetic testing in the Askenazi Jewish Population. *Am. J. Hum. Genet.* **57,** A57.

Eng, C. M., Slaugenhaupt, S. A., Blumenfeld, A., Axelrod, F. B., Gusella, J. F., and Desnick, R. J. (1995b). Prenatal diagnosis of familial dysautonomia by analysis of linked CA-repeat polymorphisms on chromosome 9q31-33. *Am. J. Med. Genet.* **59,** 349–355.

Eng, C. M., Schechter, C., Robinowitz, J., Fulop, G., Burgert, T., Levy, B., Zinberg, R., and Desnick, R. J. (1997). Prenatal genetic carrier screening: Experience with triple disease screening in the Ashkenazi Jewish population. *J. Amer. Med. Assoc.* **278,** 1268–1272.

German, J. (1995). Bloom's syndrome. *Dermatol. Clin.* **13,** 7–18.

Goldberg, J. D., Truex, J. H., and Desnick, R. J. (1997). Tay-Sachs disease: An improved fully automated method for heterozygote identification by tear β-hemosaminidase assay. *Clin. Chim. Acta* **77,** 43.

Gravel, R. A., Clarke, J. T. R., Kaback, M. M., Mahuran, D., Sandhoff, K., and Suzuki, K. (1995). The GM2 gangliosidoses. *In* "The Metabolic and Molecular Bases of Inherited Disease," 7th ed. (C. R. Scriver, A. L. Beaudet, W. S. Sly, and D. Valle, eds.), pp. 2839–2882. McGraw-Hill, New York.

Grody, W. W., Dunkel-Schetter, C., Tatsugawa, Z. H., Fox, M. A., Fang, C. Y., Cantor, R. M., Novak, J. M., Bass, H. N., and Crandall, E. F. (1998). PCR-based screening for cystic fibrosis carrier mutations in an ethnically diverse pregnant population. *Am. J. Hum. Genet.* **62,** 1252–1254.

Guardiola, P., Socie, G., Pasquini, R., Dokal, I., Ortega, J. J., Van Weel-Sipman, M., Marsh, J., Locatelli, F., Souillet, G., Cahn, J. Y., Ljungman, P., Miniero, R., Shaw, J., Vermylen, C., Archimbaud, E., Bekassy, A. N., Krivan, G., DiBartolomeo, P., Bacigalupo, A., and Gluckman, E. (1998). Allogeneic stem cell transplantation for Fanconi anaemia. Severe aplastic anaemia. Working Party of the EBMT and EUFAR. European Group for Blood and Marrow Transplantation. *Bone Marrow Transplant.* **21**(Suppl. 2), S24–S27.

Kaback, M., and Zeigler, R. (1972). Heterozygote detection in Tay-Sachs disease: A prototype community screening program for the prevention of recessive genetic disorders. *In* "Sphingolipids, Sphingolipidoses and Allied Disorders." (B. Volk and S. Aronson, eds.). Plenum, New York.

Kaback, M., Nathan, T., and Greenwald, S. (1977). Tay-Sachs disease: Heterozygote screening and prenatal diagnosis-US experience and world perspective. *In* "Tay-Sachs Disease: Screening and Prevention." (M. M. Kaback, ed.). R. Liss, New York.

Kaback, M., Lim-Steele, J., Dabholkar, D., Brown, D., Levy, N., and Zeiger, K. (1993). Tay-Sachs disease carrier screening, prenatal diagnosis, and the molecular era: An international prospective. *J. Am. Med. Assoc.* **270,** 2307–2315.

Kaul, R., Gao, P. G., Aloya, M., Balamurugan, K., Petrosky, A., Michals, K., and Matalon, R. (1994). Canavan disease: Mutations among Jewish and Non-Jewish patients. *Am. J. Hum. Genet.* **55,** 34–41.

Levran, O., Desnick, R. J., and Schuchman, E. H. (1991a). Niemann-Pick disease: A frequent missense mutation in the acid sphingomyelinase gene of Ashkenazi Jewish type A and B patients. *Proc. Natl. Acad. Sci. (USA)* **88,** 3748–3752.

Levran, O., Desnick, R. J., and Schuchman, E. H. (1991b). Niemann-Pick type B disease: Identification of a single codon deletion in the acid sphingomyelinase gene and genotype/phenotype correlations in type A and B patients. *J. Clin. Invest.* **88,** 806–810.

Levran, O., Desnick, R. J., and Schuchman, E. H. (1992). Identification and expression of a common missense mutation (L302P) in the acid sphingomyelinase gene of Ashkenazi Jewish type A Niemann-Pick disease patients. *Blood* **80,** 2081–2087.

Levran, O., Desnick, R. J., and Schuchman, E. S. (1993). Type A Niemann-Pick disease: A frameshift mutation in the acid sphingomyelinase gene (fsP330) occurs in Ashkenazi Jewish patients. *Hum. Mutat.* **2,** 317–319.

Li, L., Cagana, M., Robinowitz, J., Shabeer, J., Desnick, R. J., and Eng, C. M. (1997). Prenatal genetic screening in the Ashkenazi Jewish population: A pilot program of multiple option testing for five disorders. *Am. J. Hum. Genet.* **61,** A24.

Li, L., Eng, C. M., Desnick, R. J., German, J., and Ellis, N. A. (1998). Carrier frequency of the Bloom syndrome blmAsh mutation in the Ashkenazi Jewish population. *Mol. Genet. Metab.* **64,** 286–290.

Massarik, F., and Kaback, M. M. (1981). Being a carrier: Is there a stigma. *In* "Genetic Disease Control: A Social Psychological Approach," Sage Lib. Soc. Res. **110,** 102–117. Sage, Thousand Oaks, CA.

Matalon, R., Michale, K., and Kaul, R. (1995). Canavan disease: From spongy degeneration to molecular analysis. *J. Pediatr.* **127,** 511–517.

Mistry, P. K., and Abrahamov, A. (1997). A practical approach to the diagnosis and management of Gaucher disease. *In* "Gaucher's Disease" (A. Zimran, ed.), Billiere's Clinical Hematology.

Myerowitz, R., and Costigan, F. C. (1988). The major defect in Ashkenazi Jews with Tay-Sachs disease is an insertion in the gene for the alpha-chain of beta-hexosaminidase. *J. Biol. Chem.* **263,** 18587–18589.

National Institutes of Health (1997). Genetic testing for cyctic fibrosis. NIH Consensus Development Statement, vol. 15, no.4.

Navon, R., and Proia, R. L. (1989). The mutations in Ashkenazi Jews with adult GM2 gangliosidosis, the adult form of Tay-Sachs disease. *Science* **243,** 1471–1474.

Okada, S., and O'Brien, J. S. (1969). Tay-Sachs disease: Generalized absence of a beta-D-N-acetylhexosaminidase component. *Science* **165,** 698.

Rucknagel, D. L. (1983). A decade of screening in the hemoglobinopathies: Is a national program to prevent sickle cell anemia possible? *Am. J. Pediatr. Hematol. Oncol.* **5,** 373–379.

Schuchman, E. H., and Desnick, R. J. (1995). Niemann-Pick disease types A and B: Acid sphingomyelinase deficiencies. *In* "The Metabolic and Molecular Bases of Inherited Disease," 7th ed. (C. R. Scriver, A. L. Beaudet, W. S. Sly, and D. Valle, eds.), pp. 2601–2624. McGraw-Hill, New York.

Scriver, C. R., Bardanis, M., Cartier, L., Clow, C. L., Lancaster, G. A., and Ostrowsky, J. T. (1984). β-Thalassesmia disease prevention: Genetic medicine applied. *Am. J. Hum. Genet.* **36,** 1024–1028.

Sibille, A., Eng, C. M., Kim, S.-J., Pastores, G., and Grabowski, G. A. (1993). Phenotype/genotype correlations in Gaucher disease type 1. *Am. J. Hum. Genet.* **52,** 1094–1101.

Slaugenhaupt, S. A., Blumenfeld, A., Gill, S. P., Leyne, M., Mull, J., Cuajungco, M. P., Liebert, C. B., Chadwick, B., Idelson, M., Reznik, L., Robbins, C. M., Makalowska, I., Brownstein, M. J., Krappmann, D., Scheidereit, C., Maayan, C., Axelrod, F. B., and Gusella, J. F. (2001). Tissue-specific expression of a splicing mutation in the *IKBKAP* gene causes familial dysautonomia. *Am. J. Hum. Genet.* **68,** 598–605.

Tambor, E. S., Bernhardt, B. A., Chase, G. A., Faden, R. R., Geller, G., Hofman, K. J., and Holtzman, N. A. (1994). Offering cystic fibrois carrier screening to an HMO population: Factors associated with untilization. *Am. J. Hum. Genet.* **55,** 626–637.

Triggs-Raine, B. L., Feigenbaum, A. S., Natowicz, M., Skomorowski, M.-A., Schuster, S. M., Clarke, J. T. R., Mahuran, D. J., Kolodny, E. H., and Gravel, R. A. (1990). Screening for carriers of Tay-Sachs disease among Ashkenazi Jews. *N. Engl. J. Med.* **323,** 6–12.

Triggs-Raine, B. L., Mules, E. H., Kaback, M. M., Lim-Steele, J. S., Dowling, C. E., Akerman, B. R., Natowicz, M. R., Grebner, E. E., Navon, R., Welch, J. P., Greenberg, C. R., Thomas, G. H., and Gravel, R. A. (1992). A pseudodeficiency allele common in non-Jewish Tay-Sachs carriers: Implications for carrier screening. *Am. J. Hum. Genet.* **51,** 793–801.

Vanier, M. T., Ferlinz, K., Rousson, R., Duthel, S., Louisot, P., Sandhoff, K., and Suzuki, K. (1993). Deletion of arginine (608) in acid sphingomyelinase is the prevalent mutation among Niemann-Pick disease type B patients from Northern Africa. *Hum. Genet.* **92,** 325–330.

Verlander, P. C., Kaporis, Liu, Q., Seligsohn, U., and Auerbach, A. D. (1995). Carrier frequency of the IVS4+4 A mutation of the Fanconi anemia gene FAC in the Ashkenazi Jewish population. *Blood* **86,** 4034.

Welsh, M., Tsui, L. C., Boat, T., and Beaudet, A. L. (1995). Cystic fibrosis. *In* "The Metabolic and Molecular Bases of Inherited Disease," 7th ed. (C. R. Scriver, A. L. Beaudet, W. S. Sly, and D. Valle, eds.), pp. 3799–3878. McGraw-Hill, New York.

23 The Dor Yeshorim Story: Community-Based Carrier Screening for Tay-Sachs Disease

Josef Ekstein and Howard Katzenstein
Dor Yeshorim
The Committee for Prevention of Jewish Genetic Diseases
Brooklyn, New York/Jerusalem, Israel

   I. Introduction
  II. Understanding a Community at Risk
      A. Orthodox and Hasidic Jews: Preserving a Religion
         and Community
      B. Beliefs about Health and Medical Issues
      C. Beliefs about Procreation
      D. Courtship and Marriage: The Concern about
         Genetic Stigmatization
      E. Summary
 III. Early Efforts at Screening
  IV. Mechanics of the Premarital, Anonymous Screening
      Program
      A. Quality Control and Quality Assurance
      B. Goals and Guidelines
   V. Findings and Accomplishments
  VI. Research
 VII. Can the Dor Yeshorim Model Be Applied to Other
      Communities?
VIII. Analytical Laboratories
      References

I. INTRODUCTION

The Dor Yeshorim program, like other carrier testing programs for autosomal recessive diseases, is in the practice of helping individuals to prevent the occurrence in their families of serious human genetic diseases. This objective is being accomplished through voluntary testing of large portions of the religious Jewish population, initially in the metropolitan New York area and now worldwide, wherever concentrations of Orthodox Jews live. The Dor Yeshorim program began in 1983 with screening for Tay-Sachs disease, and has developed a novel system, which provides for anonymous testing in an effort to prevent the disease and at the same time avoid the stigma which could be attached to knowledge of an individual's carrier status. As these issues are being discussed by the genetics community, the Dor Yeshorim program is one that could be studied for new ideas about minimizing the risks associated with finding out about our genetic constitution.

In this chapter we will describe the community characteristics and dynamics, which are the context for the Dor Yeshorim program, as well as the mechanics of the program. The 17-year history and its findings and accomplishments are summarized; some interesting and surprising results are also discussed. Since genetics is fraught with difficult issues, in terms of how and when to use its powerful capabilities, we are likely to come to solutions that our society can accept if we remain open to a wide diversity of ideas. We hope that this discussion of the Dor Yeshorim program can contribute to such solution building. This program serves as a successful model of "genetic compatibility" testing and was designed to meet the special needs of the Orthodox Jewish community.

II. UNDERSTANDING A COMMUNITY AT RISK

Dor Yeshorim, the Committee for Prevention of Jewish Genetic Diseases, began within the context of the Hasidic Jewish community, then expanded to the broader Orthodox Jewish communities of the metropolitan New York area. The program expanded its services in later years to include similar communities across the United States, Canada, Europe, Israel, and Australia (Broide *et al.*, 1993; Burnett *et al.*, 1995). Dor Yeshorim's success in preventing Tay-Sachs disease in these communities is a direct result of how well the program's structure fulfills the needs of its served community. Therefore, before describing the history of Dor Yeshorim or the structure of the program, it is important to first appreciate the characteristics and dynamics of these communities.

In 1990 there were 1.6 million Jews living in the metropolitan New York area; of these, 1.1 million lived in New York City. This is a very significant concentration of Jewish people, considering that there are only 5.5 million Jews in the entire United States; in other words, 1 in 5 American Jews lives within a geographic area of 327 square miles. This physical proximity is a graphic manifestation of the powerful beliefs which especially bind traditional Jewish peoples together. So-called religious or traditional Jews can be thought of as those who adhere to similar practices and beliefs and who can be categorized into two main groups: the Orthodox and Hasidim. The "modern Orthodox" population is approximately three times the size of the traditional community; this community and the potential for approaching its citizens are discussed in the Section VII of this chapter.

A. Orthodox and Hasidic Jews: Preserving a religion and community

The Hasidic community believes very strongly that the only way they will preserve their religious identity is by maintaining the old "Shtetel" (village) ways. Thus, they seek to preserve the traditional means of communication through the Yiddish language, as well as customs and mode of dress, which can be distinguished from the rest of society and even from other Jews. However, in the Orthodox community these distinguishing characteristics are subtle.

In both of these groups, the traditional interpretations of the Torah (Old Testament) and its laws are followed. Although they may use modern technologies (fax machines, computers, cellular phones, etc.), both groups generally do not own televisions, and even radio is discouraged; the media may be viewed as providing immodest and/or immoral messages or images. There is a tremendous effort and concentration on teaching children that some of the ways of the outside society may be detrimental to a "kosher" way of life. Thus, rabbinical leaders, parents, and community leaders screen communications from the outside world.

B. Beliefs about health and medical issues

Members of the observant Jewish community believe that they have a religious obligation to care for their physical well-being. Therefore, seeking help for illness and pursuing preventive measures to guard against illness are widely practiced. They seek help first, and often exclusively within the community, and frequently consult with their rabbi about where to seek medical advice.

Because this community is distrustful, or at least skeptical, about some of the information being generated outside their community, a portion of the literature regarding medical practices and options is not acceptable to them. In particular, standard educational material about genetic diseases is typically viewed as containing biased or value-laden information inasmuch as it may present options such as abortion or other practices prohibited by Biblical law. For these reasons, only information emanating from a community-based organization with the endorsement of its community and rabbinical leaders is able to make an impact on the attitudes of this group toward any medical or public health intervention.

C. Beliefs about procreation

Like many people of various religious or philosophical perspectives, observant Jews view having children as the most fundamental expression of their humanity. The typical Hasidic or Orthodox couple is interested in producing a large family to fulfill the Biblical commandment to "be fruitful and multiply." Most families have 5 to 10 children. Jewish law does not permit birth control or abortion. However, individuals may seek special dispensation from their rabbi in extreme circumstances on a case-by-case basis. Consideration of this basic value of the absolute sanctity of human life was a founding principle for the Dor Yeshorim screening program.

D. Courtship and marriage: The concern about genetic stigmatization

Since the observant Jew believes in separation of the sexes, social dating and other such customs are unacceptable. There are no social clubs or other means of meeting a potential spouse. The couple usually come from similar backgrounds and values. Most girls marry at around 18, most boys by their early 20s. Although marriages among Hasidim can be considered to be "arranged," the young couple in question do have choice in the matter, and either one can decline a suggested partner. Within Orthodox circles, dating will occur for the purpose of seeking a marriage partner.

Since the primary purpose of the marriage is to begin a new, hopefully large family, information about the health of each partner and the potential health of future children is of great concern when considering a potential spouse; this would include information about genetic diseases in either family. Today, with almost two decades of experience with the Dor Yeshorim program, it is quite typical for a family to seek genetic counseling upon hearing of genetic disease in the family of a potential partner. Previously, it was more common that a decision would be made not to pursue this "problem family" because of the lack of accurate scientific information in hand. This practice often led community members to "hide" genetic problems for fear that the "marriageability" of their normal children and relatives would be harmed.

E. Summary

Because the Orthodox and Hasidic Jewish communities adhere to strict religious practices, are health conscious, and have strong beliefs about family, any program directed at serving these communities must accommodate these principles and practices. It was in this context of a tightly knit religious community with strong concerns about genetic stigmatization that the Dor Yeshorim screening program was conceived.

III. EARLY EFFORTS AT SCREENING

After the establishment of standard carrier screening programs in the 1970s and 1980s by Dr. Michael Kaback and others, including one in close proximity to the religious Jewish communities (at the Mount Sinai School of Medicine), it became apparent that the Orthodox and Hasidic communities were generally not participating. The birth rate of babies with Tay-Sachs disease began to decline in the 1970s, but this was not true among the children of religious Jews.

To bring preventive carrier testing to this community required the close collaboration of a motivated rabbi, an expert, dedicated physician, and two de-voted community members. In 1983, Rabbi Josef Ekstein found himself the father of yet a fourth child with Tay-Sachs disease. He elected to translate this personal tragedy of four children with Tay-Sachs disease into a mission to spare others in his community from experiencing what he realized was *preventable*. Rabbi Ekstein sought the help and advice of Dr. Robert Desnick, chairman of the Department of Human Genetics at the Mount Sinai School of Medicine, the physician who diagnosed and cared for the fourth of Rabbi Ekstein's Tay-Sachs children. With Dr. Desnick's expert advice and strong support, Rabbi Ekstein and his colleagues proposed to their community leaders that a screening program be established from within.

However, since traditional screening programs focused on testing in-dividuals who were already married, if not pregnant, the religious communities viewed them as inappropriate for people like themselves who would not consider abortion. Furthermore, there was widespread fear that carriers and their entire families would become stigmatized as undesirable marriage partners. In particular, family members of affected children were vehement in their opposition to testing. Community leaders also became concerned that the open identification of ado-lescents as carriers could cause them unnecessary stress and anxiety. In 1976, the prominent bioethicist, Dr. Fred Rosner, reviewed a plethora of opinions that had accumulated from medical and Jewish authorities (Rosner, 1976) on the pros and cons of testing young people. Indeed, Rabbi Moses Feinstein, the most notable Jewish scholar on the interpretation of the law for postwar Orthodox Jewry, wrote in 1973 against mass carrier testing of young adults for all these reasons (Feinstein, 1973).

Together with his colleagues, Mr. Elias Horowitz and Mr. Kalman Weiss, Rabbi Ekstein countered these objections to screening with a proposal for a community-based program, which could provide absolute anonymity and confi-dentiality to those tested. Because these community leaders had the commitment of a leader in the medical genetics community in Dr. Desnick, their proposal had credibility as a true collaboration between medicine and community. The novel structure of the Dor Yeshorim program allowed all young couples considering marriage to be protected from the risk of having a child with Tay-Sachs disease, while also protecting the identity of carriers in the community. Indeed, even the members of the Dor Yeshorim staff would not know the identity of carriers in this truly anonymous system. The dedicated vision of Dr. Desnick and the other founders of the Dor Yeshorim program resulted in a slow but meaningful reversal of previous objections to screening. For example, a year after the initiation of the Dor Yeshorim program in 1983, Rabbi Feinstein wrote a public letter supporting the Dor Yeshorim screening program for all Jews, since previous concerns had been addressed by the screening program.

IV. MECHANICS OF THE PREMARITAL, ANONYMOUS SCREENING PROGRAM

As young people in these communities typically marry by age 20, the Dor Yeshorim program is primarily offering testing to high school senior girls and seminary boys (90% of tests obtained through Dor Yeshorim) in order to achieve early *premarital* screening. Before a screening day, students are introduced to the clinical picture of Tay-Sachs disease and the 1-in-4 risk if two carriers marry. Students also learn that carriers are not themselves ill, and that one carrier in a couple presents no risk for affected children. This and other information is provided in a pamphlet that is given out at school, and the concepts are discussed in the classroom. The parents are encouraged to discuss testing and its merits at home and must sign a consent form for their minor children to be tested. Since the Dor Yeshorim program has been in effect for almost two decades, these young people also may have learned of the program through older siblings. Approximately 90% of students choose to be tested.

On the day of testing, each student who has a signed parental consent form is given an individual consent form and a card printed with a unique identification number. All materials including the blood collection tubes are labeled with the individual's unique identification number. A random control number is also added to each identification number to provide a quality control measure. Each student who wishes to be tested writes his or her date of birth and family's countries of origin. The consent form is signed with his or her identification number. From this point forward, each individual and blood sample can only be identified by the *anonymous* identification number. The card which each students keeps has the phone number of Dor Yeshorim on it as well as his or her preprinted identification number and a reminder to call for compatibility results in the early stages of a potential marriage relationship. Professional, licensed phlebotomists draw blood samples.

When a young couple is considering a "shidduch" (marriage match), they have the opportunity to call Dor Yeshorim, to which they can provide their two

ID numbers and dates of birth and receive a genetic compatibility assessment. The couple is told *only* whether or not they are compatible; two carriers for Tay-Sachs disease are incompatible. Two noncarriers or one carrier and one noncarrier are compatible. It is not divulged that one member is a carrier, so as to protect the carrier and his or her family from stigmatization. Since the staff at Dor Yeshorim have only the identification numbers and birthdates of the individuals in question, they are also blind as to the identity of carriers. If a couple is "potentially incompatible" (inconclusive by inconclusive or carrier by carrier or inconclusive by carrier), they are counseled regarding the disease, and the 1-in-4 risk with each pregnancy of having an affected child, if both are carriers. Blood samples are redrawn for confirmatory retesting. During retesting they are continuously counseled. If the couple is confirmed as incompatible, they are offered further genetic counseling. The choice about whether or not to pursue the marriage in light of this information is strictly up to the couple. In addition, the couple is reminded that each of them is very likely to be compatible with another individual in another potential partnering.

The Dor Yeshorim program provides hours of counseling to couples who have been found to both be carriers of Tay-Sachs disease or another Jewish genetic disease (e.g., cystic fibrosis, Canavan disease, Fanconi anemia type C), to educate them about their specific risk. Such a couple is keenly focused on this information, which pertains to them *in particular*; i.e., this is education about *themselves*. Thus, genetic counseling is efficiently focused on at-risk couples instead of on each individual carrier identified, permitting in-depth counseling and, in part, addressing the problem of the shortage of genetic counselors trained in these issues as well as the high cost of carrier counseling. Dor Yeshorim is committed to spending as much time as necessary with these at-risk couples; they can afford to do so, since the number of such couples is small compared to the number of individual carriers in the population.

A. Quality control and quality assurance

Dor Yeshorim sends blood samples to several quality control laboratories in the United States for DNA testing and/or enzyme assays. Samples are labeled only with the identification number. Results from participating laboratories are entered into a secured computer file by identification number. Entry is performed twice by different data entry personnel. The computer compares the two entries for any discrepancies and does not allow the deposition of nonduplicate data.

The use of several independent laboratories for testing is an important aspect of the quality control measures that are embedded in Dor Yeshorim's program. Both academic and commercial laboratories have provided or are currently providing testing services to the program. Each laboratory has all the basic elements required for CLIA and other required certifications as general clinical laboratories.

However, selecting a specialty genetics laboratory has required further assessment of the following aspects of laboratory practice: (1) quality control measures appropriate to DNA or enzyme analysis practiced by the laboratory; (2) turnaround time for results back to Dor Yeshorim; (3) qualifications for interpretation of results by the laboratory directors; (4) ability to manage large volumes of samples; and (5) willingness to participate in blind control sampling as part of Dor Yeshorim's quality control program. A list of the excellent laboratories that have provided testing services to Dor Yeshorim are provided in Section VIII of this chapter.

As mentioned above, Dor Yeshorim includes random control samples in most shipments of samples to each laboratory. Multiple laboratories, using all methods in practice, have analyzed each control sample so that the genotype of the sample is well documented. Indeed, as revealed in Table 23.2, control sampling and confirmatory testing at one time represented approximately 32% of all samples sent for analysis. This large quantity and frequency of control sampling has proven to be of great value in enzyme testing, since the assessment of normal, inconclusive, and carrier quantities of hexosaminidase A activity is so sensitive to minor variations in samples, patient, and assay conditions. Far fewer DNA controls are included in shipments, since mutation analysis is not sensitive to such external factors. Since any discrepant results on control samples are discussed confidentially with the laboratory, this program is typically viewed by the laboratory as a valuable external assessment of reliability and laboratory performance.

B. Goals and guidelines

The Dor Yeshorim program has two goals. The first is to provide a system whereby young men and women of marriageable age can avoid the risk of having a child with Tay-Sachs disease and other recessive diseases that are prevalent among the Ashkenazim (Jews of Eastern European descent). Importantly, the goal of the program is to provide this preventive measure without placing these young people at risk of social stigmatization and discrimination in the community or by insurance carriers or employers as a result of test results. The second goal is to allow young people with a family history of Tay-Sachs disease or other genetic diseases to participate without the fear of being identified and stigmatized.

Dor Yeshorim stresses confidentiality and anonymity very strongly. For couples who are identified to be at risk, it is essential that a confidential environment for obtaining information and assistance be guaranteed. Indeed, since its founding, Dor Yeshorim has seen a regular flow of parents, siblings, and other family members of individuals afflicted with *other* genetic diseases coming to seek information and help in preventing genetic disease. As Dor Yeshorim has come to represent a reliable mechanism for obtaining compatibility assessments, so has it become a resource for referrals and for research on many of the recessive illnesses found among Jewish peoples (see below).

V. FINDINGS AND ACCOMPLISHMENTS

Since 1983, the Dor Yeshorim program has received and processed blood samples from over 120,000 individuals (Table 23.1). The volume of testing has increased over the years, with almost 12,500 being tested in 2000. Dor Yeshorim added testing for cystic fibrosis to its program in 1993, and for Canavan disease and type C Fanconi anemia in 1995. Dor Yeshorim also facilitates access to testing for other autosomal recessive diseases upon request from families. In addition, Dor Yeshorim has explored through pilot studies the merits of screening for Gaucher disease. These new screening opportunities have provoked the retesting of an additional 18,608 individuals (Table 23.1). This expansion of the Dor Yeshorim program to other recessive disorders is discussed elsewhere (Desnick, in press). The numbers are noteworthy, since the individuals who were retested for these additional diseases did so by special request; this avid reuse of the program seems a significant affirmation of how the community has embraced the screening program.

Dor Yeshorim has had almost 60,000 requests for compatibility assessment from couples, with 7,759 of these in the last year alone (Table 23.1). These queries have resulted in finding 295 potential matches that were confirmed as incompatible for at least one of the disorders for which carrier tests were available.

To date, the program has logged Tay-Sachs carrier status results for 117,302 individuals, plus an additional 55,487 control sample results (Table 23.2). Taken together, some 172,789 tests for Tay-Sachs disease have been performed and cataloged over 17 years. Among the 117,302 tested individuals, 5,416 were found to be carriers by enzyme or DNA analysis. The overall carrier rate of all testing methods was found to be 1:22. The carrier rate from serum testing was approximately 1:20. The rate of inconclusive enzyme analysis results was 1:15.

Since 1992, there have been 39,574 individuals who also received gene-based testing for Tay-Sachs disease by analysis of the three common mutations in the Ashkenazim (see Table 23.3). Of these, 1,718 were found to be carriers of one mutation, a carrier rate of 1:23. This carrier rate, when compared with the 1:22 rate derived from total testing (serum and DNA tests), is consistent with

Table 23.1. Summary of Screening Activity at Dor Yeshorim: 1983–2000

Description	1983–2000	Period: 1/1/00–12/31/00
Total individuals tested	120,019	12,479
Total retested for additional disease or confirmation[a]	18,608	619
Total couples called to check compatibility	59,700	7,759
Total incompatible; all diseases	295	44

[a] Testing for CF was added in 1993, Canavan disease and Fanconi anemia Type C in 1995.

Table 23.2. Tay-Sachs Screening; 1983–2000

	Number of individuals tested	Carriers (rate)	Inconclusives
Total individuals sampled	118,689		
Total individuals analyzed to date[a]	117,302	5,416(1:22)[b]	8,036(1:15)
Carriers and inconclusives by serum, retested by platelet or leukocyte assay	1,227	415	98
Normal control samples tested	55,487		
Total tests	172,789		
Incompatible couples	136		

Comparison of biochemical/DNA results			
	Number of individuals tested	DNA positive (rate)	DNA negative
Individuals tested by DNA and serum	39,574	1,718(1:23)	37,856
Carriers by serum	2,037	1,602	435[c]
Inconclusives by serum	4,152	99	4,053
Negatives by serum	33,385	17[d]	33,368
Individuals tested by platelet or leukocytes and DNA	1,047	232	815
Carriers detected by platelet or leukocyte assay	239	224	15[e]
Inconclusives by above platelet or leukocyte assay	63	8	55[e]
Negatives by platelet or leukocyte assay	745	0	745
Control DNA samples	5,107		
Total DNA results	44,681		

[a]Serum and DNA together.
[b]By serum, not confirmed.
[c]Three individuals in this category were found to have the common pseudodeficiency gene.
[d]One of the individuals had a bone marrow transplant.
[e]Two individuals in each of these categories were found to have the common pseudodeficiency gene.

Table 23.3. Mutation Analysis of β-Hexosaminidase A Gene[a]

Specimens	Number of individuals	Mutation frequency
Total specimens	59,819	
Total mutations detected	2491	
1,277 insertion	2,202	88.4%
1,421 splice site	206	8.3%
G269	79	3.2%
Other	4[b]	.2%

[a]Analysis of mutations includes research. Samples.
[b] Includes one convert and three individuals with sephardic lineage.

the observation of 435 individuals who were determined to be carriers by serum testing but who were negative for all three mutations by DNA analysis. It has not been possible to retest all 435 individuals to clear the discrepant results. It is likely that the discrepant results are due to the known false positive rate associated with serum tests.

After six years of parallel Tay-Sachs testing by enzyme and mutation analyses, Dor Yeshorim concluded that mutation analysis offered close to 100% sensitivity and specificity for Ashkenazi Jewish individuals (Bach et al., 2001). DNA testing did not have the drawbacks associated with enzyme analyses, the latter having approximately 15% inconclusives as well as false positives. Of 151 Tay-Sachs disease chromosomes from patients or obligate carriers, all were identified by the analysis of the three Tay-Sachs disease mutations. Blood samples from 478 enzyme positive or inconclusive and DNA negative individuals were analyzed for 10 other Tay-Sachs disease/pseudodeficiency mutations. Five carriers of a pseudodeficiency mutation were found, as well as one carrier of the C509A mutation. Ten of the specimens that were also positive/inconclusive by leukocyte/platelet, and negative for the three common Tay-Sachs disease mutations, were analyzed for the 10 above-mentioned mutations. Two were found to be carriers of pseudodeficiency mutations. The remaining eight patients were analyzed by gene sequencing, and no Tay-Sachs disease mutation was detected.

Since the inception of Dor Yeshorim, over 135 proposed matches were found to be incompatible for Tay-Sachs disease. Based on the 1:23 carrier rate derived from the DNA analysis, one would predict that 222 incompatible couples would be identified. However, since incompatible matches were identified only upon the request of particular individuals at the time of considering a potential partner, there is a lag between when Dor Yeshorim obtains carrier results and when incompatibilities are identified. This lag is 2–3 years, because girls are screened on average 2–3 years prior to the time of peak marriage considerations. As many as 25% of the individuals in the Dor Yeshorim database have not been queried for compatibility due to this lag. Therefore, the observed incompatibility rate is expected to be lower than that theoretically predicted from the total individuals tested.

Currently, the Dor Yeshorim program relies exclusively on DNA results for these individuals of Ashkenazi Jewish descent. Individuals who are not solely of Ashkenazi descent [i.e., Sephardi Jews (Jews of Mediterranean descent) or converts] undergo enzyme and DNA analysis.

Dor Yeshorim has observed that the vast majority of at-risk, incompatible couples did not pursue marriage, based on subsequent queries with different suitors. However, a handful of these couples did proceed with the engagement, after further counseling at Dor Yeshorim and elsewhere.

In 1996, there were no children with Tay-Sachs disease in the special unit for children with inborn errors of metabolism at Kingsbrook Jewish Medical Center in Brooklyn, New York (personal communication, Dr. Larry Schneck).

This hospital has had a special ward devoted to the care of patients with Tay-Sachs disease and other inherited metabolic diseases for the past 40 years. Prior to Tay-Sachs prevention programs and the Dor Yeshorim program, the ward had as many as 16 Tay-Sachs patients and had a waiting list of affected children. The last child, the fifth affected child of a religious family, expired in March 1996. It is now a rare occurrence indeed for a child to be born with Tay-Sachs disease in the community served by the Dor Yeshorim program. Clearly, the Dor Yeshorim program is meeting its primary goal of prevention.

VI. RESEARCH

As the religious Jewish community has been greatly served by the endeavors and successes of the general research community, so has the religious Jewish community endeavored to help further advance work in the area of Jewish recessive genetic diseases. Dor Yeshorim has been able to provide hundreds of trial samples to laboratories wishing to establish testing services, as well as positive and negative control samples for laboratory standards. Furthermore, after the discovery of the β-hexosaminidase A gene, the program provided samples to help establish mutation frequencies for Tay-Sachs disease in this population as part of worldwide survey efforts. Dor Yeshorim also has assisted in the identification and validation of the genes that cause Canavan disease, Gaucher type 1 disease, and Fanconi anemia—all diseases which occur quite commonly in the Dor Yeshorim catchments population (Desnick, in press). This synergy between prevention of serious diseases and research is a valuable outcome of the firm base of trust and participation that Dor Yeshorim has established in its community. The benefits of such cooperation and mutual assistance cannot be over-estimated.

VII. CAN THE DOR YESHORIM MODEL BE APPLIED TO OTHER COMMUNITIES?

The Dor Yeshorim screening program must still reach the much larger modern Orthodox community, since this community has not yet embraced the concept of couple-based screening. However, with education and appropriate methods of practice in the program, we believe that a modified version of the Dor Yeshorim model could be applied successfully in this community.

It is for religious and social reasons that the Dor Yeshorim program is focused on testing premarital young adults. At this point in developing a relationship, genetic factors as well as values, family, and moral character are considered. In Canada there has also been a pilot program of testing high school students for Tay-Sachs disease and more recently for cystic fibrosis carrier status, with a reportedly positive impact (Zeesman *et al.*, 1984; Scriver, 1993). Furthermore, the current program of couple-based carrier screening in Britain seems to have met with a positive response in that population (Livingstone *et al.*, 1993; Brock, 1996). Studies on the perceived value that American youth might place on genetic compatibility as one important factor to consider when contemplating a potential marriage relationship might be very interesting. After all, many young couples dissolve relationships upon discovering that the partner does not share similar values with regard to child rearing, religion, or career aspirations. What would such young people do with information that they are both carriers of a severe recessive genetic disease, and have a 1-in-4 risk for an affected child? They would have to choose options including having no children vs. risking having a child with a serious untreatable disease vs. terminating a pregnancy. Perhaps these are choices that a potential couple would choose to avoid, just as many now choose to avoid the difficult choices surrounding the blending of dissonant philosophical or religious views.

VIII. ANALYTICAL LABORATORIES

Dor Yeshorim would like to acknowledge the expert services of all the laboratories which have provided Tay-Sachs carrier-testing services for the Dor Yeshorim program: Mount Sinai School of Medicine, New York, NY; Albert Einstein College of Medicine, Bronx, NY; Baylor College of Medicine Kleeberg DNA Laboratory, Houston, TX; University of Pittsburgh, Pittsburgh, PA; Kingsbrook Jewish Medical Center, Brooklyn, NY; New York University, New York, NY; Hadassah Hebrew University Hospital, Jerusalem, Israel; National Hospital, Institute of Neurology, London, England; Thomas Jefferson Hospital, Philadelphia, PA; Scripps Research Laboratory, San Diego, CA; Genzyme Genetics, Framingham, MA.

Acknowledgments

Dor Yeshorim's success (even survival) has relied heavily on the service and contributions of many people in laboratories, offices, and centers around the world. Rabbi Ekstein and the staff at Dor Yeshorim would like to particularly thank Dr. Robert Desnick, Mr. Elias Horowitz, and Mr. Kalman Weiss for their vision and support in their co-founding and guidance of the Dor Yeshorim program. In addition, Dr. Harold M. Nitowsky's ongoing commitment to Dor Yeshorim as medical director and general advisor has given the program much-needed knowledge, perspective, and professional guidance. The expertise and work of almost 20 years of Frances Berkwits, M.S., as our genetic counselor has given the program the ability to reach out to those in difficult times. We are grateful to Karen Greendale, M.A., and The New York State Department of Health for their assistance.

Dor Yeshorim is also grateful for the advice and assistance of Dr. Barbara Handelin, in particular for help in preparation of this manuscript.

Many thanks also are extended to the laboratory directors and coordinators whose dedication to quality and service are very much appreciated: Dr. Christine Eng, Dr. Ruth Kornreich, Dr. Sue Richards, Dr. John Barranger, Dr. Bernice Allito, Dr. Jean DeMarchi, Ms. Gutta Perle, Dr. Edwin Kolodny, Dr. Tomczak, Dr. E. J. Thompson, Ms. Patricia Morris, and Dr. Eugene Grebner. Special thanks especially go to Dr. S. Nikagawa and Profs. Gideon Bach and Dvorah Abeliovich, who on a daily basis provide service and support to Dor Yeshorim. Finally, Dor Yeshorim has been enabled and guided by the office of The New York State Department of Health through the advocacy and diligent efforts of Dr. Ann Willey, Dr. Kenneth Pass, Ms. Katherine Harris, and Dr. Jane Lin Fu and Dr. Margaret Lee of H.H.S. Bureau of Maternal and Child Health.

Rabbi Ekstein is grateful to have received the Robert Wood Johnson Community Health Leadership award for 1996; Dor Yeshorim is very proud to receive such external recognition and support.

References

Bach, G., Tomczak, J., Risch, N., and Ekstein, J. (2001). Tay-Sachs screening in the Jewish Ashkenazic population: DNA testing is the preferred procedure. *Am. J. Med. Genet.* **99,** 70–75.

Brock, D. J. H. (1996). Prenatal screening for cystic fibrosis: 5 years' experience reviewed. *Lancet* **347,** 148–149.

Broide, E., Zeigler, M., Ekstein, J., and Bach, G. (1993). Screening for carriers of Tay-Sachs disease in the ultra orthodox Ashkenazi Jewish community in Israel. *Am. J. Med. Gen.* **47,** 213–215.

Burnett, L., Proos, A. L., Chesher, D., Howell, V. M., Longo, L., Tedeshi, V., Yang, V. A., Stafakos, N., and Turner, G. (1995). The Tay-Sachs disease prevention program in Australia: Sydney pilot study. *Med. J. Austr.* **163,** 298–300.

Desnick, R. J., ed. (in press). *Advances in Jewish Genetic Diseases*.

Feinstein, M. (1973). Responsum on Tay-Sachs testing. *Igros Moshe. Even Ho'ezer* **IV,** 10.

Livingstone, J., Axton, R. A., Mennie, M., Gilfillan, A., and Brock, D. J. H. (1993). A preliminary trial of couple screening for cystic fibrosis: Designing an appropriate information leaflet. *Clin. Genet.* **43,** 57–62.

Rosner, F. (1976). Tay-Sachs disease: To screen or not to screen? *J. Rel. & Health* **15,** 271–280.

Scriver, C. R. (1993). Human genetics: Schoolyard experiences. *Am. J. Hum Genet.* **52,** 243–245.

Zeesman, S., Clow, C. B., Cartier, L., and Scriver, C. R. (1984). A private view of heterozygosity: Eight year follow up on carriers of Tay-Sachs gene detected in high school screening in Montreal. *Am. J. Med. Genet.* **18,** 769–778.

24

Tay-Sachs Disease and Preimplantation Genetic Diagnosis

Christoph Hansis and Jamie Grifo
Department of Obstetrics and Gynecology
Program for IVF, Reproductive Surgery and Infertility
NYU School of Medicine
New York, New York 10016

I. Tay-Sachs Disease
II. Preimplantation Genetic Diagnosis
References

I. TAY-SACHS DISEASE

According to the Online Mendelian Inheritance in Man (OMIM, 1999) database, Tay-Sachs disease is an autosomal recessive progressive neurodegenerative disorder which is characterized by a deficiency of the enzyme hexosaminidase A. In the classic infantile form it is usually fatal by age 2 or 3 years while in the juvenile form patients die by age 15. Clinical features include developmental retardation, paralysis, dementia, and blindness.

Okada and O'Brien (1969) demonstrated that the α unit, one of the two α and β units of hexosaminidase A, is disturbed in Tay-Sachs disease. While in the classic infantile form the α unit is absent, the juvenile form may suffer from a partial loss of function. The substrates of the α unit, acting synergistically with activator proteins called G_{M2}, which are sulfated substances, the most important one being the G_{M2} ganglioside. In the absence of the α unit, G_{M2} ganglioside is not degraded and accumulates in neurons.

The α unit of hexosaminidase A was finally located by Nakai *et al.* (1991) at the long arm of chromosome 15 (15q23-q24) and named *HEXA*. Proia and

Soravia (1987) showed that the *HEXA* gene is about 35 kb long and consists of 14 exons. Differential transcription of the most 3' exon generates two *HEXA* mRNAs with different 3' untranslated regions.

To date, 78 mutations of the *HEXA* gene have been described (Myerowitz, 1997), with one large and 10 smaller deletions, two small insertions, and 65 single-base substitutions. Eighty percent of the carriers of Tay-Sachs disease of Ashkenazi Jewish origin, a population with high frequency of Tay-Sachs disease, are characterized by an insertion of 4 bp in exon 11. This mutation introduces a premature termination signal, resulting in a deficiency of mRNA. The insertion is also present in some Cajun families of southwest Louisiana, but at a much lower frequency. Another affected distinct population of French-Canadian origin shows a very large deletion of 7.5 kb which affects exon 1 and neighboring sequences.

According to Petersen *et al.* (1983), the Tay-Sachs mutation in the Ashkenazi Jewish population occurred after the second Diaspora at 70 A.D. and before the major migrations to Poland and Russia at 1100 A.D. one of 31 North American Jews is a carrier of Tay-Sachs disease, 88% of them being of Ashkenazi origin. While parental consanguinity is relatively infrequent in the Jewish cases, it is frequent in non-Jewish cases.

It is still not clear whether the Tay-Sachs mutation spread as a founder effect within a relatively small group of a few thousand individuals, a genetic drift, differential immigration patterns, or heterozygote advantage. A combination of these factors seems the most likely scenario.

The diagnosis of Tay-Sachs disease is based on both DNA and enzymatic testing, including prenatal testing. While the most severe problem of enzymatic testing is the high frequency of false positive results, DNA testing may suffer from false negative results due to many rare alleles. It is therefore crucial to thoroughly examine and determine the genetic status of the parents before testing the embryo for Tay-Sachs. Genetic testing of the parents can be done in several ways, including allele-specific oligonucleotide hybridization (ASO), heteroduplex polymerase chain reaction (PCR) fragment analysis, or single-cell whole-genome preamplification.

II. PREIMPLANTATION GENETIC DIAGNOSIS

Preimplantation genetic diagnosis (PGD) is an excellent tool to prevent a future child from suffering with a severe disease. Not only can the child be protected from a genetic disease, but so are his descendants through the generations. If the parents are known carriers for a genetic disease, the mutation can be tested, affected embryos excluded, and healthy children born. Terminations of affected pregnancies can also be avoided, since the affected embryos are not transferred back. After having experienced a pregnancy or a birth of an affected child, it is often very important for many parents to have the security that the next child will not suffer from the same disease.

Tay-Sachs disease is a very likely candidate for PGD and provides excellent reasons to test the embryo. Since 78 mutations are known and well characterized, they can be examined by means of molecular biological techniques. Relatively high frequencies of the mutations in distinct populations result in many couples dealing with the disease, which leads to well-established laboratory procedures and extensive experience of the medical staff. Finally, Tay-Sachs disease means fatal consequences for the child and the parents, and should therefore be avoided.

PGD starts with an exact definition of the parents' mutations in their blood: What kind of mutations are present? What consequences would it have for the child to be homozygous for the mutations? Many Ashkenazi Jewish descendants might have the same 4-bp insertion, but due to the large number of possible mutations the parents could also have different mutations which might not lead to disease if transmitted. Only when those questions are answered should the embryo be tested in PGD. During the whole process, the couple undergoes extensive genetic counseling in order to make informed decisions.

Prior to PGD a blood sample from the couple is drawn again to repeat the mutation analysis and to prepare the embryo for testing in the *in vitro* fertilization (IVF) research laboratory. In a conventional IVF cycle, with or without injection of sperm cells (ICSI), embryos are obtained and cultured to day 3. Then, suitable embryos are chosen for possible transfer by criteria such as morphology, cleavage rate, and number of nuclei. Usually they consist of 6–8 cells at this stage. Under a micromanipulator the zona pellucida of those embryos is "drilled" by mechanical force or application of acidic tyrode's solution and one or two blastomeres are carefully sucked out. It is crucial not to disrupt the membrane of the blastomeres, since the DNA would be spread out and not be available for genetic testing.

The blastomere is then placed in equilibrated PCR lysis buffer for transportation to the molecular biological lab. According to our lab, the blastomere is lysed and a PCR is performed. In the PCR a distinct DNA fragment is amplified with specific primers which were determined by the parents' genetic analyses. The primers are usually chosen on both sides of the mutation to generate a product of 200–400 nucleotides which stretches across the mutation. In order to confirm the amplification of the right product, in many cases a "nested" PCR is performed: after the first round of PCR a small portion of the product is submitted to a second round of PCR, where the primers are located inside the first primer pair. The second primer pair should not amplify nonspecific products of the first PCR but just the specific product and therefore makes sure that the right product is amplified.

Great care should be taken to optimize the PCR conditions, such as annealing temperature, number of cycles, and primer conditions—if not, allelic dropout might occur with fatal misdiagnoses if the mutated allele is missed. The consequences are naturally more severe in dominant inherited diseases such as Huntington's disease when the mutated allele is missed. However, in recessive diseases like Tay-Sachs there is also the risk of not detecting a carrier status of the embryo, who might transmit it to the next generation.

The product of the nested PCR is then submitted to a digestion: by means of restriction endonucleases, which recognize distinct DNA sequences, the DNA is cut into smaller pieces. If chosen in the appropriate way, a restriction endonuclease might cut the mutated gene but not the nonaffected or the other way. The products of the digestion are then loaded on an agarose or a polyacrylamide gel, made visible by staining, and analyzed according to the digestion pattern.

Another possibility is the "heteroduplex analysis": in case of a carrier, both the mutated and the normal allele are amplified in a PCR. If the mutated allele bears an insertion and is therefore longer than the nonmutated, it can form a loop if the two products are hybridized. This complex usually runs slower on a gel than two hybridized normal products, which are also present in heterozygous carriers. As a result, two bands are visible on the gel instead of one band in a healthy individual.

If possible, healthy embryos are chosen for the transfer to the recipient, while additional embryos might be frozen for a later cycle. Finally, the diagnosis can be confirmed by amniocentesis or chorionic villus sampling.

In addition to DNA testing for mutations, the chromosomes can also be examined by FISH: a DNA probe binds to a specific chromosome and emits a fluorescent signal which can be caught by a fluorescent microscope. This approach is especially useful to screen for aneuploidies such as trisomy 21 (when the mother is older then 35 years) or to select the gender of the embryo to exclude X-linked diseases such as hemophilia. Since Tay-Sachs disease is an autosomal recessive disease, gender selection does not provide any advantages.

References

Myerowitz, R. (1997). Tay-Sachs disease-causing mutations and neutral polymorphisms in the Hex A gene. *Hum. Mutat.* **9**, 195–208.

Nakai, H., Byers, M. G., Nowak, N. J., and Shows, T. B. (1991). Assignment of beta-hexosaminidase A alpha-subunit to human chromosomal region 15q23-q24. *Cytogenet. Cell Genet.* **56**, 164.

Okada, S., and O'Brien, J. S. (1969). Tay-Sachs disease: Generalized absence of a beta-D-N-acetylhexosaminidase component. *Science* **165**, 698–700.

Online Mendelian Inheritance in Man, OMIM (TM) (1999). Johns Hopkins University, Baltimore, MD. MIM Number: 272800; Date last edited: 5/25/1999. World Wide Web URL: http://www.ncbi.nlm.nih.gov/omim/

Petersen, G. M., Rotter, J. I., Cantor, R. M., Field, L. L., Greenwald, S., Lim, J. S. T., Roy, C., Schoenfeld, V., Lowden, J. A., and Kaback, M. M. (1983). The Tay-Sachs disease gene in North American Jewish populations: Geographic variations and origin. *Am. J. Hum. Genet.* **35,** 1258–1269.

Proia, R. L., and Soravia, E. (1987). Organization of the gene encoding the human beta-hexosaminidase alpha-chain. *J. Biol. Chem.* **262,** 5677–5681.

25

Treatment of G$_{M2}$ Gangliosidosis: Past Experiences, Implications, and Future Prospects

Mario C. Rattazzi
Department of Human Genetics
NYS Institute for Basic Research in Developmental Disabilities
Staten Island, New York

Kostantin Dobrenis
Department of Neuroscience
Albert Einstein College of Medicine
New York, New York

I. Introduction
II. Early Enzyme Infusion Trials
III. Studies in G$_{M2}$ Gangliosidosis Cats
IV. Cell Targeting of Hexosaminidase A
V. TTC-HEX A and Neuronal Storage
VI. Implications and Open Questions
VII. Bone Marrow Transplantation and Enzyme Secretion
VIII. Delivery of Macromolecules to the Brain Parenchyma
IX. CNS Gene Therapy
X. Conclusions
   Acknowledgments
   References

I. INTRODUCTION

As highlighted elsewhere in this volume, the molecular biology advances of the past 15 years have made possible a precise understanding of the genetic defects underlying G$_{M2}$ gangliosidosis (Tay-Sachs, TSD, and Sandhoff diseases), the development of sensitive and accurate diagnostic techniques, and the establishment of experimental animal models of the human disease. In this as in other lysosomal storage diseases, research is now increasingly focused on therapy, a subject that, after the early, disappointing experimental attempts carried out in the 1970s, had been virtually ignored until very recently. The renewed, active interest in therapeutically oriented research is fueled by recent developments in molecular genetics, which have made it possible to foresee permanent cures by means of gene therapy. This exciting prospect, however, is hampered by the realization that numerous problems have to be solved before therapy becomes a practical reality. These problems fall into three broad categories: (1) strictly molecular genetics problems, common to all gene therapy endeavors, dealing with the design and implementation of efficient transduction vectors with minimal side effects, capable of appropriate insertion into the host genome, and providing stable and controllable transgene expression; (2) cell targeting problems, particularly relevant in the case of diseases of the central nervous system (CNS), where functional differentiation and complex intercellular relationships—only dimly understood at present—will probably require that therapeutic molecules be directed to a large number of diverse neural cells; and (3) organ delivery problems peculiar to the CNS, where anatomical and functional barriers must be overcome or circumvented in order to reach the target cells.

Solutions to the problems in the first category will come from a better understanding of fundamental aspects of molecular genetics, currently under active investigation in a number of laboratories. The problems in the other two categories are not new, having been considered in several early and recent studies, in the context of neural cell physiopathology and of therapeutic modalities based on delivery of macromolecules to the CNS, in particular lysosomal enzyme therapy. In this chapter, after a brief review of early experimental approaches to enzyme therapy in G_{M2} gangliosidosis, we will focus on cell targeting and brain delivery aspects, highlighting some of our work relevant to these subjects.

II. EARLY ENZYME INFUSION TRIALS

As envisioned by C. de Duve more than 30 years ago, enzyme-deficient cells exposed to normal enzyme would take it up by endocytosis; by endocellular fusion of the endocytic vesicles with other vesicles of the lysosomal system the enzyme would eventually be delivered to the sites of storage and degrade the stored substrate (de Duve, 1964). The classic experiments of cross-correction by E. Neufeld's group (Fratantoni et al., 1968) as well as numerous enzyme replacement experiments in cultured cells by other investigators showed the validity of de Duve's concept, and spurred a number of attempts at enzyme therapy in patients with diverse lysosomal storage diseases, aimed at exploring its therapeutic value (see Tager et al., 1980, for a review). In G_{M2} gangliosidosis, the first enzyme therapy attempt was that of R. Desnick and collaborators (Desnick *et al.*, 1972), who injected normal human plasma and plasma concentrate enriched in β-hexosaminidase intravenously in a 13-month-old girl with Sandhoff disease. Although the authors stated clearly that the attempt was not meant to be therapeutic, they did notice an increase of the plasma levels of glycosphingolipids (G$_{M3}$ ganglioside, globotriaosylceramide) distal to the enzyme blocks (Hex A and Hex B deficiency), and a decrease of GL4 globoside in urinary sediment, suggestive of a catabolic effect of the infused enzyme.

Shortly thereafter, R. Desnick, R. Brady, and collaborators (Johnson *et al.*, 1973), using Hex A purified from human urine by a method devised by R. Brady's group, again injected the same patient with a higher but still relatively modest amount of enzyme (0.7 mg). The plasma clearance of Hex A was rapid ($T_{1/2} = 10$ min), in contrast to that of plasma β-hexosaminidase infused previously ($T_{1/2} = 2$ h). This discrepancy was puzzling at that time, as the existence of glycosyl residue-specific, receptor-mediated endocytosis of glycoproteins was just beginning to be recognized (Ashwell and Morell, 1974). It is now known, of course, that the rapid plasma clearance of most organ-derived lysosomal enzymes is due to mannosyl-specific receptors, which cause these high mannose-type glycoprotein enzymes to be rapidly endocytosed by macrophages of the reticuloendothelial system (RES), especially Kupffer cells (KC) of liver (Schlesinger *et al.*, 1980).

In this carefully conducted study (Johnson *et al.*, 1973), which included several precisely timed brain and liver biopsies, the authors were able to document uptake of Hex A by liver, but found no evidence of enzyme in cerebrospinal fluid (CSF), nor in brain parenchyma. The hopes that β-hexosaminidase may cross or circumvent the blood–brain barrier (BBB) as do, to some extent, albumin and other serum proteins with a long plasma $T_{1/2}$, were unfulfilled. Although an increase in plasma glucosylceramide concentration suggested that the infused enzyme had been catabolically active, there was no evidence of clinical effects.

Though clinically ineffective, these early attempts at enzyme replacement in G$_{M2}$ gangliosidosis, and similar ones in other storage diseases (Tager *et al.*, 1980), contributed to the identification of the main problems of this approach: preferential uptake by KC and other RES cells, and ineffective delivery to extrahepatic target organs in general and to the CNS in particular, the latter owing to the impermeability of the BBB to proteins. The recognition of these problems, however, decreased considerably the initial enthusiasm for enzyme therapy.

The next trial took place several years later, using large amounts of human Hex A from placenta (Geiger and Arnon, 1978) both native and conjugated with polyvinylpyrrolidone (PVP) that prevented its rapid plasma clearance (Geiger *et al.*, 1977). The observation that proteins injected into the subarachnoid spaces of the CNS and into the cerebral ventricles—thus inside the BBB—could penetrate the brain parenchyma (Brightman, 1965) appeared to provide a route of access to the CNS. This prompted B. von Specht, B. Geiger, and collaborators to inject large doses of Hex A and PVP-conjugated Hex A intra-ventricularly and intracisternally into two patients with TSD (von Specht *et al.*, 1979). A total of 20–30 mg and 40–60 mg of pure enzyme were injected into the two children over a period of 14 and 42 weeks, respectively. In patient #1 after intraventricular injection of enzyme, the CSF levels of both Hex A and PVP-Hex A fell rather rapidly and the enzyme appeared in the plasma. A biopsy of brain cortex after 3 months of treatment showed persistence of massive glycol-ipid storage, consistent with the observation that the patient's neurologic status had continued to deteriorate. Of particular note was the absence of anti-Hex A antibodies in the patient's plasma (von Specht *et al.*, 1979).

In contrast to patient #1, who had overt clinical signs and symptoms when treatment was started at 14 months of age, patient #2 was clinically normal when treatment was initiated at 6 weeks of age. Despite repeated enzyme infusions for almost 1 year, however, the disease followed its typical course; as in the case of patient #1, no antibodies to Hex A were present in this child after treatment was discontinued. A third patient, reported separately, was treated in the same way as patient #2 in this study, but only 11 weeks, with no clinical improvement (Godel *et al.*, 1978). Possible explanations for the lack of success of these therapeutic trials may be found in the following considerations, which have a bearing on any therapeutic methodology involving attempts at delivering macromolecules to neural cells via the CSF.

1. The CSF is secreted by the choroid plexus and excreted into the blood through the arachnoid villi at a very high rate, rapidly diluting and removing macromolecules that may be present in the subarachnoid spaces (Wood, 1980). Active transcytosis by the choroid plexus contributes to the transfer of proteins injected into the cerebral ventricles from CSF to blood (van Deurs, 1976). Thus, the rapid decrease of Hex A activity in the CSF of patient #1 of von Specht *et al.* (1979) and its appearance in plasma shortly after intraventricular injection are not surprising.

2. Although there is an anatomical route that may allow diffusion of pro-teins into the brain parenchyma from the subarachnoid spaces, (Brightman, 1965; Brightman and Reese, 1969; Zhang *et al.*, 1990), under physiologic conditions this diffusion appears unlikely. About 20% of CSF is secreted as interstitial fluid at the level of parenchymal capillaries and flows to the surface of the brain through intercellular and perivascular spaces (Wood, 1980; Rosenberg *et al.*, 1980; Cserr, 1988). Thus it is difficult to envisage *centripetal* movement of macromolecules from CSF into perivascular and intercellular spaces, against the normal *centrifugal* flow of interstitial fluid.

3. Assuming that some Hex A did penetrate the interstitial spaces, there is no evidence that this enzyme may be internalized by the neurons via an efficient mechanism, i.e., adsorptive endocytosis. In fact, as discussed later, endocytosis of placental Hex A by neurons is minimal.

4. The same argument about lack of enzyme internalization applies *a fortiori* to PVP-Hex A, specifically designed to prevent cellular uptake; indeed, the rationale of injecting into the CSF an enzyme modified as to escape RES uptake and have a longer plasma $T_{1/2}$ is not clear. In addition, there are reasons to suspect that Hex A derivatized with PVP may not be able to degrade G_{M2} ganglioside; this was not tested by Geiger *et al.*(1977).

III. STUDIES IN G_{M2} GANGLIOSIDOSIS CATS

The identification of cats with genetic β-hexosaminidase deficiency and generalized G_{M2} gangliosidosis (Cork *et al.*, 1977)—a natural animal model of human Sandhoff disease—made it possible to study various aspects of enzyme therapy. Our studies ranged from enzymatic and immunologic characterization to enzyme therapy-oriented experiments. These included injection of highly purified human placental Hex A with depression of RES uptake and delivery of enzyme to extrahepatic tissues, delivery of Hex A to the CNS by reversible permeabilization of the BBB, and assessment of the catabolic effectiveness of the exogenous enzyme (O'Neil *et al.*, 1979; Rattazzi *et al.*, 1979a, 1979b, 1980, 1981, 1982; Rattazzi, 1983).

By competitively inhibiting the mannosyl-specific, RES clearance of Hex A by intravenous injection of yeast mannans, we could show enhanced exogenous Hex A uptake by, and glycolipid degradation in, visceral organs. In the same animals we were able to deliver Hex A to the CNS by transient permeabilization of the BBB. Assay of brain homogenates showed that the activity of exogenous Hex A reached 100% of the endogenous activity found in normal cat brain, but the amount of G_{M2} ganglioside degradation in these brains was minimal (Rattazzi *et al.*, 1982). Immunofluorescence microscopy in fact showed that human Hex A was present mainly in perivascular macrophages and pericytes, in a few astrocytes, but not in neurons (Rattazzi, unpublished observations). Thus, a critical limiting factor was a low affinity of Hex A for neurons and other neural cells.

Lysosomal enzymes extracted from tissue originate mostly from lysosomes, and therefore have lost the mannose 6-phosphate (M6P) residues present on newly synthesized enzymes (Glaser *et al.*, 1975). If these residues were present, as in the case of enzymes overexpressed and secreted by transduced cells in culture (Ioannou *et al.*, 1992; Grubb *et al.*, 1993; Kakkis *et al.*, 1994), the enzymes might be endocytosed by astrocytes, which have M6P receptors (Hill *et al.*, 1985), and possibly by some neurons, in which M6P receptors (M6PR), present early in development (Nielsen and Gammeltoft, 1990), apparently persist in postnatal life (Couce *et al.*, 1992; Kar *et al.*, 1993), albeit with reduced expression (Nissley *et al.*, 1993). In contrast, tissue-derived enzymes with M6P-less, high-mannose-type glycosyl chains will be endocytosed mainly by macrophages and related cells, through the mannosyl-specific receptor (MR) present on phagocytic cells (Pontow *et al.*, 1992). Furthermore, in the absence of acute insults to brain tissue or of cytotoxic protein markers such as those utilized in early experiments to study BBB disruption, receptor-independent, fluid-phase endocytosis of proteins by neurons and by nonphagocytic neural cells is minimal (Ohata *et al.*, 1990). Thus, appropriate cell targeting strategies are needed to obtain efficient uptake of lysosomal enzymes by neural cells.

IV. CELL TARGETING OF HEXOSAMINIDASE A

In order to enhance Hex A uptake nonspecifically by all neural cells, we coupled the enzyme with low-molecular-weight poly-L-lysine (PLL). This positively charged moiety results in adsorption of the conjugate to the cell membrane and internalization by adsorptive but ligand-independent, "constitutive" endocytosis (Shen and Ryser, 1978). When tested on cultured fibroblasts—which take up minimal amounts of native placental Hex A by fluid-phase endocytosis (Rattazzi, 1983)—the conjugate, PLL-Hex A was taken up avidly by these cells, and , as shown by subcellular fractionation, it was internalized in lysosomes (Rattazzi *et al.*, 1987).

To obtain specific uptake by neurons, we utilized a polypeptide derived from tetanus toxin (TT), a protein that has a very high affinity for neurons and binds almost exclusively to these cells (Wellhoner, 1982). This 50-kDa, nontoxic proteolytic fragment of TT, called "fragment C" (TTC), binds to the neuronal cell membrane almost as avidly as TT, is internalized, and appears to end up in the neuronal lysosomes (Bizzini *et al.*, 1977; Simpson, 1985; Schwab *et al.*, 1979). We coupled TTC to Hex A by thiolation and mixed disulfide formation (Dobrenis *et al.*, 1992; Bizzini *et al.*, 1980). The conjugate, TTC-Hex A, was active against G_{M2} ganglioside in the presence of activator protein, with approximately the same K_m as Hex A (3–5 μM G_{M2}).

We tested the uptake of Hex A and its conjugates by neural cells in neural cell cultures from embryonic rat (Dobrenis *et al.*, 1992), and neonatal cats (Dobrenis *et al.*, 1995; Dobrenis, unpublished procedure), identifying cell types by morphology and established markers. In these cultures exposed to native Hex A there was very little uptake of enzyme overall as judged by immunofluorescence microscopy. The amount of Hex A was negligible in larger neurons, somewhat greater in small neurons, and still greater in same non-neuronal "flat cells" and astrocytes. In parallel neural cell cultures exposed to Lucifer yellow (LY, a marker for fluid-phase endocytosis), the distribution of the fluorescent dye was superimposable on that of Hex A. Exposure of the cultures to PLL-Hex A at the same concentration and time resulted in a somewhat greater uptake by large and small neurons, and in a quite noticeable uptake by non-neuronal cells.

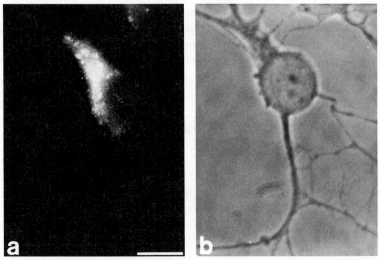

Figure 25.1. *Neuronal Uptake of TTC-Hex A.* Photomicrographs of neuron with internalized TTC-Hex A, detected with anti-Hex A antibody and fluorescein-conjugated 2nd antibody and visualized by epifluorescence (*a*) and phase-contrast (*b*) microscopy. Fifty-five-day rat brain cell culture. (Scale bar = 10 μm.)

Finally, after exposure of rat neural cultures to TTC Hex A under the same conditions, there was a remarkable uptake by both small and large, well-differentiated neurons, while uptake by non-neuronal cells was relatively much less. Immunofluorescent Hex A-positive granules were visible in neuronal perikaria and in proximal regions of neurites (Figure 25.1A).

We can draw the following conclusions from these uptake experiments.

1. Native placental Hex A apparently was taken up by neurons and most other neural cells predominantly by nonadsorptive, fluid-phase endocytosis. This contention is supported by the similarity between Hex A and LY uptake. The low uptake of Hex A by neurons and higher uptake by astrocytes is a reflection of the relative endocytic activity of these cells (Broadwell and Brightman, 1976).

2. Although generally greater, the uptake of PLL-Hex A showed the same relative proportion as that of Hex A among different cell types, because increased uptake of PLL conjugates results from their higher affinity for the cell membrane, superimposed on a normal rate of constitutive endocytosis (Li *et al.*, 1986). Thus, in the context of enzyme therapy, the PLL conjugate approach would not provide a great advantage if neurons are the target cells, given their low rate of constitutive endocytosis (Broadwell and Brightman, 1976), although it could be generally useful by enhancing uptake by all cell types.

3. TTC conjugation specifically enhanced uptake of Hex A by rat neurons. As determined in subsequent experiments (Dobrenis *et al.*, 1992), uptake

was approximately 16- to 40-fold greater than that of Hex A, and this did not result from enhanced endocytic activity triggered by TTC, but rather by the selective, high affinity of TTC for the neuronal plasma membrane. Thus in the context of enzyme therapy, TTC derivatization would effectively target a protein to all neurons.

V. TTC-HEX A AND NEURONAL STORAGE

We assessed the effectiveness of TTC-Hex A in degrading lysosomally stored G_{M2} ganglioside on mixed cultures of neural cells and neuron-enriched cultures prepared from cerebral cortex of neonatal kittens with G_{M2} gangliosidosis (Dobrenis et al., 1992). As shown in Figure 25.2, using a monoclonal antibody to G_{M2} ganglioside (Natoli et al., 1986), we were able to visualize ganglioside storage in these cells by immunofluorescence microscopy. When these cultures were exposed to Hex A or TTC-Hex A (50 μM) the binding and uptake patterns of both enzymes were similar to those observed in rat neural cells, suggesting that

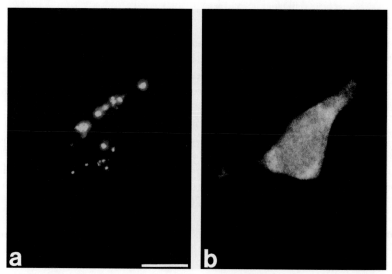

Figure 25.2. *Neuron with G_{M2} ganglioside storage*. Photomicrographs of a neuron with G_{M2} ganglioside storage detected with an anti-G_{M2} monoclonal antibody and fluorescein-conjugated 2nd antibody by epifluorescence microscopy (*a*). The cell is identified as a neuron by reaction with an antibody to neuron-specific enolase, and Texas red-conjugated 2nd antibody (*b*). Thirty-six-day cell culture from G_{M2} gangliosidosis kitten cerebral cortex. (Scale bar = 10 μm.)

neural cell endocytic mechanisms are not grossly altered by ganglioside storage. Upon overnight incubation with enzyme, TTC-Hex A reduced G_{M2} ganglioside immunoreactivity more than Hex A in cells identified as neurons by TT or TTC binding, but not in non-TT/TTC-binding cells , suggesting that enhanced uptake of the conjugated enzyme resulted in enhancement of ganglioside cleavage. By double immunolabeling for G_{M2} ganglioside and neuron-specific enolase we also observed a significant reduction of ganglioside-positive neurons in the TTC-Hex A-treated cultures, but not in the Hex A-treated ones. Finally, when we exposed the cultures to enzyme twice in a 3-day period, virtually no G_{M2} ganglioside-positive neurons were visible in the cultures treated with TTC-Hex A, whereas in those exposed to the unconjugated enzyme most neurons were still positive (Dobrenis *et al.*, 1992). To our knowledge, this is the first demonstration that exogenous Hex A can degrade stored G_{M2} ganglioside in neurons, indicating the potential usefulness of neuron-targeted TTC-Hex A for enzyme therapy of G_{M2} gangliosidosis. In principle, TTC derivatization is applicable to neuronal targeting of any protein, but it must be considered that TTC conjugates are internalized mainly to the degradative environment of the endosome–lysosome system. Thus, if the protein is not a lysosomal enzyme (Francis *et al.*, 1995), the system is likely to be much less effective, unless the conjugate is modified to obtain transfer to the cytosol (Rattazzi and Ioannou, 1996).

An issue relevant to the design of TTC derivatives obtained as chimeric proteins by recombinant DNA techniques is the state of the conjugate after internalization. Our experiments demonstrate that *enzyme activity* was delivered to the secondary lysosomes, the site of ganglioside storage, but we have no evidence that the enzyme present in these organelles was still TTC-bound, although the presence of enzyme adhering to the neuronal plasma membrane even after a prolonged chase argues for the persistence of intact conjugate (Dobrenis *et al.*, 1992). Most of the ganglioside-bound conjugate may have remained in the endosome-plasma membrane recirculating ganglioside pool, only a small fraction being slowly routed to lysosomes (Sofer *et al.*, 1996). Alternatively, Hex A routing to the lysosomes might have been contingent on reductive cleavage in the endosome (Shen *et al.*, 1985) of the disulfide bond to TTC; if this were the case, it might be necessary to introduce appropriate cleavable sequences in TTC-containing chimeric proteins that may be produced by recombinant DNA techniques to obtain routing of the active moiety to the lysosome.

This issue is also relevant to another advantage, that TTC-Hex A and similar conjugates might offer in the context of CNS-directed enzyme therapy. It is known that TT and TTC are taken up at the axon terminal and are transported retrogradely to the neuronal perikaryon, and that TT can undergo transneuronal, transsynaptic transport, with uptake by adjacent neurons (Schwab *et al.*, 1979; Fishman and Carrigan, 1987; Cabot *et al.*, 1991). If TTC conjugates share these properties, they might spread transcellularly into the CNS after gaining entry to the CNS. The persistence of intact conjugates would then be essential for transcellular transport.

VI. IMPLICATIONS AND OPEN QUESTIONS

Although the practical applicability of neuronal targeting by TTC derivatization depends on solving additional problems, such as for instance TTC immunogenicity (Andersen-Beckh et al., 1989), our cell culture experiments demonstrate its effectiveness. In addition to enzyme therapy applications, the concept of TTC-mediated targeting can be extended to various CNS-directed gene therapy modalities (Ioannou and Rattazzi, 1996). For instance, constructs resulting in secretion of TTC-chimeric, neuron-targeted enzymes could be used for transduction of bone marrow-derived progenitor cells in gene therapy *ex vivo;* similar constructs could be used for transduction *in vivo* of neural cells that may be physically more accessible, more easily targeted, or more endocytically active than neurons; and finally, TTC derivatization of synthetic transgene vectors could be useful for direct transduction of neurons (Rattazzi and Ioannou, 1996). Recent data on transfection of neuronal cells in culture using TTC-derivatized PLL as DNA carrier (Knight et al., 1999) support this contention.

In our experiments, we used a nonphysiologic ligand to explore the feasibility and effects of neuronal targeting; of course, more physiologic ligand–receptor systems should also be considered, provided that the receptor undergoes lysosome-directed endocytosis. An attractive candidate for lysosomal enzyme therapy is the M6PR, although its downregulation in postnatal neurons poses a problem. This obstacle might be circumvented by administration of recombinantly produced, M6P-rich enzyme (Ioannou et al., 1992) in the immediate postnatal period. Indeed, the results of enzyme therapy experiments by infusion of recombinant β-glucuronidase in 1-day-old mice with mucopolysaccharidosis (MPS) VII (Vogler et al., 1993; 1996; Sands et al., 1994; Vogler et al., 1996) appear to bear this concept out: a significant reduction in neuronal storage was observed in these animals, as opposed to mice in which enzyme infusion was started later in life.

The application of the same concept (i.e., early postnatal intervention) to bone marrow transplantation would require the assumption that graft-derived, brain perivascular macrophages and microglia—the cells most likely to enter the host's brain in BMT (Perry, 1994; Walkley et al., 1996) and play a major role in providing normal enzyme to the surrounding enzyme-deficient neural cells—secrete lysosomal enzymes bearing M6P residues. Early data on M6P-inhibitable uptake by cultured fibroblasts of β-hexosaminidase secreted by mouse peritoneal macrophages (Jessup and Dean, 1982) lend some support to this hypothesis, but a more direct test is needed.

VII. BONE MARROW TRANSPLANTATION AND ENZYME SECRETION

Attempts at therapy of G_{M2} gangliosidosis by BMT have given disappointing results both in human patients (Hoogerbrugge et al., 1995) and in the feline model (Walkley et al., 1994a, 1996). Some insight as to possible reasons for the apparent ineffectiveness of the approach may be gained from a comparison between the therapeutically unsuccessful BMT experiments in feline G_{M2} gangliosidosis and the analogous but successful ones in feline β-mannosidosis, in which BMT virtually eliminated the clinical neurologic symptoms. Brain tissue sections from BMT-treated β-mannosidosis cats showed the presence of numerous perivascular macrophages and of some parenchymal macrophages and microglia with apparently normal β-mannosidase activity, thus most likely derived from the graft. Enzyme activity was also detectable in neurons, and clearing of storage in neuronal lysosomes was evident (Walkley et al., 1994b). In BMT-treated G_{M2} gangliosidosis kittens—which did not show any clinical improvement—enzymatic assay of brain homogenates showed exogenous β-hexosaminidase activity equivalent to 30% of normal. Brain tissue sections showed in fact the presence of a large number of strongly enzyme-positive, perivascular and parenchymal macrophages, most likely of graft origin. However, no enzyme-positive neuron was found, and no significant depletion of neuronal G_{M2} ganglioside storage was visible (Walkley et al., 1994a, 1996).

The results of cell culture experiments using purified microglia from cat cerebral cortex provided an explanation. Enzyme activity accumulating in the serum-free culture medium over a 1- to 6-day period was significantly different for different lysosomal enzymes. The activity of β-mannosidase was very high, that of β-hexosaminidase and β-galactosidase, very low; for example, after 1 day it was 300, 40, and 4 nmol substrate cleaved/h/1 × 10^6 cells, respectively. Extracellular activity was equivalent to 30% of intracellular activity for β-mannosidase, but only to <1% of intracellular activity for β-hexosaminidase and β-galactosidase (Dobrenis et al., 1994, 1996). Thus, at least in so far as the cell culture conditions reflect brain tissue environment in vivo, one can infer from these observations that a difference in the amount of macrophage/microglia-secreted, active enzyme available for uptake by neurons was most probably a critical factor in determining the different outcome of BMT in the two different feline diseases.

It is of interest that in cultures of mouse microglia, prepared using the same protocol as the cat cells, the relative extracellular and intracellular activities of α-mannosidase and β-galactosidase are similar to those observed in the feline cultures; in contrast, the extracellular activity of β-hexosaminidase is much higher, being for instance equivalent to 30% of intracellular activity after a 1-day incubation (Dobrenis et al., 1996). Thus, not only the success of BMT in a given enzyme/disease system does not guarantee a similar success in a different system, but also extrapolation of results to a different species within the same system may not be appropriate. An extensive discussion of BMT for the treatment of lysosomal diseases with neurologic involvement is beyond the scope of this chapter. In general, however, these observations suggest that overexpression and secretion by graft-derived cells of enzymes in natural or engineered, neuron-targeted forms would probably improve significantly the outcome of this procedure. BMT experiments in Hex B knockout, Sandhoff disease mice (Norflus *et al.*, 1998), which have yielded more encouraging but equally puzzling results, are relevant in that respect. The life span of the BMT-treated mice was doubled, and their neurological deterioration progressed at a slower pace as compared to affected controls. However, in these animals' brains neither the levels of G_{M2} ganglioside and GA2 glycolipid, nor the histopathological picture were significantly different from those in untreated mice. A detailed histochemical study (Oya *et al.*, 2000) showed a very low number of Hex–positive perivascular macrophages and microglia, and no enzyme-positive neurons in the treated mice; the majority of Hex-positive cells were found in the leptomeninges and in the choroid plexus. Insight into the mechanism underlying clinical improvement has come from a recent study on neural pathogenesis in these mice (Wada *et al.*, 2000). A cDNA microarray analysis of neural tissue indicated the existence of an inflammatory process and of microglia activation, which was followed by neuronal apoptosis. After infusion of bone marrow from normal mice into γ-irradiated Sandhoff mice, inflammatory processes, microglia activation, and apoptosis were suppressed. Remarkably, there was no appreciable change in G_{M2} ganglioside levels in the neural tissue, suggesting that exogenous β-hexosaminidase had not played a significant role and that neuronal storage of ganglioside was not the immediate cause of apoptosis. The authors propose that neurons are in some way damaged by ganglioside storage, and are phagocytosed by microglia. These cells, however, are also unable to digest the ganglioside and are activated by its accumulation, recruiting inflammatory cells and secreting neuronotoxic factors that induce neuronal apoptosis. According to the authors' model, irradiation and BMT replaced enzyme-deficient microglia with normal cells, which could degrade G_{M2} ganglioside and did not become activated, thus interrupting the neurodegenerative cascade. The authors point out that therapy based on anti-inflammatory agents alone would not be an answer, since the initial, storage-induced neuronal damage would still occur, and they suggest that substrate depletion by inhibition of ganglioside synthesis combined with BMT could be beneficial. Recent experiments along this line in Sandhoff disease mice (Jeyakumar *et al.*, 2001) appear to support this concept. It is doubtful that this experimental procedure, involving total body irradiation, can easily be applied as such to human patients. However, these experiments should induce clinicians and investigators to look at BMT in neurodegenerative storage diseases from a different angle and may provide a blueprint for more effective therapeutic attempts.

VIII. DELIVERY OF MACROMOLECULES TO THE BRAIN PARENCHYMA

In the experiments in newborn MPS VII mice (Vogler *et al.*, 1993, 1996; Sands *et al.*, 1994; Vogler *et al.*, 1996) the amelioration of neuronal pathology was obtained by intravenous injection of enzyme, without specific attempts to induce BBB disruption. The results are puzzling in the sense that the commonly held contention of "immaturity" of the BBB in fetal and possibly early postnatal life—proposed as a possible explanation by the authors—is questionable, at least in terms of permeability to proteins. In fact, a critical review of the experiments on which this contention is based indicates that in each case the apparent BBB permeability most probably resulted from inappropriate methodologies or experimental design, such as injection of marker dye in excess of plasma albumin–binding capacity, toxicity to the vascular endothelium of marker protein infused in excessive amounts, or retrograde axonal transport from outside the BBB owing to protracted marker protein infusion (Møllgard and Saunders, 1986; Moos and Møllgard, 1993). Furthermore, ultrastructural data indicate that interendothelial tight junctions—the key structural components of the BBB—are well established at or about embryonic day 14 in the mouse, rat, and chick, and by the 10th gestational week in the human fetus (Møllgard and Saunders, 1986). In view of these observations, and of the methodology used in these experiments (Sands and Barker, 1999), one wonders whether the intravenous injection into newborn mice of volumes of enzyme solution approximately equal to the total blood volume of these animals, rather than BBB immaturity, may have been responsible for enzyme extravasation into the brain parenchyma.

The apparently encouraging results of enzyme infusion in newborn mice (Vogler *et al.*, 1993, 1996; Sands *et al.*, 1994) are contrasted by the disappointing results of early postnatal enzyme infusion without BBB disruption in a human infant with neuronopathic Gaucher disease type II (Bove *et al.*, 1995). Injection of glucocerebrosidase was started at 4 days of age and continued until shortly before the patient's death at age 15 months. Although the progression of neurologic deterioration was judged to be slower than usual, postmortem examination of brain showed the typical neuropathologic picture of this disease (Bove *et al.*, 1995). The issue here is complicated by the fact that the glycosyl chains of the therapeutic preparations of glucocerebrosidase are modified to favor recognition by mannosyl-specific receptors on RES cells (Grabowski *et al.*, 1995); thus, the disappointing outcome of this trial may not be attributable solely to BBB impermeability, but may reflect also enzyme clearance by the RES and neural cell uptake problems. Clearly, however, the issue of BBB permeability in the perinatal period in experimental animals and in humans needs clarification. If it exists, it may provide a relatively simple therapeutic methodology, particularly useful if it is found that the neurologic manifestations of a given storage disease are not reversible after depletion of neural cell storage by enzyme therapy at a later age.

In general, the impermeability of the BBB to proteins is still the main obstacle to the therapy of neurodegenerative disorders. The osmotic BBB disruption pioneered by S. Rapoport (Rapoport, 1976, 2000) and applied to human patients for the treatment of brain tumors (Gumerlock et al., 1992; Doolittle et al., 2000) appears to provide both a global access to brain parenchyma and the capacity to deliver the relatively large amounts of enzyme protein needed for therapy of storage diseases affecting the entire brain. The long-term consequences of repeated BBB disruption, however, need a better evaluation than that obtainable from brain tumor patients (Roman-Goldstein et al., 1995; McAllister et al., 2000). Delayed immunologic complications are of particular concern, as it is now clear that the brain is an immunologically quiescent rather than immunologically privileged site (Hickey, 1991; Cserr and Knopf, 1992; Gordon et al., 1992; Fabry et al., 1994), and breakdown of the BBB may result in exposure of immunocompetent cells to antigens both locally and systemically.

Alternative CNS delivery approaches that circumvent the BBB—thus decreasing the danger of breaching the relative immunologic isolation of the brain—are possible. Convection-enhanced intraparenchymal infusion (Bobo et al., 1994) appears promising, and has already been applied to human patients for the experimental treatment of brain tumors (Laske et al., 1997), although at present it seems useful mainly for regional distribution of macromolecules (Lieberman et al., 1995). Plasma osmolarity-driven convection from subarachnoid spaces (Rattazzi and Dobrenis, 1991) requires further development and evaluation. Finally, advances in the understanding of the molecular structure and function of tight junctions (Hirase et al., 1997; Rubin and Staddon, 1999) may uncover pharmacologic means to modulate BBB permeability in a more controlled and selective way than that provided by the osmotic disruption method.

IX. CNS GENE THERAPY

Enzyme therapy requires repeated infusions of enzyme to keep pace with a continuing production and storage of substrate. In contrast, gene therapy, by introducing into enzyme-deficient cells a cDNA capable of sustained expression of normal enzyme, in theory offers the possibility of a long-lasting treatment, particularly desirable in the case of the CNS. A detailed discussion of gene therapy is beyond the scope of this paper, but it is appropriate to highlight some of the issues relevant to CNS.

Gene therapy directed to the CNS presents problems in common with enzyme therapy. As in the case of exogenous enzyme protein, DNA access to the brain is prevented by the BBB. The methodologies to overcome the BBB outlined previously are applicable to cDNA delivery with an important limitation, however, the 20 nm width of the interendothelial gaps obtainable by osmotic disruption (Vorbrodt et al., 1993) and of the interstitial spaces through which parenchymal neural cells can be reached (Brightman, 1965; Brightman and Reese, 1969). This size limitation reduces significantly the usefulness of most viral vectors (see for instance Doran et al., 1995), which are a most popular tool in gene therapy, owing to their effectiveness in cell transfection and transduction. In addition, the problem of possible inflammatory and immunologic reactions to these vectors is particularly worrisome in the case of the CNS. In contrast, the nonpathogenic adeno-associated virus (AAV) has a diameter of ~15 nm and can diffuse into the brain parenchyma through an osmotically disrupted BBB (Rattazzi and Friedrich, unpublished observations). For this and other reasons, AAV is an attractive candidate for neural gene therapy (Xiao et al., 1997). It is effective in transducing neural cells after direct injection into the CNS (Kaplitt et al., 1994; McCown et al., 1996), and its neural cell tropism, efficiency of transfection, and possibly duration of transgene expression may be optimized by the choice of an appropriate serotype (Davidson et al., 2000). If a multiplicity of infection comparable to that obtained by direct injection can be attained by the vascular route, AAV vectors have a great potential for CNS-directed gene transfer.

The proponents of nonviral, synthetic vectors also face a size problem, as the size of a 5-10 Kb plasmid is in excess of 200 nm (Dunlap et al., 1997). DNA, however, can be compacted to particles <20 nm in diameter, by complexing it with polycations such as poly-L-lysine (PLL) or salmon sperm protamine (SSP) under carefully controlled conditions (Perales et al., 1997). As shown by recent experiments (Rattazzi et al., 1997; Ioannou, 2000), these compacted cDNA-protein complexes are amenable to delivery to the CNS by osmotic BBB disruption. Our experiments were carried out in 28 adult rats, including 7 controls, using a 5 Kb plasmid encoding A. victoria green fluorescent protein (GFP), reacted with PLL or SSP to obtain toroidal complexes 10–20 nm in diameter, as assessed by electron microscopy. The DNA suspensions (100 μl) were infused into the terminal internal carotid artery of rats after unilateral BBB disruption obtained by infusion of a 100 μl bolus of 1.7 molal arabinose into the same artery. We detected plasmid DNA by the polymerase chain reaction (PCR) in homogenates of brain hemispheres and cerebellum from 17 rats sacrificed and perfused at 2 h. PCR product was present in 16 rats, in areas showing Evans blue-albumin (EBA) extravasation, indicating BBB disruption. We could separate plasmid DNA from endothelial cells and plasma membranes by a novel capillary depletion method (modified from Triguero et al., 1990); this indicated that most of the plasmid DNA had entered the brain parenchyma. Little or no PCR product was detectable in the brain of 4 rats infused with compacted DNA in the absence of BBB disruption, which indicated minimal binding or internalization by vascular endothelial cell. As expected, none was detected after infusion of uncomplexed

DNA with BBB disruption. In 4 rats sacrificed at 5 days, we assessed transduction of neural cell by PLL-cDNA resulting in GFP expression by fluorescence- and confocal microscopy. Paravascular and parenchymal cells exhibiting green cytoplasmic fluorescence were visible in diverse brain regions, usually coincident with red-fluorescing, extravasated EBA. Our protein-DNA complexes were not derivatized for uptake by specific neural cells; thus, as we expected most of the transduced cells were macrophages and microglia, but we also observed occasional astrocytes and putative neurons showing GFP expression. Furthermore, GFP expression, apparently below the limit of detection by fluorescence microscopy, was detectable by immunocytochemistry in numerous putative neurons (Rattazzi and Friedrich, unpublished data). From these initial experiments one can conclude that transduction of neural cells can be obtained by delivery of compacted DNA-protein complexes to the brain parenchyma across a reversibly permeabilized BBB.

In addition to small size, artificial vectors have to fulfill equally important prerequisites to approach the efficiency of viral vectors. The addition of moieties for cell recognition and uptake discussed earlier for protein enzymes applies equally to protein-DNA complexes, with the caveat that internalization to the endosome-lysosome system, desirable in lysosomal enzyme therapy, is instead a problem in gene therapy. DNA degradation in lysosomes is a major contributor to the low transduction efficiency of nonviral vectors, as shown by the improvement provided by lysosomotropic drugs that depress lysosomal enzyme activity (Erbacher et al., 1996). Therefore, it is necessary to add to the DNA carrier a plasma membrane-destabilizing moiety to effect rapid translocation to the cytoplasm (Wyman et al., 1997), and a nuclear localization signal (Boulikas, 1993) to obtain transport of the cDNA to the nucleus for transcription. Finally, the problem of cell-appropriate, modulated, sustained expression should be considered by choosing appropriate promoters for the cDNA.

X. CONCLUSIONS

The effectiveness of lysosomal enzyme therapy has been convincingly established by the successful treatment of patients with Gaucher disease type I (Grabowski et al., 1995). Together with the ease of production of recombinant enzymes, this is generating renewed interest in this therapeutic approach. Basic problems underlying the lack of success of the early trials are now better understood, and rational solutions to these problems have been or can be devised. Application of this therapeutic modality to neurodegenerative lysosomal diseases, however, still requires refinement of neural cell targeting and CNS delivery methodologies. Efforts in this direction are likely to facilitate as well the development of gene therapy, in which cell targeting and brain delivery are also critical issues. The power of molecular genetics and the availability of well-characterized, natural and genetically engineered animal models should make it possible to develop and test rational approaches to these problems, and to assess their clinical effectiveness, eventually leading to therapy of G$_{M2}$ gangliosidosis and other storage diseases with neurologic involvement. In the case of Tay-Sachs disease, until such therapy can be implemented, prevention through screening should continue to provide an effective way to reduce its incidence, while we look forward with considerable confidence to the development of valid therapeutic approaches.

Acknowledgments

The authors' work whose results are discussed in this paper was supported in part by NIH grants NS 13677, NS 21404, and NS 32169. The authors are grateful to Claudia Papka Rattazzi, Ph.D., for a critical reading of the manuscript. This paper is dedicated to her memory.

References

Andersen-Beckh, B., Binz, T., Kurazono, H., Mayer, T., Eisel, U., and Niemann, P. (1989). Expression of tetanus toxin subfragments in vitro and characterization of epitopes. *Infect. Immun.* **57,** 3498–3505.

Anderson, J. M., and Van Itallie, C. M. (1995). Tight junctions and the molecular basis for regulation of paracellular permeability. *Am. J. Physiol.* **269,** G467–G475.

Ashwell, G., and Morell, A. (1974). Role of surface carbohydrates in the hepatic recognition and transport of circulating glycoproteins. *Adv. Enzymol.* **41,** 99–128.

Bizzini, B., Stoeckel, K., and Schwab, M. (1977). An antigenic polypeptide fragment isolated from Tetanus toxin: Chemical characterization, binding to ganglioside and retrograde axonal transport in various nervous systems. *J. Neurochem.* **28,** 529–542.

Bizzini, B., Akert, R., Glicksman, M., and Grob, P. (1980). Preparation of conjugates using two tetanus toxin derived fragments: Their binding to gangliosides and isolated synaptic membranes and their immunological properties. *Toxicon* **18,** 561–572.

Bobo, R. H., Laske, D. W., Akbasak, A., Morrison, P. F., Dedrick, R. L., and Oldfield, E. H. (1994). Convection-enhanced delivery of macromolecules into the brain. *Proc. Natl. Acad. Sci. (USA)* **91,** 2076–2080.

Boulikas, T. (1993). Nuclear localization signals (NLS). *Crit. Rev. Eukaryot. Gene Expr.* **3,** 193–227.

Bove, K. E., Daugherty, C., and Grabowski, G. A. (1995). Pathological findings in Gaucher disease type 2 patients following enzyme therapy. *Hum. Pathol.* **26,** 193–207.

Brightman, M. W. (1965). The distribution within the brain of ferritin injected into cerebral fluid compartments. *Am. J. Anat.* **117,** 193–220.

Brightman, M. W., and Reese, T. S. (1969). Junctions between intimately apposed cell membranes in the vertebrate brain. *J. Cell Biol.* **40,** 618–677.

Broadwell, R. D., and Brightman, M. W. (1976). Entry of peroxidase into neurons of the central and peripheral nervous system from extracerebral and cerebral blood. *J. Comp. Neurol.* **166,** 257–284.

Cabot, J. B., Menneone, A., Bogan, N., Carroll, J., Evinger, C., and Erichsen, J. T. (1991). Retrograde, transsynaptic and transneuronal transport offragment C of tetanus toxin by sympathetic preganglionic neurons. *Neuroscience* **40,** 805–823.

Cork, L. C., Munnell, J. F., Lorenz, M. D., Murphy, J. W., Baker, W., and Rattazzi, M. C. (1977). G$_{M2}$ ganglioside storage disease in cats with β-hexosaminidase deficiency. *Science* **196,** 1014–1017.

Couce, M. E., Weatherington, A. J., and McGinty, J. F. (1992). Expression of insulin-like growth factor-II (IGF-II) and IGF-II/mannose-6-phosphate receptor in the rat hippocampus: An in situ hybridization and immunocytochemical study. *Endocrinology* **131**, 1636–1642.

Cserr, H. F. (1988). Role of secretion and bulk flow of brain interstitial fluid in brain volume regulation. *Ann. N.Y. Acad. Sci.* **529**, 9–20.

Cserr, H. F., and Knopf, P. M. (1992). Cervical lymphatics, the blood–brain barrier and the immunore-activity of the brain: A new view. *Immunol. Today* **13**, 507–512.

Davidson, B. L., Stein, C. S., Heth, J. A., Martins, I., Kotin, R. M., Derksen, T. A., Zabner, J., Ghodsi, A., and Chiorini, J. A. (2000). Recombinant adeno-associated virus type 2, 4, and 5 vectors: Transduction of variant cell types and regions in the mammalian central nervous system. *Proc. Natl. Acad. Sci. U.S.A.* **97**, 3428–3432.

de Duve, C (1964). From cytases to lysosomes. *Fed. Proc.* **23**, 1045–1049.

Dejana, E., Corada, M., and Lampugnani, M. G. (1995). Endothelial cell-to-cell junctions. *FASEB J.*, 910–918.

Desnick, N., Snyder, P. D., Desnick, S. J., Krivit, W., and Sharp, H. L. (1972). Sandhoff's disease: Ultrastructural and biochemical studies. *In* "Sphingolipid, Sphingolipidosis and Allied Disorders" (B. W. Volk and S. M. Aronson, eds.), pp. 351–371. New York: Plenum.

Dobrenis, K., and Rattazzi, M. C. (1987). Enhanced uptake of human β-hexosaminidase A (Hex A) by CNS neurons in vitro. *Soc. Neurosci. Abst.* **13**(3), 1685.

Dobrenis, K., and Rattazzi, M. C. (1988). Enzyme replacement in cultured neurons with tetanus toxin fragment C (TTC)-β-hexosaminidase A (Hex A) conjugate. *Pediatr. Res.* **23**, 551A.

Dobrenis, K., Schwartz, P., Joseph, A., and Rattazzi, M. C. (1987). Neuronal enzyme replacement in vitro using tetanus toxin fragment C (TTC)-conjugated β-hexosaminidase A (Hex A). *Am. J. Hum. Genet.* **41**, A5.

Dobrenis, K., Joseph, A., and Rattazzi, M. C. (1992). Neuronal lysosomal enzyme replacement using fragment C of tetanus toxin. *Proc. Natl. Acad. Sci. (USA)* **89**, 2297–2301.

Dobrenis, K., Wenger, D. A., and Walkley, S. U. (1994). Extracellular release of lysosomal glycosidases in cultures of cat microglia. *Mol. Biol. Cell.* **(Suppl.)** **5**, 113A.

Dobrenis, K., Makman, M. H., and Stefano, G. B. (1995). Occurrence of the opiate alkaloid m_3 receptor in mammalian microglia, astrocytes and Kupffer cells. *Brain Res.* **686**, 239–248.

Dobrenis, K., Finamore, P. S., Masui, R., and Walkley, S. U. (1996). Secretion of lysosomal glycosidases by microglia in culture. *Mol. Biol. Cell* **(Suppl.)7**, 325A.

Doolittle, N. D., Miner, M. E., Hall, W. A., Siegal, T., Jerome, E., Osztie, E., McAllister, L. D., Bubalo, J. S., Kraemer, D. F., Fortin, D., Nixon, R., Muldoon, L. L., and Neuwelt, E. A. (2000). Safety and efficacy of a multicenter study using intraarterial chemotherapy in conjunction with osmotic opening of the blood-brain barrier for the treatment of patients with malignant brain tumors. *Cancer* **88**, 637–647.

Doran, S. E., Ren, X. D., Betz, A. L., Pagel, A. L., Neuwelt, E. A., Roessler, B. J., and Davidson, B. L. (1995). Gene expression from recombinant viral vectors in the central nervous system after blood-brain barrier disruption. *Neurosurgery* **36**, 965–970.

Dunlap, D. D., Maggi, A., Soria, M. R., and Monaco, L. (1997). Nanoscopic structure of DNA condensed for gene delivery. *Nucl. Acid Res.* **25**, 3095–3101.

Erbacher, P., Roche, A. C., Monsigny, M., and Midoux, P. (1996). Putative role of chloroquine in gene transfer into a human hepatoma cell line by DNA/lactosylated polylysine complexes. *Exp. Cell. Res.* **225**, 186–194.

Fabry, Z., Raine, C. S., and Hart, M. C. (1994). Nervous tissue as an immune compartment: The dialect of the immune response in the CNS. *Immunol. Today* **15**, 218–224.

Fishman, P. S., and Carrigan, D. R. (1987). Retrograde transneuronal transfer of the C-fragment of tetanus toxin. *Brain Res.* **406**, 275–279.

Francis, J. W., Hosler, B. A., Brown, R. H., Jr, and Fishman, P. S. (1995). CuZn superoxide dismutase

(SOD-1): Tetanus toxin fragment C hybrid protein for targeted delivery of SOD-1 to neuronal cells. *J. Biol. Chem.* **270,** 15434–154342.

Fratantoni, J. C., Hall, C. W., and Neufeld, E. F. (1968). Hurler and Hunter syndromes. Mutual correction of the defect in cultured fibroblasts. *Science* **162,** 570–571.

Geiger, B., and Arnon, R. (1978). Hexosaminidases A and B from human placenta. *Meth. Enzymol.* **50(C),** 547–555.

Geiger, B., von Specht, B. U., and Arnon, R. (1977). Stabilization of human β-D-N-Acetyl hexosaminidase A towards proteolytic inactivation by coupling it to poly (N-vinyl pyrrolidone). *Eur. J. Biochem.* **73,** 141–147.

Glaser, J., Rosen, K. J., Brut, P. E., and Sly, W. S. (1975). Multiple isoelectric and recognition forms of human beta glucuronidase activity. *Arch. Biochem. Biophys.* **166,** 536–542.

Godel, V., Bluhmenthal, M., Goldman, B., Keven, G., and Padeh, B. (1978). Visual functions in Tay-Sachs diseased patients following enzyme replacement therapy. *Metab. Ophthalmol.* **2,** 27–32.

Gordon, L. B., Knopf, P. M., and Cserr, H. F. (1992). Ovalbumin is more immunogenic when introduced into brain or cerebrospinal fluid than into extracerebral sites. *J. Neuroimmunol.* **40,** 81–87.

Grabowski, G. A., Barton, N. W., Pastores, G., Dambrosia, J. M., Banerjee, T. K., McKee, M. A., Parker, C., Schiffmann, R., Hill, S. C., and Brady, R. O. (1995). Enzyme therapy in type 1 Gaucher disease: Comparative efficacy of mannose-terminated glucocerebrosidase from natural and recombinant sources. *Ann. Intern. Med.* **122,** 33–39.

Grubb, J. H., Kyle, J. W., Cody, L. B., and Sly, W. S. (1993). Large scale purification of phosphorylated recombinant human b-glucuronidase from overexpressing mouse L cells. *FASEB J.* **7,** A1255.

Gumerlock, M. K., Belshe, B. D., Madsen, R., and Watts, C. (1992). Osmotic blood-brain barrier disruption and chemotherapy in the treatment of high grade malignant glioma: Patient series and literature review. *J. Neurooncol.* **12,** 33–46.

Hickey, W. F. (1991). Migration of hematogenous cells through the blood-brain barrier and the initiation of CNS inflammation. *Brain Pathol.* **1,** 97–105.

Hill, D. F., Bullock, P. N., Chiappelli, F., and Rome, L. H. (1985). Binding and internalization of lysosomal enzymes by primary cultures of rat glia. *J. Neurosci. Res.* **14,** 35–47.

Hirase, T., Staddon, J. M., Saitou, M., Ando-Akatsuka, Itoh, M., Furuse, M., Fujimoto, K., Tsukita, S., and Rubin, L. L. (1997). Occludin as a possible determinant of tight junction permeability in endothelial cells. *J. Cell Sci.* **110,** 1603–1610.

Hoogerbrugge, P. M., Brouwer, O. F., Bordigoni, P., Ringden, O., Kapaun, P., Ortega, J. J., O'Meara, A., Cornu, G., Souillet, G., Frappaz, D., *et al.* (1995). Allogeneic bone marrow transplantation for lysosomal storage diseases. The European Group for Bone Marrow Transplantation. *Lancet* **345**(8962), 1398–1402.

Ioannou, Y. A. (2000). Gene therapy for lysosomal storage disorders with neuropathology. *J. Am. Soc. Nephro.* **11,** 1542–1547.

Ioannou, Y. A., Bishop, D. F., and Desnick, R. J. (1992). Overexpression of human alpha-galactosidase A results in its intracellular aggregation, crystallization in lysosomes, and selective secretion. *J. Cell. Biol.* **119,** 1137–1150.

Jessup, W., and Dean, R. T. (1982). Secretion by mononuclear phagocytes of lysosomal hydrolases bearing ligands for the mannose-6-phosphate receptor system of fibroblasts: Evidence for a second mechanism of spontaneous secretion? *Biochem. Biophys. Res. Commun.* **105,** 922–927.

Jeyakumar, M., Norflus, F., Tifft, C. J., Cortina-Borja, M., Butters, T. D., Proia, R. L., Perry, V. H., Dwek, R. A., and Platt, F. M. (2001). Enhanced survival in Sandhoff disease mice receiving a combination of substrate deprivation therapy and bone marrow transplantation. *Blood* **97,** 327–9.

Johnson, W. A., Desnick, R. J., Lang, D. M., Sharp, H. L., Krivit, W., Brady, R., and Brady, R. O. (1973). Intravenous injection of purified hexosaminidase A into a patient with Tay-Sachs disease. *In* "Enzyme Therapy in Genetic Diseases" (D. Bergsma, ed.), pp. 120–124. Williams &

Wilkins, Baltimore, for the National Foundation-March of Dimes, Birth Defects: Original Article Series IX(2).

Kakkis, E. D., Matynia, A., Jonas, A. J., and Neufeld, E. F. (1993). Overexpression of the human lysosomal enzyme alpha-L-iduronidase in Chinese hamster ovary cells. *Protein Exp. Purif.* **5**, 225–232.

Kaplitt, M. G., Leone, P., Samulski, R. J., Xiao, X., Pfaff, D. W., O'Malley, K. L., and During, M. J. (1994). Long-term gene expression and phenotypic correction using adeno-associated virus vectors in mammalian brain. *Nature Genetics* **8**, 148–153.

Kaplitt, M. G., and Makimura, H. (1997). Defective viral vectors as agents for gene transfer in the nervous system. *J. Neurosci. Methods* **71**, 125–132.

Kar, S., Chabot, J. G., and Quirion, R. (1993). Quantitative autoradiographic localization of [125I] insulin-like growth factor I, [125I] insulin-like growth factor II, and [125I] insulin receptor binding sites in developing and adult rat brain. *J. Comp. Neurol.* **333**, 375–397.

Knight, A., Carvajal, J., Schneider, H., Coutelle, C., Chamberlain, S., and Fairweather, N. (1999). Non-viral neuronal gene delivery mediated by the HC fragment of tetanus toxin. *Eur. J. Biochem.* **259**, 762–769.

Kusiak, J. W., Toney, J. M., Quirk, J. M., and Brady, R. O. (1979). Specific binding of 125 I-labeled b-hexosaminidase A to rat brain synaptosomes. *Proc. Natl. Acad. Sci. (USA)* **76**, 982–985.

Laske, D. W., Youle, R. J., and Oldfield, E. H. (1997). Tumor regression with regional distribution of the targeted toxin TF-CRM107 in patients with malignant brain tumors. *Nature Medicine* **3**, 1362–1368.

Li, W., Ryser, H. G. P., and Shen, W. C. (1986). Altered endocytosis in a mutant of LM fibroblasts defective in cell-cell fusion. *J. Cell Physiol.* **126**, 161–166.

Lieberman, D. M., Laske, D. W., Morrison, P. F., Bankiewicz, K. S., and Oldfield, E. H. (1995). Convection-enhanced distribution of large molecules in gray matter during interstitial drug infusion. *J. Neurosurg.* **82**, 1021–102.

McAllister, L. D., Doolittle, N. D., Guastadisegni, P. E., Kraemer, D. F., Lacy, C. A., Crossen, J. R., and Neuwelt, E. A. (2000). Cognitive outcome and long-term follow-up results after enhanced chemotherapy delivery for primary central nervous system lymphoma. *Neurosurgery* **46**, 51–60.

McCown, T. J., Xiao, X., Li, J., Breese, G. R., and Samulski, R. J. (1996). Differential and persistent expression patterns of CNS gene transfer by an adeno-associated virus (AAV) vector. *Brain Res.* **713**, 99–107.

Møllgard, K., and Saunders, N. R. (1986). The development of the human blood-brain and blood-CSF barrier. *Neuropathol. Appl. Neurobiol.* **12**, 337–358.

Moos, T., and Møllgard, K. (1993). Cerebrovascular permeability to azo dyes and plasma proteins in rodents of different ages. *Neuropathol. Appl. Neurobiol.* **19**, 120–127.

Natoli, E. J., Livingston, P. O., Pukel, C. S., Lloyd, K. O., Wiegandt, H., Szalsy, J., Oettgen, H. F., and Old, L. J. (1986). A murine monoclonal antibody detecting N-acetyl-and N-glycolyl-G_{M2}: Characterization of cell surface reactivity. *Cancer Res.* **46**, 4116–4120.

Nielsen, F. C., and Gammeltoft, S. (1990). Mannose-6-phosphate stimulates proliferation of neuronal precursor cells. *FEBS Lett.* **262**, 142–144.

Nissley, P., Kiess, W., and Sklar, M. (1993). Developmental expression of the IGF-II/mannose 6-phosphate receptor. *Mol. Reprod. Dev.* **35**, 408–413.

Norflus, F., Tifft, C., McDonald, M. P., Goldstein, G., Crawley, J. N., Hoffmann, A., Sandhoff, K., Suzuki, K., and Proia, R. L. (1998). Bone marrow transplantation prolongs life span and ameliorates neurologic manifestations in Sandhoff disease mice. *J. Clin. Invest.* **101**, 1881–1888.

Ohata, K., Marmarou, A., and Povlishock, J. T. (1990). An immunocytochemical study of protein clearance in brain infusion edema. *Acta. Neuropathol. (Berl.)* **81**, 162–177.

O'Neil, D. C., Bartholomew, W. R., and Rattazzi, M. C. (1979). Antigenic homology of feline and human β-hexosaminidase. *Biochim. Biophys. Acta* **580**, 1–9.

Oya, Y., Proia, R. L., Norflus, F., Tifft, C. J., Langaman, C., and Suzuki, K. (2000). Distribution of enzyme-bearing cells in GM2 gangliosidosis mice: regionally specific pattern of cellular infiltration following bone marrow transplantation. *Acta Neuropathol. (Berl)* **99**, 161–168.

Perales, J. C., Grossmann, G. A., Molas, M., Liu, G., Ferkol, T., Harpst, J., Oda, H., and Hanson, R. W. (1997). Biochemical and functional characterization of DNA complexes capable of targeting genes to hepatocytes via the asialoglycoprotein receptor. *J. Biol. Chem.* **272**, 7398–7407.

Perry, V. H. (1994). "Macrophages and the Nervous System," pp. 6–26. Landes, Austin, TX.

Pontow, S. E., Kery, V., and Stahl, P. D. (1992). Mannose receptor. *Int. Rev. Cytol.* **137B**, 221–244.

Rapoport, S. I. (1976). "Blood-Brain Barrier in Physiology and Medicine." Raven, New York.

Rapoport, S. I. (2000). Osmotic opening of the blood-brain barrier: principles, mechanisms and therapeutic applications. *Cell Mol. Neurobiol.* **20**, 217–230.

Rattazzi, M. C. (1983). β-Hexosaminidase isozymes and replacement therapy in G_{M2} gangliosidosis. *Isozymes: Curr. Top. Biol. Med. Res.* **11**, 65–81.

Rattazzi, M., and Dobrenis, K. (1991). Enzyme replacement: Overview and prospects. *In* "Therapy of Genetic Diseases" (R. J. Desnick, ed.), pp. 131–152, *Churchill-Livingstone,* New York.

Rattazzi, M. C., and Ioannou, Y. A. (1996). Fragile X syndrome: Molecular genetic approachesto therapy. *In* "Fragile X Syndrome: Diagnosis, Treatment and Research," 2nd ed. (R. Hagerman and A. Cronister, eds.), pp. 412–452. Johns Hopkins Univ. Press, Baltimore.

Rattazzi, M. C., Baker, H. J., Cork, L. C., Cox, N. R., Lanse, S. B., McCullough, R. A., and Munnell, J. F. (1979a). The domestic cat as a model for human G_{M2} gangliosidosis. Pathogenetic and therapeutic aspects. *In* "Models for the Study of Inborn Errors of Metabolism" (F. Hommes, ed.), pp. 57–72. Elsevier, Amsterdam.

Rattazzi, M. C., McCullough, R. A., Downing, C. J., and Kung, M. P. (1979b). Toward enzyme therapy in G_{M2} gangliosidosis: β-Hexosaminidase infusion in normal cats. *Pediatr. Res.* **13**, 916–923.

Rattazzi, M. C., Lanse, S. B., McCullough, R. A., Nester, J. A., and Jacobs, E. A. (1980). Towards enzyme therapy in G_{M2} gangliosidosis: Organ disposition and induced central nervous system uptake of human β-hexosaminidase in the cat. *In* "Enzyme Therapy in Genetic Diseases: 2" (R. J. Desnick, ed.), pp. 179–193. A. R. Liss, New York.

Rattazzi, M. C., Appel, A. M., Baker, H. J., and Nester, J. (1981). Toward Enzyme replacement in G_{M2} gangliosidosis: Inhibition of hepatic uptake of human β-hexosaminidase in the cat. *In* "Lysosomes and Lysosomal Storage Diseases" (J. W. Callahan, and J. A. Lowden, eds.), pp. 405–424. Raven, New York.

Rattazzi, M. C., Appel, A. M., and Baker, H. J. (1982). Enzyme replacement in feline G_{M2} gangliosidosis: Catabolic effects of human β-hexosaminidase A. *In* "Animal Models of Inherited Metabolic Diseases" (R. J. Desnick, D. E. Patterson, and D. F. Scarpelli, eds.), pp. 213–220. A. R. Liss, New York.

Rattazzi, M. C., Dobrenis, K., Joseph, A., and Schwartz, P. (1987). Modified β-D-N-Acetyl-hexosaminidase isozymes for enzyme replacement in G_{M2} gangliosidosis. *Isozymes: Curr. Top. Biol. Med.Res.* **16**, 49–65.

Rattazzi, M. C., Tsuda, T., Benjamin, C. J., Gordon, R. E., Friedrich, V. L., Desnick, R. J., and Ioannou, Y. A. (1997). Delivery of DNA across the blood-brain barrier (BBB). *Am. J. Hum. Genet.* **61**(4), A358.

Roman-Goldstein, S., Mitchell, P., Crossen, J. R., Williams, P. C., Tindall, A., and Neuwelt, E. A. (1995). MR and cognitive testing of patients undergoing osmotic blood-brain barrier disruption with intraarterial chemotherapy. *Am. J. Neuroradiol.* **16**, 543–553.

Rosenberg, G. A., Kyner, W. T., and Estrada, E. (1980). Bulk flow of brain interstitial fluid under normal and hyperosmolar conditions. *Am. J. Physiol.* **283**, F42–F49.

Rubin, L. L., and Staddon, J. M. (1999). The cell biology of the blood-brain barrier. *Ann. Rev. Neurosci.* **22**, 11–28.

Sands, M., and Barker, J. E. (1999). Percutaneous intravenous injection in neonatal mice. *Lab. Animal Sci.* **49**, 328–330.

Sands, M. S., Vogler, C., Kyle, J. W., Grubb, J. H., Levy, B., Galvin, N., Sly, W. S., and Birkenmeier, E. H. (1994). Enzyme replacement therapy for murine mucopolysaccharidosis type VII. *J. Clin. Invest.* **93**, 2324–2331.

Schlesinger, P. M., Rodman, J. S., Doebber, T. W., Stahl, P. D., Lee, Y. C., Stowell, C. P., and Kuhlenschmidt, T. B. (1980). The role of extrahepatic tissues in the receptor-mediated plasma clearance of glycoproteins terminated by mannose or N-acetylglucosamine. *Biochem. J.* **192**, 596–606.

Schuchman, E. H., Ioannou, A. Y., Rattazzi, M. C., and Desnick, R. J. (1995). Neural gene therapy for inherited diseases with mental retardation: Principles and prospects. *Ment. Ret. Dev. Disab.Res. Rev.* **1**, 39–48.

Schwab, M. E., Suda, K., and Thoenen, H. (1979). Selective retrograde transsynaptic transfer of a protein, tetanus toxin, subsequent to its retrograde axonal transport. *J. Cell Biol.* **82**, 798–810.

Shen, W. C., and Ryser, H. J. P. (1978). Conjugation of poly-L-lysine to albumin and horseradish peroxidase: A novel method of enhancing the cellular uptake of proteins. *Proc. Natl. Acad. Sci. (USA)* **75**, 1872–1876.

Shen, W.-C., Ryser, H. J.-P., and Lamanna, L. (1985). Disulfide spacer between methotrexate and poly (D-lysine). A probe for exploring the reductive process inendocytosis. *J. Biol. Chem.* **260**, 10905–10908.

Simpson, L. (1985). Pharmacologic experiments on the binding and internalization of the 50,000 dalton carboxyterminus of tetanus toxin at the cholinergic neuromuscular junction. *J. Pharmacol. Exp. Ther.* **234**, 100–105.

Sofer, A., Schwarzmann, G., and Futerman, A. H. (1996). The internalization of short acyl chain analogue of ganglioside GM1 in polarized neurons. *J. Cell Sci.* **109**, 2111–2119.

Staddon, J. M., Herrenknecht, K., Schulze, C., Smales, C., and Rubin, L. L. (1995). Signal transduction at the blood-brain barrier. *Biochem. Soc. Trans.* **23**, 475–479.

Tager, J. M., Hamers, M. N., Schram, A. W., van den Berg, F. A., Rietra, P. J., Loonen, C., Koster, J. F., and Slee, R. (1980). An appraisal of human trials in enzyme replacement therapy of genetic diseases. *In* "Enzyme Therapy in Genetic Diseases: 2" (R. J. Desnick, ed.), pp. 343–359. A. R. Liss, New York, for the March of Dimes Birth Defect Foundation, Birth Defects: Original Article Series XVI(1).

Triguero, D., Buciak, J., and Pardridge, W. M. (1990). Capillary depletion method for quantification of blood-brain barrier transport of circulating peptides and plasma proteins, *J. Neurochem.* **54**, 1882–1888.

van Deurs, B. (1976). Choroid plexus absorption of horseradish peroxidasefrom the cerebral ventricles. *J. Ultrastruct. Res.* **55**, 400–416.

Vogler, C., Sands, M., Higgins, A., Levy, B., Grubb, J., Birkenmeier, E. H., and Sly, W. S. (1993). Enzyme replacement with recombinant beta-glucuronidase in the newborn mucopolysaccharidosis type VII mouse. *Pediatr. Res.* **34**, 837–840.

Vogler, C., Sands, M., Levy, B., Galvin, N., Birkenmeier, E. H., and Sly, W. S. (1996). Enzyme replacement with recombinant beta-glucuronidase in murine mucopolysaccharidosis type VII: Impact of therapy during the first six weeks of life on subsequent lysosomal storage, growth, and survival. *Pediatr. Res.* **39**, 1050–1054.

von Specht, B. U., Geiger, B., Arnon, R., Passwell, J., Keven, G., Goldman, B., and Padeh, B. (1979). Enzyme replacement in Tay-Sachs disease. *Neurology* **29**, 848–854.

Vorbrodt, A. W., Lossinsky, A. S., Dobrogowska, D. H., and Wisniewski, H. M. (1993). Cellular mechanisms of the blood-brain barrier (BBB) opening albumin-gold complex. *Histol. Histopathol.* **8**, 51–61.

Wada, R., Tifft, C. J., and Proia, R. L. (2000). Microglial activation precedes acute neurodegeneration in Sandhoff disease and is suppressed by bone marrow transplantation. *Proc. Natl. Acad. Sci. U.S.A.* **97**, 10954–10959.

Walkley, S. U., Thrall, M. A., Dobrenis, K., March, P. A., and Wurzelmann, S. (1994a). Bone marrow transplantation in neuronal storage disorders. *Brain Pathol.* **4,** 376A.

Walkley, S. U., Thrall, M. A., Dobrenis, K., Huang, M., March, P. A., Siegel, D. A., and Wurzelmann, S. (1994b). Bone marrow transplantation corrects the enzyme defect in neurons of the central nervous system in a lysosomal storage disease. *Proc. Natl. Acad. Sci. (USA)* **91,** 2970–2974.

Walkley, S. U., Thrall, M. A., and Dobrenis, K. (1996). Targeting gene products to the brain and neurons using bone marrow transplantation: A cell-mediated delivery system for therapy of inherited metabolic human disease. *In* "Protocols for Gene Transfer in Neuroscience: Towards Gene Therapy of Neurological Disorders" (P. R. Lowenstein and L. W. Enquist, eds.), pp. 275–302. Wiley, London and New York.

Wellhoner, N. M. (1982). Tetanus neurotoxin. *Rev. Physiol.Biochem. Pharmacol.* **93,** 1–68.

Wood, J. M. (1980). Physiology, pharmacology, and dynamics of cerebrospinal fluid. *In* "Neurobiology of Cerebrospinal Fluid." (J. M. Wood, ed.), pp. 1–15. Plenum, New York.

Xiao, X., Li, J., McCown, T. J., and Samulski, R. J. (1997). Gene transfer by adeno-associated virus vectors into the central nervous system. *Exper. Neurol.* **144,** 113–124.

Zhang, E. T., Inman, C. B., and Weller, R. O. (1990). Interrelationships of the pia mater and the perivascular (Virchow-Robin) spaces in the human cerebrum. *J. Anat.* **170,** 111–123.

26

Tay-Sachs Disease: Psychologic Care of Carriers and Affected Families

Leslie Schweitzer-Miller
New York University Psychoanalytic Institute
New York University Medical School
New York, New York 10016

A century ago the door to understanding the unseen workings of the human body and mind began to swing open, heralding the extraordinary advances in science we are witnessing today. At approximately the same time as Dr. Bernard Sachs began to publish his findings, Dr. Sigmund Freud, also a neurologist, turned his attention from the study of the brain and neurologic system to unraveling the mysteries of the human mind. Based on his clinical observations, he began to elucidate mental phenomena, as distinct from brain functions. He conceptualized the structure of the mind, the existence of an unconscious, and the barriers the mind erects to prevent what is stored in the unconscious from reaching awareness, which he termed the defenses. He was able to demonstrate conclusively that, for a host of reasons, the personally unacceptable wishes, fantasies, dreams, and fears that people inevitably harbor remain locked in their unconscious, out of awareness. These thoughts and feelings may become overtly conscious, but often they masquerade, leaving only traces that allow their presence to be deduced. To provide appropriate and useful care to the carriers and families affected by Tay-Sachs disease, we must first understand what they may not be able to tell us, but rely on us to know.

When geneticists speak of "the affected" they are referring to the proband, the one with the disease, in this case, the infant who will ultimately die of Tay-Sachs. But the long-lasting, devastating mental effect of Tay-Sachs disease is on the living, making the carriers, carrier couples, and families—parents, grandparents, and siblings—of Tay-Sachs children the truly "living" affected. It is they who require sophisticated understanding and psychologic support, which is recognized and mandated by both American and Canadian clinical genetics accrediting bodies. However, this is easier said than done, as the extent and quality of the impact may remain hidden from the patient as well as from the professional.

Clinicians in Tay-Sachs disease centers around the world (medical geneticists and genetic counselors) have now had vast experience in screening, identifying more than 36,000 people as carriers since 1974. Similar overt psychologic findings are reported. Whereas most people accept and incorporate the information without visible psychologic difficulties, many initially experience transient anxiety when they are notified. Although some of the anxiety may be a result of misunderstanding the implications and risks, it is more likely related to the shock of receiving unexpected and unwanted news, a manifestation of "how could this be true, there's never been a problem in my family" or "why did this happen to me?" I say more likely because even well-informed carriers, who comprehend the lack of threat, experience this initial anxiety. It has also been reported that although people identified as carriers do not differ from controls or those identified as noncarriers in their perception of their past or current health, they view their future health with less optimism than the other two groups. In fact, in a prospective psychologic study carried out in conjunction with the original Minnesota Tay-Sachs screening program, it was clearly demonstrated that those people identified as carriers "forgot" (repressed) and/or ignored (denied) that they and a carrier spouse might have special limitations to consider when planning a family.

Human beings develop a self-image, a familiar, fixed perception of themselves, which can be disrupted when they learn they have a "defective" gene, regardless of the implications. In the vernacular, it is a blow to the ego, albeit minor, which, when it occurs, is dealt with by the defenses. The ego has many defenses at its disposal to cope with the blow, but carriers often use a combination of unconscious mechanisms that robs the distressing news of its affective component and diminishes its importance. The immediate emotional reaction, the shock, anger, fear, or anxiety, becomes disassociated from the information in this maneuver, allowing it to be unemotionally incorporated into the self-representation. Thus, the long-term impact is usually subtle, and a minor price to pay for the real benefit that is gained.

Clinical experience verifies the above phenomena. Anyone who has ever carefully explained something to a patient who then returns the following week asking the same question can attest to the reality of denial and/or repression. Often, people do not mentally register what they do not want to know, for fear of what it will mean to them and how it will make them feel. That is why every person informed of being a carrier should receive a follow-up phone call several weeks after the initial contact. The caller should:

Inquire about the individual's general state of mind and health

Be ready to answer any questions

Be alert to any unstated signs of depression, such as change in appetite, sleep patterns, or mood, decreased sex drive or irritability, and/or a hopeless view of the future

Investigate and pursue an abrupt change in social or work relationships

Be prepared to make a referral to a psychiatrist who is not only familiar with genetics, but also interested in trying to understand what is causing distress in this particular individual.

In rare instances, a newly identified heterozygote, whether unmarried or married to a noncarrier, responds to the information with prolonged fear, anxiety, or other severe psychologic symptoms. Whereas this may seem totally irrational to a scientist, it must be remembered that the unconscious harbors many irrational thoughts and feelings that therefore cannot be reasoned away by using "facts." As message bearers, we are obliged to spend time and human resources helping these people find an appropriate way to become comfortable. We may, for example:

Provide them an opportunity to explain what they understand about recessive genes, so that any misconceptions, and what is causing them, may be sensitively addressed

Schedule additional meetings to go over the same ground as often as necessary while any confusion remains

Inform them that we all carry numbers of such genes, i.e., that it is "normal"

Explain the concept of gene "activity," as opposed to an all-or-nothing apparatus—they are not totally missing something essential

Reinforce the zero risk to themselves or any offspring, if married to a noncarrier

Transform what may be perceived as a weakness (a defective gene) into a strength by helping them to feel courageous and pleased with themselves for being tested

Reinforce that being a carrier is nothing to be ashamed or embarrassed about

Refer them to psychotherapy to further explore and attempt to resolve the underlying issues if psychologic symptoms persist

When it is necessary to inform partners that they both carry the Tay-Sachs gene—as has been the case with over 1100 couples—we must remember we are telling them they may be forced to participate, either actively or passively, in the loss of a child. We can help them choose to have only healthy children by offering pre-embryo biopsy or some form of prenatal diagnosis and abortion of an affected fetus. Alternatively, if they choose to (child's) take an affected pregnancy to term, their "bad" genes will cause the child's demise. Therefore they may be helplessly propelled into playing a kind of Russian roulette in each attempted pregnancy, with the risk of becoming an unwilling accomplice to the destruction of a life. It is easy to forget that alongside the good news that we can help couples to have only normal babies, is also potentially devastating trauma. In my experience with these couples, their fantasies verify that they do, indeed, feel they have been potentially doomed to destroy their own child. The couple may not disclose these thoughts and related feelings to the geneticist, first because each is ashamed to have such fantasies and, in addition, because each imagines being viewed by the geneticist as foolish and ungrateful.

In a psychotherapy setting, these couples often talk about their enormous guilt in bringing the "bad" gene to the marriage, feeling a loss of control over their lives, being put in a position they did not anticipate and do not want to be in, and enormous anger at their bad luck. They often start to wonder what they did to deserve being punished. If either partner's guilt is too much to bear, the tendency is to start to project it onto the other partner as blame. Every couple I have seen has reported initially wanting to blame the other, saying things like, "If I hadn't married you I wouldn't have to go through any of this." They feel trapped and tortured, and often this issue dominates their relationship and all their life plans.

Testing both members of a couple concurrently can help minimize the anxiety inherent in waiting when the results identify the first as a carrier. It also relieves the emotional burden placed singularly on either the husband or wife, making the problem to be faced one that is shared by the couple, rather than a source of antagonism between them. It also minimizes the risk of any laboratory errors resulting in unknown affected pregnancies, when one spouse is mistakenly identified as a noncarrier and the other subsequently feels secure in not getting tested. When tested together, if one member of the couple is identified as carrier, the one identified as a noncarrier should have his or her results verified.

People inevitably feel as they do, and simply telling someone "don't feel that way" never changes anything. If being a carrier couple is causing problems in the marriage, or if either partner becomes intolerant of socializing with people who have children, or develops new health concerns, a referral for psychologic help should be considered. If such a couple is faced with the decision to abort an affected fetus, it can be useful to encourage them to talk to a psychotherapist who is independent of the genetics center. This gives them the opportunity to speak about their guilt, the sadness and despair over the loss of what might have been, their anger at the entire situation, and, at times, anyone associated with it. The temptation to shoot the messenger is not uncommon, and requires patient and diplomatic intervention from a neutral source in a setting of privacy and trust.

Of the "living affected," it is the family members of a child afflicted with Tay-Sachs disease who suffer the greatest trauma. Each parent believes he or she gave "the wrong gene," and often develops quasi-superstitions, such as there is a right time of day, position, vitamin, or exercise that will allow one to give the right gene next time. They often develop what I term a Job syndrome, believing they were singled out as a bad couple and are doomed to be tested and suffer, not only by watching the fatal deterioration of the child they love while providing it with tedious, daily care, but also by having more "bad" babies. Their sense of dread about their personal future and life in general often inflicts itself on all their relationships, on their ability to work successfully, and on the stability of the marriage. It is not uncommon for them to lose the capacity to derive pleasure out of any aspect of their lives. Overwhelming rage that this happened to them, and guilt at producing such a child, can easily be transformed into blame and projected onto a physician, an institution, society, or their parents. The rage can be overpowering and difficult to contain, providing some with excuses for acting on sadistic or masochistic impulses.

Ordinary people who can tolerate knowing things about themselves they do not like, at one time or another may be aware of a wish to get away from or get rid of their child. It is not unusual even in the most loving parents. Children place unexpected burdens on parents that can occasionally and transiently result in that wish. However, the wish can pass in the moment it takes for a child to smile, and does not move into the realm of action, as it ultimately does for the parents who have passively given the lethal gene to their Tay-Sachs child. In and of itself, this can cause unbearable guilt along with pain. In addition, as the child becomes more difficult to manage, develops seizures, dysphagia, and creates financial and emotional havoc, parents often have fantasies of murdering the child, their once-desired and beloved baby. This serves to increase their guilt, to which they may respond in a variety of ways, including hating themselves, each other, or overextending themselves on behalf of the declining child. The entire focus of the family, including any siblings, may come to revolve around this child in a pathologic overattachment which defends against the forbidden wish.

The siblings of these children may have difficulties similar to those experienced by the parents. Wishing a sibling would disappear or die is considered a normal part of ordinary sibling rivalry, and no cause for alarm or guilt—that is, not unless it comes true. Most children wish to be the center of their parents' universe, the one and only, and it is not unusual to hear an older child request to send the new addition back. When it really happens, the child can feel frighteningly powerful, as well as guilty for having survived. Such children may be terrified, imagining they will be punished and die like the sibling. They may become depressed, phobic, and/or have psychologic sequelae affecting their ability to learn and function normally in the future. These children should have the opportunity to talk about these feelings, if they can, or reveal their fears through play and art to a professional who will understand and, over time, help them come to some satisfactory resolution.

If these issues are not dealt with in a meaningful way before the death of the child, or the birth of another, it may well result in pathologic or unresolved mourning. Ordinary mourning is the normal process of saying good-bye, which goes through stages including guilt and rage but culminates in remembering— what was loved and hated—allowing the mourner to go on. Since a Tay-Sachs– affected child brings something bad to the family, it can easily be secretly hated, unconsciously, and never allow the parents to go on in a normal way. A subsequent child may be irrationally resented for replacing the lost one, despite the joy and relief its normal development brings.

The psychologic manifestation I have outlined are meant to further delineate the nature of the trauma experienced by this group. Understanding their inner emotional world provides the rational not only for management of psychologic issues, but also informs the approach to imparting necessary concrete information. Planning a course of action with a patient, discussing amniocentesis versus CVS, the availability and ramifications of preimplantation genetics, how to proceed if an affected child is conceived, and whether, where, and when to place an affected child, helps restore a much needed sense of control, particularly at a time there is so much that must be passively accepted. Although the options we can offer patients are ever increasing, they do not ameliorate the psychologic nature of the trauma to any individual. In these vulnerable situations, the patient gives extraordinary weight to the recommendations of the professional. Therefore, with an appreciation of what often remains unstated by patients, the geneticist can strengthen and increase his or her therapeutic armamentarium, continuing in the great tradition of Dr. Sachs and Dr. Freud to look for answers in what cannot be seen.

Bibliography

Borreani, C., and Gangeri, L. (1996). Genetic counseling: Communication and psychosocial aspects. *Tumori* **82**(2), 147–150.

Burnett, L., Proos, A. L., Howell, V. M., Longo, L., Tedeschi, V., Yang, V. A., Siafakas, N., and Turner, G. (1995). The Tay-Sachs disease prevention program in Australia: Sydney pilot study. *Med. J. Austral.* **163**(6), 298–300.

Desnick, R. J. (1983). The Minnesota Tay-Sachs carrier screening program—A prospective Study. Personal Communication.

Freud, A. (1926). Inhibitions, symptoms.

Freud, A. (1936). "The Ego and the Mechanisms of Defense. Writings 2." International Univ. Press, New York, 1966.

Freud, S. (1894). The neuro-psychoses of defense. S E 3.

Freud, S. (1917). Mourning and melancholia. S E 14.

Freud, S. (1923). The ego and the id. S E 19. and anxiety. S E 20.

Garber, A. P., Platt, L. D., Wang, S. J., Jam, K., Carlson, D. E., and Rotter, J. I. (1993). Determinants of utilization of Tay-Sachs screening. *Obstet. Gynecol.* **82**(3), 460–643.

Gibbons, W. E., Gitlin, S. A., Lanzendorf, S. E., Kaufmann, R. A., Slotnick, R. N., and Hodgen, G. D. (1995). Preimplantation genetic diagnosis for Tay-Sachs disease: Successful pregnancy after pre-embryo biopsy and gene amplification by polymerase chain reaction. *Fertil. Steril.* **63**(4), 723–728.

Hechtman, P., and Kaplan, F. (1993). Tay-Sachs disease screening and diagnosis: Evolving technologies. *DNA Cell Biol.* **12**(8), 651–665.

Kaback, M., Lim-Steele, J., Dabholkar, D., Brown, D., Levy, N., and Zeiger, K. (1993). Tay-Sachs disease—Carrier screening, prenatal diagnosis, and the molecular era. An international perspective, 1970–1993. The International TSD Data Collection Network. *J. Am. Med. Assoc.* **270**(19), 2307–2315.

Marteau, T. M., van Duijn, M., and Ellis, I. (1992). Effects of genetic screening on perception of health: A pilot study. *J. Med. Genet.* **29**(1), 24–26.

Natowicz, M. R., and Alper, J. S. (1991). Genetic screening: Triumphs, problems, and controversies. *J. public Health Policy* **12**(4), 475–491.

Sharpe, N. F. (1994). Psychological aspects of genetic counseling: A legal perspective. *Am. J. Med. Genet.* **50**(3), 234–238.

27

Future Perspectives for Tay-Sachs Disease

Robert J. Desnick
Department of Human Genetics
Mount Sinai School of Medicine of New York University
New York, New York 10029

Michael M. Kaback
Departments of Pediatrics and Reproductive Medicine
University of California, San Diego School of Medicine
San Diego, California 92123

I. Introduction
II. Substrate Deprivation
III. Chemical Chaperones
IV. Stem Cells
V. Oligonucleotide Recombination
VI. Genetic Counseling and Psychosocial Support
VII. Prevention
    Acknowledgments
    References

I. INTRODUCTION

As we enter the new millennium, it is easy to recount the accomplishments of the past, but difficult to predict the revolutionary breakthroughs of the future. It is anticipated that the rapid pace of biomedical scientific discovery will change the practice of medicine. For example, the recent deciphering of the human genome and the translation of its information should lead to the understanding of common and rare diseases, and the means to develop prediction/prevention programs, novel

Copyright © 2001 by Academic Press
0065-2660/01 $35.00

treatments, and cures. But what will be the impact of these future advances on Tay-Sachs disease? Will the paradigm of prevention change in the anticipated era of predictive medicine? Will there be novel and effective treatments or cures for neurodegenerative diseases such as Tay-Sachs disease?

Informed optimists take the position that we are on the threshold of remarkable advances in molecular and cellular biology that will change current diagnostic and therapeutic paradigms. History documents unexpected breakthroughs that have dramatically changed the unlikely into reality. The informed pragmatist is optimistic, but cautious, about the likelihood of future therapeutic breakthroughs. For example, it took decades to develop enzyme replacement therapy for Gaucher disease, but it has little benefit for the neurologic forms of the disease. Similarly, the clinical benefit of bone marrow transplantation (BMT) to treat neurodegenerative lysosomal storage diseases is still controversial, particularly for rapidly progressive disorders. BMT of mice with Sandhoff disease (the total β-hexosaminidase-deficient variant of Tay-Sachs disease) prolonged their life-span, but they ultimately experienced the typical neurologic deterioration of un-treated mice (Norflus et al., 1998). Although the expectations for gene therapy are high, this strategy is still in an early developmental phase, and the ability to deliver therapeutic genes to neurons throughout the central nervous system is an enormous, if not daunting, challenge (Guidotti et al., 1998, 1999).

Within that context, researchers are currently investigating a variety of intriguing experimental approaches to treat neurodegenerative lysosomal storage disorders like Tay-Sachs disease. These include "substrate deprivation," "chemical chaperones," neural stem cell transplantation, and in vivo treatment with chimeric oligonucleotides. The recent availability of mouse models of Tay-Sachs disease permits biochemical evaluation of these strategies as discussed below (Yamanaka et al., 1994; Cohen-Tannoudji et al., 1995; Sango et al., 1995; Phaneuf et al., 1996; Huang et al., 1997; Liu et al., 1997; Tifft and Proia, 1997; Chavany and Jendoubi, 1998).

II. SUBSTRATE DEPRIVATION

Substrate deprivation therapy is designed to decrease the rate of substrate accumulation by inhibition of an early enzyme in the synthesis of the glycolipid precursors of GM2 ganglioside (Platt and Butters, 1998). This approach can decrease the rate of substrate synthesis, but cannot degrade already accumulated GM2 ganglioside. However, it is unlikely that they will be effective in the rapidly progressive infantile form of Tay-Sachs disease. In fact, such a therapeutic approach may only decrease the rate of neurologic disease progression, thereby prolonging the life of the affected child, but not altering the ultimate fatal outcome. For example, a pilot trial of substrate derivation in asymptomatic Tay-Sachs knockout mice was recently performed using N-butyldeoxynojirimycin (NB-DNJ), a potent inhibitor of glucosylceramide synthase (Platt *et al.*, 1997). After four weeks of oral treatment, the rate of accumulation of neural GM2 ganglioside was decreased, indicating that the small inhibitor molecule could reach the neuronal sites of pathology and decrease glycolipid synthesis, and, therefore, the subsequent progressive accumulation of the toxic GM2 ganglioside. Subsequent studies in knockout mice with Sandhoff disease (total β-hexosaminidase deficiency) were designed to determine if substrate derivation could alter the clinical course of their neurodegenerative disease. Mice treated from 3 or 6 weeks of life had delayed onset of symptoms and lived 40% longer than untreated mice (170 vs 125 days), but then experienced severe neurodegenerative symptoms and were euthanized (Jeyakumar *et al.*, 1999). When substrate deprivation therapy was initiated at a late presymptomatic stage (11 weeks of life), no difference in the disease course was observed. However, if such compounds can effectively reduce the rate of substrate accumulation without toxic or developmental side effects, they may be useful for the treatment of patients with the adult-onset form of GM2 gangliosidosis.

III. CHEMICAL CHAPERONES

Another potential therapeutic approach for the chronic and adult-onset patients with GM2 gangliosidosis involves the use of small inhibitors that act as "chemical chaperones" for unstable enzymes with residual activity (Fan *et al.*, 1999). When used in low doses, these reversible competitive inhibitors bind the enzyme's active site and stabilize the mutant enzyme for transport to the lysosome. As a prototype, this approach has been shown to increase the residual α-galactosidase A activity in patients with the mild cardiac variant of Fabry disease (Fan *et al.*, 1999). Although β-hexosaminidase A has not been crystallized, candidate active site regions have been identified (Tse *et al.*, 1996; Fernandes *et al.*, 1997; Wright *et al.*, 2000; Mark *et al.*, 2001) and a homologous chitobiase has been crystallized at 1.9 Å (Tews *et al.*, 1996), permitting the rational design of active site reversible competitive inhibitors that could be evaluated as potential "chemical chaperones." Such small molecule competitive inhibitors that bind to the active site of the β-hexosaminidase α-chain could be tested in cultured cells from adult-onset patients with the G269S mutation to evaluate the *in vitro* feasibility of this approach. If effective *in vitro*, then studies in Tay-Sachs knockout mice that are transgenic for the human G269S mutation would assess the *in vivo* effectiveness of these inhibitors in reversing, stabilizing, or delaying the progressive clinical manifestations of these disorders.

IV. STEM CELLS

Experimental transplantation of neural stem cells into the central nervous system of animal models with various lysosomal diseases is currently underway in several laboratories (Lacorazza *et al.*, 1996; Flax *et al.*, 1998; Ourednik *et al.*, 1999). It has been shown that murine and human neural stem cells can differentiate into several neural cell types that multiply and migrate throughout the brain of mice after intracerebral injection (Yandava *et al.*, 1999). Thus, they may provide a global source of β-hexosaminidase A capable of degrading accumulated GM2 ganglioside. These mutant cells also may serve as a delivery system for neural gene therapy since they can be transduced by several viral vectors (Snyder *et al.*, 1995). Another intriguing cell type that is currently being evaluated for neural cell transplantation and gene therapy is the mesenchymal bone marrow cell. Such cells can be easily obtained autologously, and can differentiate and migrate throughout the brain after intracerebral injection (Kopen *et al.*, 1999). These cells may be used for global neural gene therapy for lysosomal neurodegenerative diseases.

V. OLIGONUCLEOTIDE RECOMBINATION

Another experimental approach involves the use of synthetic oligonucleotides which are designed to recombine with the region of a mutation in genomic DNA and replace the mutation with the normal sequence. Studies using this approach in cell culture and in animal models of human diseases have provided "proof of concept" (Cole-Strauss *et al.*, 1996; Yoon *et al.*, 1996; Kren *et al.*, 1999). Thus, it may be feasible in the future to deliver a therapeutic oligonucleotide designed to exchange with the region of the chromosomal DNA containing a specific β-hexosaminidase α-chain mutation so that a normal allele may be reconstituted *in vivo*. If the exchange can be induced to occur frequently enough to be therapeutic, then the remaining obstacle will be the global delivery and uptake of these oligonucleotides into neurons throughout the brain.

All in all, these approaches are not yet reality, and the pragmatist would argue that the likelihood is low for developing a therapeutic strategy that will significantly alter the course of a rapidly progressive neurodegenerative disease and significantly improve the quality of the patient's life. However, one must remain optimistic; breakthroughs happen and miracle drugs have been discovered.

VI. GENETIC COUNSELING AND PSYCHOSOCIAL SUPPORT

While these therapeutic possibilities offer hope to the afflicted family, caregivers should appreciate the tragedy of this disease and its devastating effects on the parents, the parents' marriage, other children, and the extended family. Physicians should recognize that Tay-Sachs disease is as much a "disease of the parents" as it is of the affected child. Parents experience the burden of the disease and will grasp at any possible treatment, no matter how experimental or risky. The parents of a Tay-Sachs baby poignantly described the pain they experienced: "For the parents, the child dies three times. First, when the diagnosis is made, usually at 6 to 10 months of life, when they learn that the child will experience an unrelenting neurodegenerative course with death by 5 years; second, when the parents can no longer provide the needed nursing care, and placement for their child in a care facility must be considered; and third, when the child actually dies. The loss is forever in their thoughts."

Psychological support of the parents and their relationship with each other, the affected child, and their normal children, is needed throughout the life, and after the loss of a Tay-Sachs disease child. Understanding and supporting their needs, medical and emotional, is the cornerstone of medical management for many families who have an affected child, particularly if they have no normal children. It should be emphasized that the planning for and birth of a "normal" baby can provide a positive, reassuring, and emotionally important focus for them.

VII. PREVENTION

Our expectation for an effective treatment or cure is high. We anticipate important breakthroughs and novel therapeutic approaches beyond our current imagination. However, prevention is the key in the interim (Kaback, 2000). In the high-risk Ashkenazi Jewish population, Tay-Sachs disease carrier detection and prevention programs have proven dramatically effective and have established the paradigm for the prevention of recessive diseases (Kaback et al., 1998; Eng et al., 1997). The extension of prenatal carrier screening to premarital carrier screening has also proven effective and has met the special cultural needs of the religious community (Ekstein and Katzenstein, this volume; Vought et al., 1986; Bach et al., 2001). It is likely that the future use of multiple mutation detection methods, for example, DNA chip and bioarray technologies, will permit testing in the general population for multiple severe genetic disorders like Tay-Sachs disease. One can imagine that such testing will be generally available in the future, and that a prospective couple could premaritally or preconceptually determine which disease genes they carry, even for very rare, but devastating traits. Various options would exist to avoid the birth of an affected child, including novel preimplantation procedures that permit the selection of only "normal" embryos for in vitro fertilization. This technique has been employed by at-risk parents who have successfully given birth to unaffected babies (Snabes et al., 1994; Gibbons et al., 1995). Such is the goal of predictive medicine, and the experience with Tay-Sachs disease has provided the guidance and experience for its future success.

Acknowledgments

This work was supported in part by grants to R.J.D. from the National Institutes of Health, including a Merit Award (5 R37 DK34045), a grant (5 M01 RR00071) for the Mount Sinai General Clinical Research Center, a grant (5 P30 HD28822) for the Mount Sinai Child Health Research Center, in part by a contract to M.M.K. from the Maternal and Child Health Section, Genetic Disease Branch, Department of Health Services, State of California, and a grant from the National Tay-Sachs Disease and Allied Disorders Association.

References

Bach, G., Tomczak, J., Risch, N., and Ekstein, J. (2001). Tay-Sachs screening in the Jewish Ashkenazi population: DNA testing is the preferred procedure. *Am. J. Med. Genet.* **99,** 70–75.

Chavany, C., and Jendoubi, M. (1998). Biology and potential strategies for the treatment of GM2 gangliosidoses. *Mol. Med. Today.* **4,** 158–165.

Cohen-Tannoudji, M., Marchand, P., Akli, S., Sheardown, S. A., Puech, J. P., Kress, C., Gressens, P., Nassogne, M. C., Beccari, T., and Muggleton-Harris, A. L., *et al.* (1995). Disruption of murine Hexa gene leads to enzymatic deficiency and to neuronal lysosomal storage, similar to that observed in Tay-Sachs disease. *Mamm. Genome.* **6,** 844–849.

Cole-Strauss, A., Yoon, K., Xiang, Y., Byrne, B. C., Rice, M. C., Gryn, J., Holloman, W. K., and Kmiec, E. B. (1996). Correction of the mutation responsible for sickle cell anemia by an RNA-DNA oligonucleotide. *Science* **273,** 1386–1389.

Eng, C. M., Schechter, C., Robinowitz, J., Fulop, G., Burgert, T., Levy, B., Zinberg, R., and Desnick, R. J. (1997). Prenatal genetic carrier testing using triple disease screening. *JAMA* **278,** 1268–1272.

Fan, J. Q., Ishii, S., Asano, N., and Suzuki, Y. (1999). Accelerated transport and maturation of lysosomal α-galactosidase A in Fabry lymphoblasts by an enzyme inhibitor. *Nat. Med.* **5,** 112–115.

Fernandes, M. J., Yew, S., Leclerc, D., Henrissat, B., Vorgias, C. E., Gravel, R. A., Hechtman, P., and Kaplan, F. (1997). Identification of candidate active site residues in lysosomal β-hexosaminidase. A. *J. Biol. Chem.* **272,** 814–820.

Flax, J. D., Aurora, S., Yang, C., Simonin, C., Wills, A. M., Billinghurst, L. L., Jendoubi, M., Sidman, R. L., Wolfe, J. H., Kim, S. U., and Snyder, E. Y. (1998). Engraftable human neural stem cells respond to developmental cues, replace neurons, and express foreign genes. *Nat. Biotechnol.* **16,** 1033–1039.

Gibbons, W. E., Gitlin, S. A., Lanzendorf, S. E., Kaufmann, R. A., Slotnick, R. N., and Hodgen, G. D. (1995). Preimplantation genetic diagnosis for Tay-Sachs disease: Successful pregnancy after preembryo biopsy and gene amplification by polymerase chain reaction. *Fertil. Steril.* **63,** 723–728.

Guidotti, J., Akli, S., Castelnau-Ptakhine, L., Kahn, A., and Poenaru, L. (1998). Retrovirus-mediated enzymatic correction of Tay-Sachs defect in transduced and non-transduced cells. *Hum. Mol. Genet.* **7,** 831–838.

Guidotti, J. E., Mignon, A., Haase, G., Caillaud, C., McDonell, N., Kahn, A., and Poenaru, L. (1999). Adenoviral gene therapy of the Tay-Sachs disease in hexosaminidase A-deficient knock-out mice. *Hum. Mol. Genet.* **8,** 831–838.

Huang, J. Q., Trasler, J. M., Igdoura, S., Michaud, J., Hanal, N., and Gravel, R. A. (1997). Apoptotic cell death in mouse models of GM2 gangliosidosis and observations on human Tay-Sachs and Sandhoff diseases. *Hum. Mol. Genet.* **6,** 1879–1885.

Jeyakumar, M., Butters, T. D., Cortina-Borja, M., Hunnam, V., Proia, R. L., Perry, V. H., Dwek, R. A., and Platt, F. M. (1999). Delayed symptom onset and increased life expectancy in Sandhoff disease mice treated with N-butyldeoxynojirimycin. *Proc. Natl. Acad. Sci. USA* **96,** 6388–6393.

Kaback, M. M. (2000). Population-based genetic screening for reproductive counseling: the Tay-Sachs disease model. *Eur. J. Pediatr.* **159,** S192–195.

Kaback, M., Lim-Steele, J., Dabholkar, D., Brown, D., Levy, N., and Zeiger, K. (1993). Tay-Sachs disease–carrier screening, prenatal diagnosis, and the molecular era. An international perspective, 1970 to 1993. The International TSD Data Collection Network. *JAMA* **270,** 2307–2315.

Kopen, G. C., Prockop, D. J., and Phinney, D. G. (1999). Marrow stromal cells migrate throughout forebrain and cerebellum, and they differentiate into astrocytes after injection into neonatal mouse brains. *Proc. Natl. Acad. Sci. USA* **96,** 10711–10716.

Kren, B. T., Parashar, B., Bandyopadhyay, P., Chowdhury, N. R., Chowdhury, J. R., and Steer, C. J. (1999). Correction of the UDP-glucuronosyltransferase gene defect in the Gunn rat model of Crigler-Najjar syndrome type 1 with a chimeric oligonucleotide. *Proc. Natl. Acad. Sci. USA* **96,** 10349–10354.

Lacorazza, H. D., Flax, J. D., Snyder, E. Y., and Jendoubi, M. (1996). Expression of human β-hexosaminidase α-subunit gene (the gene defect of Tay-Sachs disease) in mouse brains upon engraftment of transduced progenitor cells. *Nat. Med.* **2,** 424–429.

Liu, Y., Hoffmann, A., Grinberg, A., Westphal, H., McDonald, M. P., Miller, K. M., Crawley, J. N., Sandhoff, K., Suzuki, K., and Proia, R. L. (1997). Mouse model of GM2 activator deficiency manifests cerebellar pathology and motor impairment. *Proc. Natl. Acad. Sci. USA* **94,** 8138–8143.

Mark, B. L., Vocadlo, D. J., Knapp, S., Triggs-Raine, B. L., Withers, S. G., and James, M. N. (2001). Crystallographic evidence for substrate-assisted catalysis in a bacterial beta-hexosaminidase. *J. Biol. Chem.* **276,** 10330–10337.

Norflus, F., Tifft, C. J., McDonald, M. P., Goldstein, G., Crawley, J. N., Hoffmann, A., Sandhoff, K., Suzuki, K., and Proia, R. L. (1998). Bone marrow transplantation prolongs life span and ameliorates neurologic manifestations on Sandhoff disease mice. *J. Clin. Invest.* **101,** 1881–1888.

Ourednik, V., Ourednik, J., Park, K. I., and Snyder, E. Y. (1999). Neural stem cells – a versatile tool for cell replacement and gene therapy in the central nervous system. *Clin. Genet.* **56,** 267–278.

Phaneuf, D., Wakamatsu, N., Huang, J. Q., Borowski, A., Peterson, A. C., Fortunato, S. R., Ritter, G., Igdoura, S. A., Morales, C. R., and Benoit, G. *et al.* (1996). Dramatically different phenotypes in mouse models of human Tay-Sachs and Sandhoff diseases. *Hum. Mol. Genet.* **5,** 1–14.

Platt, F. M., and Butters, T. D. (1998). New therapeutic prospects for the glycosphingolipid lysosomal storage diseases. *Biochem. Pharmacol.* **56,** 421–430.

Platt, F. M., Neises, G. R., Reinkensmeier, G., Townsend, M. J., Perry, V. H., Proia, R. L., Winchester, B., Dwek, R. A., and Butters, T. D. (1997). Prevention of lysosomal storage in Tay-Sachs mice treated with N-butyldeoxynojirimycin. *Science* **276,** 428–431.

Sango, K., Yamanaka, S., Hoffmann, A., Okuda, Y., Grinberg, A., Westphal, H., McDonald, M. P., Crawley, J. N., Sandhoff, K., and Suzuki, K. *et al.* (1995). Mouse models of Tay-Sachs and Sandhoff diseases differ in neurologic phenotype and ganglioside metabolism. *Nat. Genet.* **11,** 170–176.

Snabes, M. C., Chong, S. S., Subramanian, S. B., Kristjansson, K., DiSepio, D., and Hughes, M. R. (1994). Preimplantation single-cell analysis of multiple genetic loci by whole-genome amplification. *Proc. Natl. Acad. Sci. USA* **91,** 6181–6185.

Snyder, E. Y., Taylor, R. M., and Wolfe, J. H. (1995). Neural progenitor cell engraftment corrects lysosomal storage throughout the MPS VII mouse brain. *Nature* **374,** 367–370.

Tews, I., Perrakis, A., Oppenheim, A., Dauter, Z., Wilson, K. S., and Vorgias, C. E. (1996). Bacterial chitobiase structure provides insight into catalytic mechanism and the basis of Tay-Sachs disease. *Nat. Struct. Biol.* **3,** 638–648.

Tifft, C. J., and Proia, R. L. (1997). The β-hexosaminidase deficiency disorders: Development of a clinical paradigm in the mouse. *Ann. Med.* **29,** 557–561.

Tse, R., Vavougios, G., Hou, Y., and Mahuran, D. J. (1996). Identification of an active acidic residue in the catalytic site of β-hexosaminidase. *Biochemistry* **35,** 7599–7607.

Vought, L., Schuette, J., Bach, G., Grabowski, G. A., and Desnick, R. J. (1986). Tay-Sachs disease: Compatability screening in the Orthodox Jewish community. *Am. J. Hum. Genet.* **39,** A183.

Wright, C. S., Li, S. C., and Rastinejad, F. (2000). Crystal structure of human Gm2-activator protein with a novel beta-cup topology. *J. Mol. Biol.* **304,** 411–22.

Yamanaka, S., Johnson, M. D., Grinberg, A., Westphal, H., Crawley, J. N., Taniike, M., Suzuki, K., and Proia, R. L. (1994). Targeted disruption of the Hexa gene results in mice with biochemical and pathologic features of Tay-Sachs disease. *Proc. Natl. Acad. Sci. USA* **91,** 9975–9979.

Yandava, B. D., Billinghurst, L. L., and Snyder, E. Y. (1999). "Global" cell replacement is feasible via neural stem cell transplantation: Evidence from the dysmyelinated shiverer mouse brain. *Proc. Natl. Acad. Sci. USA* **96,** 7029–7034.

Yoon, K., Cole-Strauss, A., and Kmiec, E. B. (1996). Targeted gene correction of episomal DNA in mammalian cells mediated by a chimeric RNA-DNA oligonucleotide. *Proc. Natl. Acad. Sci. USA* **93,** 2071–2076.

Index

A

β-D-N-Acetylgalactosamine, 63
N-Acetylgalactosamine, 34–35
N-Acetyl-[³H]-D-galactosamine, 56–57
N-Acetylgalactosaminidase, 54
N-Acetyl-β-D-galactosaminidase, 63
N-Acetylgalactosaminyl group, 56–57
β-D-N-Acetylglucosamine, 63
N-Acetyl-β-glucosaminide, 46
N-Acetylmannosamine, 54–55
N-Acetylneuraminic acid, 54–55
Acid sphingomyelinase, 242, 280
Adler, I., 13
AGG, see G$_{M2}$-gangliosidosis, adult-onset
Amaurotic idiocy, 67–69
Ancestry, in TSD, 28–29
Animal models, for G$_{M2}$-gangliosidosis,
    120–121
Antibodies, against hexosaminidase,
    95–96
Antidepressant drugs, 191
ARMS test, 111
Aryl sulfatase, 45
Ashkenazim
    AGG, 190–191
    Bloom syndrome, 280
    Canavan disease, 279
    carrier frequency, 286
    CD detection and carrier frequency, 291–292
    common recessive diseases, 277–281
    confidentiality issues, 290
    cystic fibrosis, 278–279
    demographic history, 244–246
    educational intervention, 287–288
    enzymatic and DNA-based screening,
        281–282
    enzyme and DNA testing, 285
    FACC, 280
    familial dysautonomia, 280–281
    future prospects, 290–291
    Gaucher disease, 240–242
    Gaucher disease type 1, 279
    genetic diseases, 243
    genetic testing demographics, 285–286
    group counseling, 288
    MLIV, 243
    multiple-option carrier screening, 283–285,
        292–293
    NPD, 242–243
    NPD type A, 291
    NPD types A and B, 280
    prenatal carrier screening, 282–283
    prenatal diagnosis, 287
    prenatal screening, 289–290
    screening anxiety, 288–289
    TSD, 137–143, 278
        epidemiology, 233–236
        mutation frequency, 236–240
        PGD for, 313
        statistical modeling, 246–248
ASM, see Acid sphingomyelinase

B

Biochemical genetics, hexosaminidases, 104
Biosynthesis
    β-hexosaminidases, 165–169
    mutant β-hexosaminidase, 169–170
BLM, 280
Bloom syndrome, 280
BMT, see Bone marrow transplantation
Bone marrow transplantation, 327–328, 350
Brain
    with amaurotic idiocy, 68–69
    in G$_{M2}$-gangliosidosis, 329–330
    monosialoganglioside, 39
N-Butyldeoxynojirimycin, 350

C

Canavan disease, 279, 291–292

Carrier screening
  ARMS, 111
  CD detection, 291–292
  in Dor Yeshorim, 302–304
  enzymatic and DNA-based, 281–282
  in Jewish communities, 301–302
  multiple-option rationale, 283–284
  multiple-option strategy, 284–285
  multiple-option type, 292–293
  mutation frequency, 286
  for new mutations, 111–112
  NPD type A frequency, 291
  prenatal, 282–283, 289–290
  for TSD, 257–259, 271–272, 341–346
  in TSD avoidance, 270–271
Catalytic sites, hexosaminidases, 155–156
CCM, see Chemical cleavage of
    mismatch analysis
CD, see Canavan disease
Cell targeting, Hex A, 322–324
Cerebrospinal fluid, 319–321
CF, see Cystic fibrosis
CFTR, see Cystic fibrosis transmembrane
    regulator
Charcot, Jean Martin, 12–13
Chemical chaperones, TSD treatment, 351
Chemical cleavage of mismatch analysis, 112
Chromatography
  gangliosides, 36–38
  hexosaminidases, 148
Cloning
  hexosaminidase genes, 113–115, 127–133,
    225–230
  in molecular biology, 105–107
Community genetics, 271–272
Courtship, in Jewish communities, 300
CSF, see Cerebrospinal fluid
Cystic fibrosis, in Ashkenazim, 278–279,
    292–293
Cystic fibrosis transmembrane regulator,
    278–279

D

Deletions, in G_{M2} activator, 86
Demographics, Ashkenazim, 244–246, 285–286
Differential assay, hexosaminidases, 47
DNA
  analysis, 107–108
  Ashkenazim genetic diseases, 285

composition, 104–105
  hexosaminidases, 148–149, 225–230
  reagents for TSD diagnosis, 269–270
Dor Yeshorim program
  accomplishments, 305–308
  analytical laboratories, 309
  community context, 298–299
  and courtship and marriage, 300
  and health and medical issues, 299
  Orthodox and Hasidic communities,
    299
  for other communities, 308–309
  and procreation beliefs, 300
  research, 308
  screening, 301–304
Dot blotting, in mutation detection,
    110–111
Drugs, antidepressant, 191

E

Education, for Ashkenazim genetic diseases,
    287–288
Emborgenic stem cells, 120–121
Enzymatic defects, in TSD
  enzyme replacement trials, 58–59
  first tests, 51–52
  G_{M2} heterogeneity, 83–85
  and hexosaminidases, 52
  labeled G_{M2}, 54–58
  labeled sphingomyelin studies, 52–54
  patient diagnosis, 58
  search for, 71–76
Enzyme therapy
  Ashkenazim genetic diseases, 285
  for G_{M2}-gangliosidosis, 120–121, 318–321,
    327–328
  in TSD, 58–59
Enzymology, variant B1, 177–178
Epidemiology, in Ashkenazim
  Gaucher disease, 240–242
  genetic diseases, 243
  HEXB mutations, 214–215
  MLIV, 243
  NPD, 242–243
  TSD, 209–210, 233–240, 246–248
ES cells, see Emborgenic stem cells
Evolution
  molecular biology, 104–107
  variant B1, 176–180

F

Fabry disease, 53–54
FACC, *see* Fanconi anemia complementation
    group C
Familial dysautonomia, 280–281
Fanconi anemia complementation group C, 280
Fatty acids, in ganglioside, 36
Fluorigenic substrates, 44–45
French Canadians, and TSD, 138

G

G_{M2}-activator protein
    gene cloning, 116
    gene mutations, 215
    specificity, 86
    and TSD variant AB, 76–80
Galactocerebroside, in early TSD tests, 51
Galactose, tolerance test in TSD, 51
Gangliosides
    N-acetylgalactosamine component, 34–35
    chromatographic separation, 36–37
    as storage compounds, 68–69
    Substance X as, 33–34
    TLC studies, 37–38
    from TSD brain, 38–39
Gangliosidoses
    G_{A2}-gangliosidosis, 76–80
    G_{M1}-gangliosidosis, 62–63, 68–69
    G_{M2}-gangliosidosis, *see* G_{M2}-gangliosidosis
Gaucher disease
    in Ashkenazim, 240–242, 279, 292–293
    and early TSD tests, 51–52
    enzymatic deficiencies, 53–54
    in TSD studies, 3
GD, *see* Gaucher disease
Genes
    ASM, 242
    BLM, 280
    CFTR, 278–279
    familial dysautonomia, 281
    GM2A, 116, 215
    HEX, 127–133
    HEXA, 113–116, 201, 207–210, 225–230,
        236–240
    HEXB, 115–116, 210–215, 225–230
Gene therapy, for G_{M2}-gangliosidosis, 120–121
Genetic counseling, for TSD, 4, 270–271,
    352–353

Genetic diseases, in Ashkenazim
    carrier frequency, 286
    carrier screening sensitivity, 281–282
    confidentiality issues, 290
    educational intervention, 287–288
    enzyme and DNA testing, 285
    future prospects, 290–291
    group counseling, 288
    multiple-option carrier screening, 283–285
    prenatal carrier screening, 282–283
    prenatal diagnosis, 287
    prenatal screening, 289–290
    screening anxiety, 288–289
    testing demographics, 285–286
    types, 243
Genetic engineering, 120–121
Genetic heterogeneity, 174–175
Genetic lesion, 137–143
Genetics
    biochemical, 104
    community, 271–272
    molecular, 175–176, 178–180
Genotype–phenotype relationships
    HEXA mutations, 201, 207–209
    HEXB mutations, 211–214
    β-hexosaminidase α-chain, 180–181
Genotyping, TSD, 194–195
G_{M2}-gangliosidosis
    adult-onset
        antidepressant drug effects, 191
        biochemical studies, 187–188
        clinical evaluation, 189–190
        molecular basis, 190–191
        mutations, 118
        overview 185–187, 191–192
        pathology, 188–189
    animal models, 120–121
    β chain mutations, 119–120
    classic infantile, 117
    heterogeneity in, 83–85
    insoluble TSD α-subunit, 118–119
    labeling for TSD studies, 54–58
    major classes, 174–175
    pathogenesis, 86
    as storage compounds, 68–69
    treatment
        by BMT, 327–328
        cat studies, 321–322
        early enzyme infusion trials, 318–321
        by enzyme secretion, 327–328

by Hex A cell targeting, 322–324
with macromolecules to brain, 329–330
by TTC–Hex A and neuronal storage,
    324–326
variant AB, 76–80
variant B1, *see* Variant B1
Globoside, 70
Globus, Joseph, 18
Glucocerebroside, 51–52
α-Glucosidase, 46
β-Glucosidase, 46
β-Glucuronidase, 45
Glycolipids, 68–70
Glycolytic processing, 153–154
Glycosidases, 45–46
GM2A, 116, 215
GM2 ganglioside, *see* Tay–Sachs disease
Group counseling, 288

H

Hasidic Jewish community, 299–302
Health issues, and Dor Yeshorim, 299
Heat inactivation method, 185–186
Heteroduplex analysis, 111–112
Heterozygote advantage, 28–29
Heterozygotes, in TSD, 64–66
Hexosamine, 34–35
Hexosaminidase A
    from allelic mutations, 83–85
    antibodies against, 95–96
    assay development, 187–188
    cell targeting, 322–324
    α-chain gene isolation, 113–115
    deficient adults, 189–190
    differential assay, 47
    in early G_{M2}-gangliosidosis studies,
        319–321
    encoding gene
        cloning, 225–230
        genotype–phenotype relationship, 201,
            207–209
        HEXB homology, 115–116
        isolation, 113–115
        mutation frequency, 236–240
        and TSD epidemiology, 209–210
    and HexB, 48, 97–99
    HIM for, 185–186
    structure, 146–148
    subunit structure, 96–97

in TSD, 3–4, 72–76
in TSD avoidance strategies, 270–271
–TTC, and neuronal storage, 324–326
variant AB, 76–80
variant B1, 80–82
Hexosaminidase B
    from allelic mutations, 83–85
    antibodies against, 95–96
    differential assay, 47
    encoding gene
        cloning, 225–230
        epidemiology, 214–215
        genotype–phenotype relationships,
            211–214
        HEXA homology, 115–116
        isolation, 115
        mutations, 210–211
    heat-labile, 194–195
    and HexA, 48, 97–99
    mutations, 119–120
    structure, 146–148
    subunit structure, 96–97
    in TSD, 72–76
    in TSD variant AB, 76–80
Hexosaminidase S, 74
Hexosaminidases
    α and β subunits, 148–149, 154–157,
        175–176
    antibodies against, 95–96
    biochemical genetics, 104
    chromatography, 52
    function and specificity, 85
    gene cloning, 127–133, 225–230
    genotype–phenotype relationships,
        180–181
    mutant, biosynthesis, 169–170
    normal biosynthetic pathway, 165–169
    pre-pro-α- and pre-pro-β-chains,
        149–154
    purification, 72–76
    structure–function relationships,
        215–216
    in TSD, 5–6
    various studies, 46–47
Hexoses, 36
HIM, *see* Heat inactivation method
Homology
    HEXA and HEXB, 115–116
    hexosaminidases, 149
7-Hydroxy-coumarin glucuronide, 45

I

Immunology, in hexosaminidases, 96–97
Inheritance, in TSD, 2–3
Insertion mutations, 140–143
Intracytoplasmic membranous bodies, 3

J

Jackson, J. Hughlings, 13
Jewish Chronic Disease Hospital, 25–26
Jewish community
    Ashkenazi, *see* Ashkenazim
    courtship and marriage, 300
    and Dor Yeshorim, 299
    early carrier screening, 301–302
    Moroccans, 192–194
    procreation beliefs, 300

K

Knapp, Herman, 13–14

L

Lecithinoid, 17
Lesion, genetic, 137–143
Lysosomes, 166–167

M

Macromolecules, 329–330
Mannose-6-phosphate, 168
Marriage, in Jewish communities, 300
Medical issues, and Dor Yeshorim, 299
4-Methylumbelliferyl, 44–46
4-Methylumbelliferyl-2-acetamido-2-deoxy-
    6-sulfoglucopyranoside, 81
4-Methylumbelliferyl-β-D-
    N-acetylglucosaminide, 58
Meynert, Theodore, 12, 16
Microscopy, in TSD research, 18–20
Missense mutations, 108–109
MLIV, *see* Mucolipidosis type IV
Models, for TSD, 120–121, 246–248
Molecular biology
    DNA composition, 104–105
    restriction enzymes and cloning, 105–107
Molecular genetics
    Hex α and β subunits, 175–176
    variant B1, 178–180

Molecular heterogeneity, TSD and Sandhoff
    disease, 157–160
Monosialoganglioside, 39
Montreal program, 271
Moroccan Jewish population,
    192–194
Mount Sinai Hospital, 15–16
4-MU, *see* 4-Methylumbelliferyl
Mucolipidosis type IV, 243
Mutations
    in AGG, 118, 190–191
    in Ashkenazim, 140–143, 236–240, 278
    in Ashkenazim FACC, 280
    in Ashkenazim NPD, 280
    classic infantile TSD, 117
    classification, 108–109
    detection, 5–6, 110–111
    in G_{M2} activator, 86
    Gaucher disease, 240–242
    GM2A, 215
    HEXA, 201, 207–210
    in Hex A and B, 83–85
    HEXB, 210–215
    β-hexosaminidase, 119–120,
        169–170
    insertion, 140–143
    insoluble TSD α-subunit, 118–119
    missense, 108–109
    new, screening, 111–112
    nonsense, 109
    in NPD, 242–243
    variant B1, 118

N

National Tay–Sachs Association, 26
National Tay–Sachs Disease and Allied
    Disorders Association, 257–259
NBDM, *see* N-Butyldeoxynojirimycin
Neuraminic acid, 35–36
Neuronal storage, 3, 324–326
Niemann–Pick disease
    in Ashkenazim, 242–243, 280, 291–293
    enzymatic deficiencies, 53–54
    in TSD studies, 3
p-Nitrophenyl-β-D-N-acetylgalactosaminide, 70
p-Nitrophenyl-β-D-N-acetylglucosaminide, 70
Nitrophenyl-β-D-galactosylpyranoside, 54
Nonsense mutations, 109
NPD, *see* Niemann-Pick disease

O

Oligonucleotide recombination, TSD treatment, 352
OMIM, *see* Online Mendelian Inheritance in Man
Online Mendelian Inheritance in Man, 311–312
Opthamology, in TSD, 14–15
Orthodox Jewish community, 299–302

P

Parenchyma, G_{M2}-gangliosidosis, 329–330
Pathology
    AGG, 188–189
    G_{M2}-gangliosidosis, 86
    β-hexosaminidases system, 167–168
    TSD, 2–3
PCR, *see* Polymerase chain reaction
PGD, *see* Preimplantation genetic diagnosis
Phenotype, *see* Genotype–phenotype relationships
PLL, *see* Poly-L-lysine
Poly-L-lysine, 322–324
Polymerase chain reaction
    in mutation detection, 110, 140–143
    in PGD, 313–314
Polyvinylpyrrolidone, 319–320
Preimplantation genetic diagnosis, 312–314
Prevention, TSD, 4, 257–259, 353
Procreation, in Jewish communities, 300
Proteolytic processing, in hexosaminidase pathway, 149–153, 168–169
Psychological care, TSD families, 341–346, 352–353
PVP, *see* Polyvinylpyrrolidone

Q

Quality assurance, in Dor Yeshorim screening, 303–304
Quality control, in Dor Yeshorim screening, 303–304

R

RDS, *see* Riley–Day syndrome
Recessive disease, in Ashkenazim, 277–281

Restriction enzymes
    digestion, 111, 138–139
    in molecular biology, 105–107
Riley–Day syndrome, 28

S

Sachs, Bernard
    clinical training, 12–13
    early years, 11
    education, 12
    familial nature of TSD, 17
    and first TSD case, 13–14
    with Herman Knapp, 13–14
    hospital appointments, 15
    with I. Adler, 13
    with Ira van Gieson, 14
    with Isadore Strauss, 18
    with J. Hughlings Jackson, 13
    with Jean Martin Charcot, 12–13
    with Joseph Globus, 18
    with Karl Schaefer, 17
    retirement, 15–16
    with Theodore Meynert, 12, 16
    A Treatise on the Nervous Diseases of Children, 17
    TSD understanding, 268–270
    vs. Warren Tay, 15
Sandhoff disease, 157–160, 210–215
Schaefer, Karl, 17
Screening, *see* Carrier screening
Single-strand conformational polymorphism analysis, 112
Slot blotting, 110–111
Sphingolipidosis Registry, 27–29
Sphingomyelin, 52–54
Sphingosine, 36
SSCP, *see* Single-strand conformational polymorphism analysis
Statistical modeling, TSD in Ashkenazim, 246–248
Stem cells
    emborgenic, 120–121
    TSD treatment, 351–352
Strandin, 35–36
Strauss, Isadore, 18
Structure–function relationships, hexosaminidases, 154–157, 215–216
Substance X, 33–34

Substrate–binding sites, hexosaminidases, 156–157
Substrate deprivation therapy, TSD, 350–351

T

Tay, Warren, 15
Tay–Sachs disease, *see also* G$_{M2}$-gangliosidosis
amaurotic idiocy as, 67–68
ancestry, 28–29
in Ashkenazim, 137–143, 233–240, 246–248, 278, 292–293
avoidance strategies, 270–271
basics, 311–312
and Bernard Sachs, 13–14, 268–270
carrier testing, 271–272
classic infantile, 117
deficient activities, 3–4
enzymatic defects, 51–59, 71–76
epidemiology, 209–210
fluorigenic substrates usage, 44–45
future perspectives, 349–353
gangliosides, 38–39
genotyping, 194–195
G$_{M2}$ heterogeneity, 83–85
heterozygote detection, 64–66
HEXA mutations, 201, 207–209
hexosaminidase antibodies, 95–96
hexosaminidase gene cloning, 127–133
international symposiums, 30
at Jewish Chronic Disease Hospital, 25–26
molecular identification, 5–6
in Moroccan Jews, 192–194
National Tay Sachs Association birth, 26
pathogenesis, 86
PGD, 312–314
psychological care, 341–346
and RDS, 28
research advances, 2
and Sandhoff disease, 157–160
Sphingolipidosis Registry, 27–28
strandin isolation, 35–36
and tuberculosis, 29–30
variant AB, 76–80
variant B1, *see* Variant B1
variant O, 70–71
with visceral involvement, 70–71
Tay–Sachs program
community-based screening, 257
conception, 256–257
individual programs, 261–265
key people, 254–256
origins, 253–254
program results, 257–259
Testing, *see* Carrier screening
Tetanus toxin, 322
Tetanus toxin fragment C, 322–326
TGN, *see* Trans-Golgi network
Thymidine kinase, 121
TK, *see* Thymidine kinase
Transgenic animals, 120
Trans-Golgi network, 168
A Treatise on the Nervous Diseases of Children, 17
Treatment
G$_{M2}$-gangliosidosis, 329–330
by BMT, 327–328
cat studies, 321–322
early enzyme infusion trials, 318–321
by enzyme secretion, 327–328
by Hex A cell targeting, 322–324
with macromolecules to brain, 329–330
by TTC–Hex A and neuronal storage, 324–326
TSD, 6, 349–353
TSD, *see* Tay–Sachs disease
TT, *see* Tetanus toxin
TTC, *see* Tetanus toxin fragment C
Tuberculosis, 29–30

V

van Gieson, Ira, 14
Variant AB, 76–80
Variant B1
enzymology, 177–178
history, 176–177
molecular genetics, 178–180
mutations, 118
TSD basics, 80–82
Variant O, 70–71

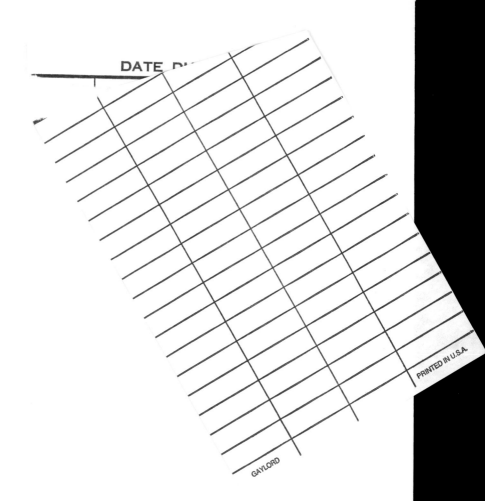

DATE D

PRINTED IN U.S.A.

GAYLORD